Handbook of Wind Energy Aerodynamics

Handbook of Wind Energy Aerodynamics

Handbook of Wind Energy Aerodynamics

Edited by Olive Murphy

SYRAWOOD
PUBLISHING HOUSE
New York

Published by Syrawood Publishing House,
750 Third Avenue, 9th Floor,
New York, NY 10017, USA
www.syrawoodpublishinghouse.com

Handbook of Wind Energy Aerodynamics
Edited by Olive Murphy

International Standard Book Number: 978-1-64740-438-3 (Hardback)

Cataloging-in-publication Data

Handbook of wind energy aerodynamics / edited by Olive Murphy.
p. cm.
Includes bibliographical references and index.
ISBN 978-1-64740-438-3
1. Wind power. 2. Aerodynamics. 3. Wind turbines--Aerodynamics.
4. Wind energy conversion systems. I. Murphy, Olive.
TJ820 .H36 2023
621.312 136--dc23

TABLE OF CONTENTS

Preface VII

Chapter 1 Development of a CFD-Based Wind Turbine Rotor Optimization Tool in Considering Wake Effects 1
Jiufa Cao, Weijun Zhu, Wenzhong Shen, Jens Nørkær Sørensen and Tongguang Wang

Chapter 2 Comparative Study of Dynamic Stall under Pitch Oscillation and Oscillating Freestream on Wind Turbine Airfoil and Blade 22
Chengyong Zhu and Tongguang Wang

Chapter 3 Development of an Efficient Numerical Method for Wind Turbine Flow, Sound Generation and Propagation under Multi-Wake Conditions 42
Zhenye Sun, Wei Jun Zhu, Wen Zhong Shen, Emre Barlas, Jens Nørkær Sørensen, Jiufa Cao and Hua Yang

Chapter 4 Aerodynamic Force and Comprehensive Mechanical Performance of a Large Wind Turbine during a Typhoon based on WRF/CFD Nesting 64
Shitang Ke, Wenlin Yu, Jiufa Cao and Tongguang Wang

Chapter 5 Experimental Study on Aerodynamic Characteristics of a Gurney Flap on a Wind Turbine Airfoil under High Turbulent Flow Condition 90
Junwei Yang, Hua Yang, Weijun Zhu, Nailu Li and Yiping Yuan

Chapter 6 CFD Simulations of Flows in a Wind Farm in Complex Terrain and Comparisons to Measurements 111
Matias Sessarego, Wen Zhong Shen, Maarten Paul van der Laan, Kurt Schaldemose Hansen and Wei Jun Zhu

Chapter 7 Variable Pitch Approach for Performance Improving of Straight-Bladed VAWT at Rated Tip Speed Ratio 132
Zhenzhou Zhao, Ruixin Wang, Wenzhong Shen, Tongguang Wang, Bofeng Xu, Yuan Zheng and Siyue Qian

Chapter 8 Wind Turbine Optimization for Minimum Cost of Energy in Low Wind Speed Areas Considering Blade Length and Hub Height 155
Han Yang, Jin Chen and Xiaoping Pang

Chapter 9 Development of an Advanced Fluid-Structure-Acoustics Framework for Predicting and Controlling the Noise Emission from a Wind Turbine under Wind Shear and Yaw 173
Mingyue Zhou, Matias Sessarego, Hua Yang and Wen Zhong Shen

Chapter 10 Comparison of the Blade Element Momentum Theory with Computational Fluid Dynamics for Wind Turbine Simulations in Turbulent Inflow 192
Sebastian Ehrich, Carl Michael Schwarz, Hamid Rahimi, Bernhard Stoevesandt and Joachim Peinke

Chapter 11 **Aerodynamics and Motion Performance of the H-Type Floating Vertical Axis Wind Turbine** 207
Ying Guo, Liqin Liu, Xifeng Gao and Wanhai Xu

Chapter 12 **Aerodynamic Performance of Wind Turbine Airfoil DU 91-W2-250 under Dynamic Stall** 224
Shuang Li, Lei Zhang, Ke Yang, Jin Xu and Xue Li

Permissions

List of Contributors

Index

PREFACE

This book has been an outcome of determined endeavour from a group of educationists in the field. The primary objective was to involve a broad spectrum of professionals from diverse cultural background involved in the field for developing new researches. The book not only targets students but also scholars pursuing higher research for further enhancement of the theoretical and practical applications of the subject.

A wind turbine refers to a type of machine that converts kinetic energy into electrical energy. It converts wind energy into electricity by utilizing the aerodynamic forces, which involve lift and drag. These forces are generated by the rotor blades, which function similar to rotor blade of helicopter or wing of an airplane. Wind turbines are a significant source of intermittent renewable energy, which can help in decreasing the reliance on fossil fuels and thereby reducing the energy costs. Smaller wind turbines are employed for tasks such as powering traffic warning signs and battery charging for caravans or boats. Larger turbines can aid household power supply and their extra energy be added back to the electrical grid. This book aims to shed light on some of the unexplored aspects of wind energy aerodynamics and the recent researches on this topic. It contains a detailed explanation of the aerodynamics utilized by wind turbines. The book will help the readers in keeping pace with the rapid changes in this area of renewable energy.

It was an honour to edit such a profound book and also a challenging task to compile and examine all the relevant data for accuracy and originality. I wish to acknowledge the efforts of the contributors for submitting such brilliant and diverse chapters in the field and for endlessly working for the completion of the book. Last, but not the least; I thank my family for being a constant source of support in all my research endeavours.

Editor

1

Development of a CFD-Based Wind Turbine Rotor Optimization Tool in Considering Wake Effects

Jiufa Cao [1], Weijun Zhu [1,2,*], Wenzhong Shen [2], Jens Nørkær Sørensen [2] and Tongguang Wang [3]

1 School of Hydraulic Energy and Power Engineering, Yangzhou University, Yangzhou 225009, China; jfcao@yzu.edu.cn
2 Department of Wind Energy, Technical University of Denmark, 2800 Kgs. Lyngby, Denmark; wzsh@dtu.dk (W.S.); jnso@dtu.dk (J.N.S.)
3 Jiangsu Key Laboratory of Hi-Tech Research for Wind Turbine Design, Nanjing University of Aeronautics and Astronautics, Nanjing 211100, China; tgwang@nuaa.edu.cn
* Correspondence: wjzhu@yzu.edu.cn

Abstract: In the present study, a computational fluid dynamic (CFD)-based blade optimization algorithm is introduced for designing single or multiple wind turbine rotors. It is shown that the CFD methods provide more detailed aerodynamics features during the design process. Because high computational cost limits the conventional CFD applications in particular for rotor optimization purposes, in the current paper, a CFD-based 2D Actuator Disc (AD) model is used to represent turbulent flows over wind turbine rotors. With the ideal case of axisymmetric flows, the simulation time is significantly reduced with the 2D method. The design variables are the shape parameters comprising the chord, twist, and relative thickness of the wind turbine rotor blades as well as the rotational speed. Due to the wake effects, the optimized blade shapes are different for the upstream and downstream turbines. The comparative aerodynamic performance is analyzed between the original and optimized reference wind turbine rotor. The results show that the present numerical optimization algorithm for multiple turbines is efficient and more advanced than conventional methods. The current method achieves the same accuracy as 3D CFD simulations, and the computational efficiency is not significantly higher than the Blade Element Momentum (BEM) theory. The paper shows that CFD for rotor design is possible using a high-performance single personal computer with multiple cores.

Keywords: wind turbine blade optimization; computational fluid dynamic; actuator disc; wake effect; Non-dominated Sorting Genetic Algorithm (NSGA-II)

---

1. Introduction

To reduce the cost of material and improve the aerodynamic efficiency of wind energy conversion systems, aerodynamic optimization of wind turbine rotors is important. Modern horizontal axis wind turbine (HAWT) rotors are aerodynamically optimized to efficiently extract wind energy. The physical limit to extract energy from wind is known as the Betz limit. The wind energy extraction process is through the reduction of wind speed passing the wind turbine rotor. The zero wake flow scenario will yield full energy absorption from wind, which certainly cannot be achieved. Since a wake always exists behind any wind turbine, and wind turbines are mostly clustered in a wind farm, the design and optimization of wind turbines in the wake situation are indeed necessary to represent some real-world wind farm cases.

From the theoretical to practical rotor aerodynamic design, the final products suffer from an accumulation of many aerodynamic and mechanical losses, such as tip loss, root loss, blade surface roughness change loss, blade shape manufacturing accuracy loss, drive train efficiency loss and wake effect loss. Despite of mechanical and manufacturing imperfections, most of aerodynamic optimizations have considered all the above-mentioned facts. The effect of wind turbine wake has been a very popular topic which is essentially important for wind farm layout optimization. However, in terms of rotor design itself, the wake effect is seldom included due to the inherent property of aerodynamic design tools. For example, the Blade Element Momentum (BEM) method is fast for optimization of single rotors, but it cannot deal with wake flow problems; the CFD-based methods are more accurate but practically not used for rotor optimization purposes.

In the field of wind turbine design and optimization, various simulation methods are involved for single rotors and for wind farms. For single rotor aerodynamic force calculations, the wind turbine aerodynamic modeling relies mainly on the following approaches: BEM [1–6]-based methods, Vortex Wake (VW)-based methods [7–12], and CFD-based methods [13–16]. The CFD methods either apply body-fitted grid or AD/AL/AS (Actuator Disc, Actuator Line, Actuator Surface) techniques. BEM has been the most efficient method which has been widely used to solve engineering problems. As the most efficient tool, BEM coupling with an optimization algorithm is mainly adopted as the method of wind turbine rotor design [13,17,18]. The VW methods are more flexible in the sense that it deals with the turbulent wake state that cannot be done with BEM. A study of BEM and VW methods are seen in [19] where general agreements are achieved for most of the considered cases but not for the high-thrust cases. However, VW methods are still not efficient for optimization purposes, and the accuracy in wake flow predictions affects the wind turbine performance to some extent. It is worthwhile to point out that the method to be used is relied on the matter of interest. From the multi-wind turbine optimization point of view, it is important to focus on both blade aerodynamics and turbulent wake. Therefore, BEM, VM and CFD based on curvilinear grid are commonly used for load calculations. However, rotor optimization requires a high computational demand, and so far CFDs are popularly used for wind turbine aerodynamic analysis. On the other hand, rotor aerodynamics has never been isolated with wake dynamics. However, existing models are individually used either for rotor design or wake study. Due to the inherent limitations of the BEM method, it cannot be directly applied to study wind turbine wakes. The more accurate methods, such as VM and CFD methods can be good choices to investigate both rotor aerodynamics and wake dynamics, but VM is not designed for long distance wake simulation and CFD is very expansive to study long range wake propagations, such that it is very hard to combine rotor aerodynamics with wake effects for rotor optimization purpose.

For a wind turbine cluster or wind farm, the velocity deficit causes a momentum loss in its wake before reaching the downstream wind turbines. Optimal design of the most upstream turbine will certainly increase the power product of the first turbine or perhaps the entire first row of a wind farm. However, for the second and third turbines downwind in the wake of the first turbine, the original design may not be the best configuration. Although this problem is realized through wind farm wake studies, but a practical design of wind turbines with wake effect is difficult either due to the model accuracy or high computational demand. The current work relates the rotor aerodynamics to the wake dynamics of multi-wind turbines. Therefore, modeling the velocity deficit is significantly important. Numerous studies were carried out in connection of large wind farm developments [20,21]. More recent work of Göçmen et al. [22] presented a good overview of various wake models that were developed at DTU. Some selected wake modelling approaches are listed in Table 1. The methods are divided into four groups according to their computational time demand. The model of Jensen [23] and Larsen [24,25] belong to class 1 that are the fastest engineering models. Some modifications of class 1 models become necessary to better represent the flow physics. For example, Tian et al. [26,27] proposed a new Jensen model that uses a cosine wake shape instead of the top-hat shape for the velocity deficit. The VW models and its modifications are adopted from aircraft aerodynamics and well-suited for wind turbine aerodynamics in terms of accuracy and efficiency [8,9,28]. Efforts were also made to reduce the computational time using CFD-based approaches [29,30], where the source equations are highly simplified. From the CFD point of view, many turbulence models are investigated and improved either for the purpose of Atmospheric Boundary Layer (ABL) simulations or of wake dynamics. Direct Numerical Method (DNS) is not yet applicable for rotor simulations, instead, the Reynolds Averaged Navier-Stokes/Large Eddy Simulation/Detached Eddy Simulation (RANS/LES/DES) and the hybrid turbulence models are widely used [31–34]. For wake investigations, in particular wake meandering and wake stability analysis, the traditional CFD with body-fitted mesh is very expensive even for one- or two-rotor analysis. So far, the best configuration to achieve a good quality and simulation speed is the so-called AL/AD techniques [35–37] which still apply the above-mentioned turbulence models but relatively faster. Using the similar mesh topology as the AL/AD techniques, the Lattice Boltzmann Method (LBM) [38] and Immersed Boundary Method (IBM) [39,40] have an advantage of reducing the computational time, but still demand a large grid number for high Reynolds number flows.

Table 1. Selected wake models.

Class	Methodology	References
Class 1	Jensen Dynamic Wake Meandering	Jensen (1983) Larsen (2007, 2009)
Class 2	Vortex Wake Method	Bareiss (1993), Voutsinas (1993), Wang (1998)
Class 3	Fuga Eddy Viscosity Model	Ott (2011) Ainslie (1988)
Class 4	RANS/LES + curvilinear RANS/LES + AD/AL/AS LBM/IBM	Sørensen (2014), Thé (2017), Allah (2017), Abdulqadir (2017) Sørensen, Shen et al. (1998, 2002, 2009, 2010, 2012), Troldborg (2014) Deiterding (2015), Angelidis (2015), Zhu (2013)

The main purpose of the current paper is to develop a wind turbine rotor optimization tool that has a numerical accuracy of Class 4 and has a computational efficiency better than Class 2. More specifically, the current study will not only focus on single wind turbine rotor optimization but also for two or more aligned wind turbine rotors. In this scenario, the 3D CFD with RANS/AL/AD approaches are suitable methods to perform rotor aerodynamic analysis, but too computationally heavy to perform aerodynamic optimization. Benefiting from the axisymmetric characteristic of a HAWT, the 3D RANS/AD equations can be reduced into a 2D formulation. The axisymmetric boundary condition applied for a 2D simulation is a good assumption in particular for rotor optimization of ideal flow cases, such as no wind shear, no yawing effects. The novelty of such method is to achieve the same accuracy as using 3D CFD simulation with a significantly higher efficiency. Then, based on the multi-objective and multi-variable optimization algorithm, the design variables are the shape parameters comprising the chord, twist, and relative thickness of wind turbine rotor blade. Finally, the blade optimization is achieved based on an in-house developed flow solver combining with an optimization algorithm. A multi-objective optimization model is proposed with maximizing the overall power efficiency for multi-wind turbines. The non-dominated sorting genetic algorithm-II (NSGA-II) is used to handle the complex process. The novel part of the present study is to combine the CFD solver with an efficient optimization algorithm, such that the single or multiple wind turbine rotors are optimized at the same time, which finally obtain a global high aerodynamic efficiency.

The paper is organized as follows: Section 2 describes the governing equations of the aerodynamic models; Section 3 presents numerical validations of the current wake model; in Section 4, the aerodynamic methods are applied for rotor optimization. Conclusions are given in the final section.

2. Numerical Methods

The recent research development on wind turbine flow modeling has largely focused on CFD approaches. With the assumption of incompressible flows over rotor blades, the NS equations are solved together with RANS, LES, DES etc. The rotor blades can either be modeled as a solid body surface or with AD/AL/AS techniques. In addition to that, more numerical issues are involved such as inflow turbulence, and atmospheric boundary layer which lead to possible modifications of turbulence models.

2.1. Governing Equations

The incompressible RANS equations in 2D form reads

$$\frac{\partial u_i}{\partial x_i} = 0 \ \ (i = 1,2) \tag{1}$$

$$\frac{\partial u_i}{\partial t} + u_j \frac{\partial u_i}{\partial x_j} = -\frac{1}{\rho}\frac{\partial P}{\partial x_i} + \frac{\partial}{\partial x_j}(\nu \frac{\partial u_i}{\partial x_j} - \overline{u'_i u'_j}) + f_{ext} \tag{2}$$

$$-\rho\overline{u'_i u'_j} = \mu_t \left(\frac{\partial u_i}{\partial x_j} + \frac{\partial u_j}{\partial x_i}\right) - \frac{2}{3}\rho k \delta_{ij} \tag{3}$$

where x_i and u_i are the displacement and velocity components in the inertial coordinate system. t is the time, ρ is the density of fluid, P is the static pressure, v is the kinematic viscosity coefficient, μ_t is the turbulent eddy viscosity δ_{ij} is the Kronecker delta function, k is the turbulent kinetic energy, f_{ext} is the momentum source terms representing possible external forces.

2.2. Turbulence Model

Herein the turbulence model is discussed with the aim to model the nonlinear Reynolds stresses terms in Equations (2) and (3). Accurate predictions of turbulent kinetic energy and dissipation are the key factors in turbulent wake modelling. In the present work, the modified k-ω turbulence model of Menter [41] is used. Menter's baseline model combines the k-ω model by Wilcox [42] in the inner region of the boundary layer with the standard k-ε model in the outer region and the free steam outside the boundary layer. The modified transport equations are written as follows:

$$\frac{\partial k}{\partial t} + \frac{\partial(u_j k)}{\partial x_j} = \frac{1}{\rho}p_k + \frac{1}{\rho}\frac{\partial}{\partial x_j}[(\mu + \sigma_k \mu_t)\frac{\partial k}{\partial x_j}] - \beta^* k\omega \tag{4}$$

$$\frac{\partial \omega}{\partial t} + \frac{\partial(u_j \omega)}{\partial x_j} = \frac{\gamma}{\rho v_t}p_k + \frac{1}{\rho}\frac{\partial}{\partial x_j}[(\mu + \sigma_\omega \mu_t)\frac{\partial \omega}{\partial x_j}] + 2(1 - F_1)\frac{\rho\sigma_{\omega 2}}{\rho\omega}\frac{\partial k}{\partial x_j}\frac{\partial \omega}{\partial x_j} - \beta\omega^2 \tag{5}$$

where the original model constants for the inner region are:

$$\sigma_{k1} = 0.5,\ \sigma_{\omega 1} = 0.5,\ \beta_1 = 0.075,\ \beta^* = 0.09,\ \kappa = 0.41,\ \gamma_1 = 0.55$$

and for the outer region

$$\sigma_{k2} = 1.0,\ \sigma_{\omega 2} = 0.856,\ \beta_1 = 0.0828,\ \beta^* = 0.09,\ \kappa = 0.41,\ \gamma_2 = 0.44 \tag{6}$$

The re-formulation of the original turbulence model contains two parts: (1) change of model constants; (2) added the source terms to maintain the turbulence level.

(1) The modified model constants are based on the work of Prospathopoulos et al. [43] in which most of the constants remain the same except for the following ones

$$\beta_1 = 0.033,\ \beta^* = 0.025,\ \gamma_1 = 0.37$$

The new constants were suggested according to measurements. With the decrease of dissipation effect, the kinetic energy towards the rotor is effectively increased, which improves the numerical accuracy for wind turbine wake flow modeling [43].

(2) For RANS solvers, the turbulence level is usually specified from the inlet of a computational domain. For the k-ω models, it is the initial k and ω values that are responsible for the turbulence level

$$k = \frac{3}{2}(u_0 I_0)^2 \tag{7}$$

$$\omega = \frac{k}{\mu}\left(\frac{\mu_t}{\mu}\right)^{-1} \tag{8}$$

where I_0 represents the ambient turbulence level and μ_t/μ is the eddy viscosity ratio. A similar approach is implemented in some commercial CFD programs such as the ANSYS Fluent software. In another approach proposed by Richards and Hoxey [44], the boundary values of u_0, k and ω are basically a function of an aerodynamic roughness height corresponding to a specific site. This approach does not apply to the current work where an axisymmetric flow is assumed, and no wall boundary is specified. No matter what boundary values are given at inlet, the free decay of turbulence intensity cannot be avoided with the current set of RANS equations. In the free-stream flow, the mean velocity gradient is zero such that the turbulence quantities decay. Therefore, the production terms and the diffusion terms in Equations (4) and (5) can be neglected. The k and ω equations are reduced to

$$\frac{\partial(u_j k)}{\partial x_j} = -\beta^* k\omega \tag{9}$$

$$\frac{\partial(u_j \omega)}{\partial x_j} = -\beta\omega^2 \tag{10}$$

Solving the partial differential equations for k and ω, the turbulence decay in the free-stream condition is obtained as a function of downstream distance

$$k = k_{in}\left(1 + \frac{\omega_{in}\beta x}{u_0}\right)^{-\beta^*/\beta} \tag{11}$$

$$\omega = \omega_{in}\left(1 + \frac{\omega_{in}\beta x}{u_0}\right)^{-1} \tag{12}$$

where k_{in} and ω_{in} are the initial values and x is the downstream distance. It is desired that these turbulence quantities specified at inflow boundary can be maintained without unphysical decay. In the work of Spalart and Rumsey [45], measurements of turbulent airfoil flows in wind tunnel were numerically reproduced with RANS computations. Even though the turbulence intensity is very low in the wind tunnel, the numerical simulations were not in perfect agreement with measurements. Inspired from the experimental and numerical work done in [45], the idea of maintaining turbulence quantities is extended to the current external wind turbine flows. The modified transport equations are

$$\frac{\partial k}{\partial t} + \frac{\partial(u_j k)}{\partial x_j} = \frac{1}{\rho}p_k + \frac{1}{\rho}\frac{\partial}{\partial x_j}[(\mu + \sigma_k\mu_t)\frac{\partial k}{\partial x_j}] - \beta^* k\omega + \beta^* k_{amb}\omega_{amb} \tag{13}$$

$$\frac{\partial \omega}{\partial t} + \frac{\partial(u_j \omega)}{\partial x_j} = \frac{\gamma}{\rho \upsilon_t}p_k + \frac{1}{\rho}\frac{\partial}{\partial x_j}[(\mu + \sigma_\omega\mu_t)\frac{\partial \omega}{\partial x_j}] + 2(1 - F_1)\frac{\rho\sigma_{\omega 2}}{\rho\omega}\frac{\partial k}{\partial x_j}\frac{\partial \omega}{\partial x_j} - \beta\omega^2 + \beta\omega_{amb}^2 \tag{14}$$

Different from the original model, the added source terms are $\beta^* k_{amb}\omega_{amb}$ and $\beta\omega_{amb}^2$ with the subscript "*amb*" indicates the ambient values [46]. It is quite straight forward to see that the destruction terms can be exactly cancelled if the inflow values of k and ω are set equal to the ambient values, which means in the free-stream flow case, turbulence quantities will remain unchanged throughout the computational domain. An example is given in Figure 1 that clearly shows the effect of using the added source terms. In the figure, the downstream distance x is non-dimensionalized to the rotor size D. In the free stream condition, the decay of turbulent kinetic energy using different approaches is compared. The original k-ω turbulence model agrees well with the theory (Equation (10)), where the modified model maintains the turbulent kinetic energy everywhere in the computational domain.

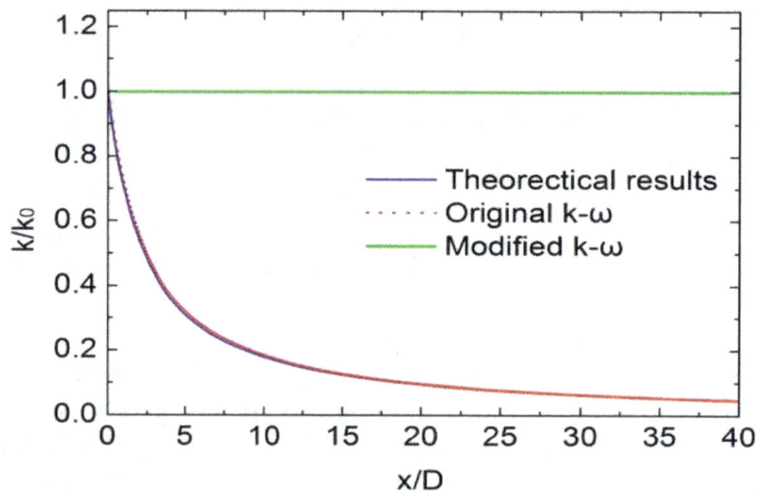

Figure 1. Comparisons of the normalized turbulence kinetic energy.

2.3. Actuator Disc

As the disc is assumed permeable, the mass conservation equation remains unchanged. In Equation (2), an external force f_{AD} is added that represents aerodynamic loads of wind turbine blades. To avoid the numerical singularity, each volume force needs to be smoothly re-distributed. A practical way of smearing the force is to use the 1D Gaussian function such that the actual forces in the disc is redistributed along one direction. In the current model, each force element is smeared locally in the blade normal direction with a distance d away from the disc using the convolution and the smeared force f' is computed as [14,47]

$$f' = f_{AD} \otimes \eta^{1D} \tag{15}$$

$$\eta^{1D}(d) = 1/(\varepsilon\sqrt{\pi}) \exp\left[-(d/\varepsilon)^2\right] \tag{16}$$

The parameter use in the smearing function is $\varepsilon = 3\Delta z$ where Δz is the reference grid size in the axis direction. Based on the hypothesis of blade elements, the body forces of rotor blades can be achieved through airfoil data and the flow field information.

The actuator disc solves the entire axisymmetric flow field and the induced velocity is naturally included into the formulation i.e., the relative velocity V_{rel} and flow angle ϕ are determined from

$$\phi = \tan^{-1}\left(\frac{V_0 - W_z}{\Omega r + W_\theta}\right),\ V_{rel}^2 = (V_0 - W_z)^2 + (\Omega r + W_\theta)^2 \tag{17}$$

As shown in Figure 2, the z-axis and θ-axis represent the rotor normal and rotational directions, respectively. The relative velocity V_{rel} seen by the blade element is a combination of axial velocity and rotational velocity, as shown in Equation (17). The lift force L and the drag force D are perpendicular and parallel to V_{rel}, respectively, as sketched in Figure 2. The local angle of attack is given by $\alpha = \phi - \gamma$, where γ is the local pitch and twist angle, the axial velocity $(V_0 - W_z)$, where V_0 is the inflow wind speed and W_z is the axial induced velocity, the tangential velocity $(\Omega r + W_\theta)$, where Ω is the rotational speed and r is the radius of wind turbine and W_θ is the tangential induced velocity. Lift and drag forces per spanwise length are found from tabulated airfoil data and the velocity triangles shown in Figure 2.

$$L = \frac{1}{2}\rho V_{rel}^2 cBC_l,\ D = \frac{1}{2}\rho V_{rel}^2 cBC_d \tag{18}$$

Therefore, projecting the force in the axial and tangential directions to the rotor gives the aerodynamic force components

$$F_z = L\cos\phi + D\sin\phi,\ F_\theta = L\sin\phi - D\cos\phi \tag{19}$$

The forces are calculated in every time-step such that the momentum equation produces a new velocity field which updates the flow angle of attack as shown in Figure 2.

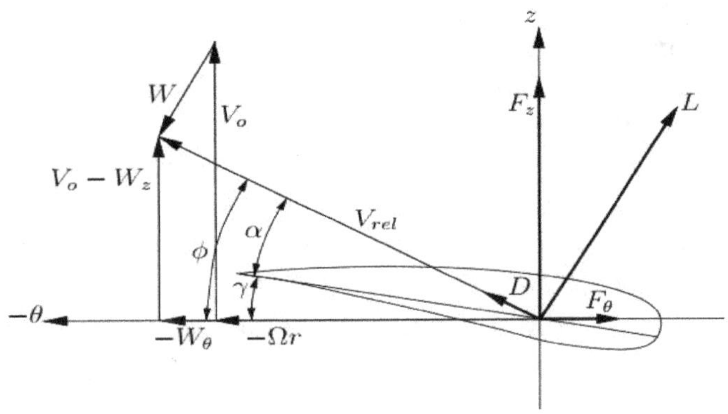

Figure 2. Sketch of velocity triangles of a blade element.

2.4. General Flow Solver

The RANS equations are solved by the *EllipSys* code [48,49], which is a general-purpose incompressible Navier-Stokes code based on a second-order multi-block finite volume method. The code solves the velocity-pressure coupled flow equations employing the SIMPLE/SIMPLEC/PISO (Semi-Implicit Method for Pressure-Linked Equations/SIMPLE Consistent/Pressure Implicit with Split Operator) methods and a multi-grid strategy. The momentum equations are first solved with a guessed pressure as a predictor. Next, the continuity equation is used as a constraint to obtain an equation for the pressure correction. In the predictor step, the momentum equations are solved by a second-order accurate backward differential scheme in time. For the spatial discretization a central difference scheme is employed for the diffusive terms and the QUICK (Quadratic Upstream Interpolation for Convective Kinematics) upwind scheme is utilized for the convective terms. In the corrector step, an improved Rhie-Chow interpolation technique is applied to suppress numerical oscillations from the velocity-pressure decoupling [50]. Furthermore, an improved SIMPLEC scheme developed for collocated grids is used to ensure that the solution does not depend on the value of relaxation parameters and the time-step [51]. The Navier-Stokes equations are discretized using a finite-volume scheme and all the information from the grid geometry is directly transferred into the discretized terms.

3. Results and Discussion

One of the major difficulties for engineering wind turbine wake models is to simulate the wake superposition effect. In this section, the 2D axisymmetric flows over two rotors are numerical validated against experimental data or LES results. The Danish Nibe wind turbines and Horns Rev wind turbines are selected for the numerical study.

3.1. Computational Setup

As sketched in Figure 3a, the two rotors are placed in tandem. The uniform inflow is applied at inlet, and the distance to the first rotor is about 10D (rotor diameter). The distance between downstream wind turbine and outlet is about 30D. The inlet and outlet width are chosen as 10D. The distance between the two rotors is not necessarily fixed which provides flexibility to be further optimized.

With a non-dimensional form, the uniform flow at the inlet and far field is Vo = 1. The change of wind turbine operational condition is through the Tip Speed Ratio (TSR). For the turbulence inflow conditions, the initial k and ω are computed through Equations (6) and (7) which are determined by the selected turbulence intensity. The mesh points are clustered around the rotor as seen in Figure 3b where only half the symmetrical plan is displayed. The mesh is block-structured with a total number of 14 blocks and 64×64 points in each block. The thick white lines indicate the two rotors whereas the mesh lines are clustered towards the rotors. With an OpenMPI (Message Passing Interface), the parallel simulation (on a Dell work station with Intel Xeon CPU E5 series) takes about 30 s for each simulation depending on the convergence rate. The computational time varies according to different TSRs as well as initial k and ω values. The simulation time using an AD method on a full 3D mesh will be more than hundred times slower than the 2D axisymmetric solver. On the other hand, the 2D CFD method is still slower than the classical steady BEM, but the 2D CFD method provides detailed flow field around the wind turbines.

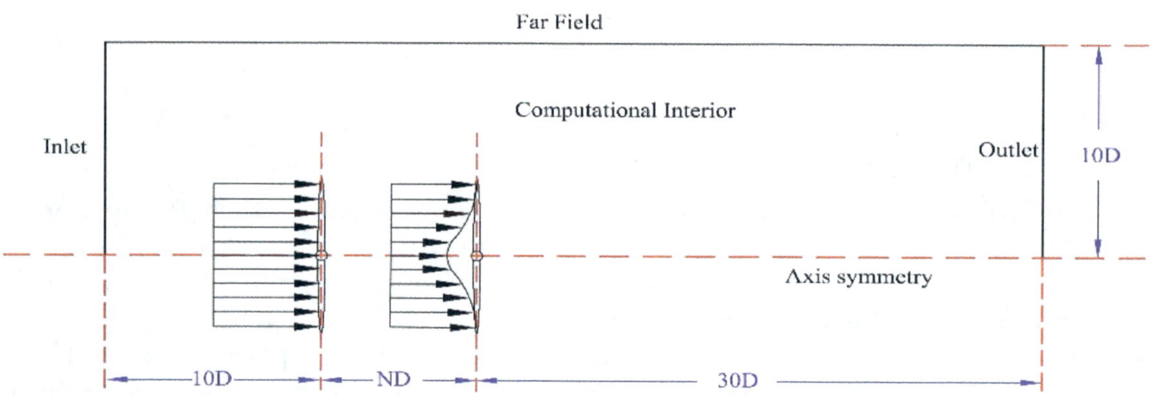

(a)

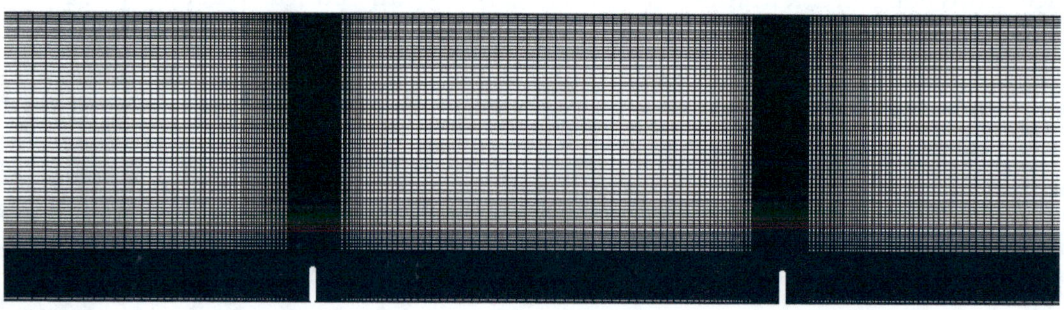

(b)

Figure 3. (**a**) Sketch of the boundary conditions and turbine positions; (**b**) The mesh configuration. The horizontal and vertical directions are defined as x and y for later reference.

3.2. Nibe Wind Turbine Case

To validate the aerodynamic model accuracy, the full-scale measurements [52] of the Nibe wind turbines are shown in Figure 4 where the two turbines are seperated with an axial distance of 5D. The Nibe wind turbines operate at a rated power of 630 KW, and the wind turbine has a rotor diameter of D = 40 m, and the hub height is H = 45 m. The simulation is performed at an inflow wind speed of 8.55 m/s and a turbulence intensity of 10%.

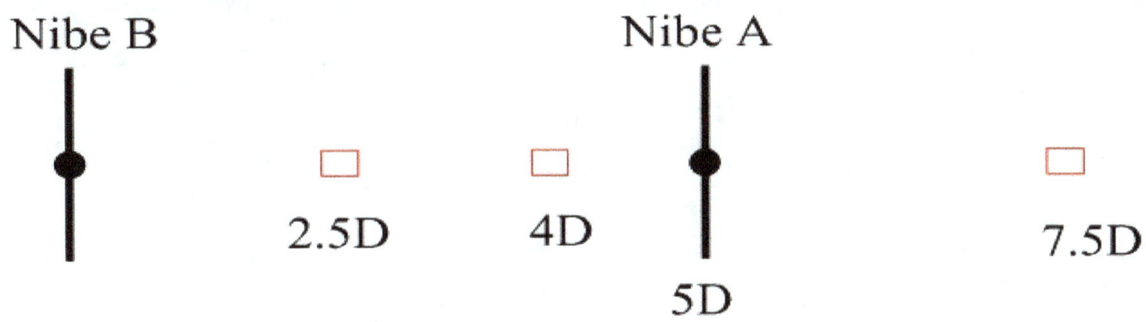

Figure 4. Sketch of the two Nibe wind turbines and the wake measurement locations.

The comparisons between the measured and simulated wake velocity profiles are shown in Figure 5. The measurement data are avialable for a single wind turbine (Nibe B) at down-stream positions of 2.5D, 4.0D and 7.5D as depicted in Figure 4. It is worth noting that for the present single wake study, Nibe A is not operating during the measurement period. The detailed wake deficit is clearly observed from Figure 5a–c which ranges from the near wake to the far wake region. In terms of maximum wake deficit and turbulence intensity, the new model shown relatively better agreement with the measured data. At the near wake region, x = 2.5D, the new model does not produce a better velocity deficit than the original model. At x = 4D and 7.5D, the new model shows better results, which is also the normal spacing between wind turbines.

For the double wake case, the streamwise velocity contour plot is presented in Figure 6. The axial velocity is largely reduced as the flow first passes through the Nibe B wind turbine and than the Nibe A wind turbine. The comparisons between the measured and simulated wake profiles for the two wind turbines at down-stream positions of 2.5D, 4.0D and 7.5D are shown in Figure 7. In this case, both Nibe A and Nibe B operate at a wind speed of 8.55 m/s. At the downstream locations 2.5D and 4D, the wake profiles are expected to be similar as that shown in Figure 6. As expected, only tiny differences are seen from the simulated results, which are due to the flow feedback from the Nibe A wind turbine, but the effect is very small. On the other hand, the experimental data showed a wider spread at 2.5D and 4D which is caused by different measurement conditions, Figure 7c shows the more complicated double wake case where the wake flow of two wind turbines are superimposed. At x = 7.5D, wake data are both measured behind the Nibe B and Nibe A wind turbines. The velocity deficit is seen to be relatively small as previously observed in Figure 5c. However, when the Nibe A wind turbine is operating, the x = 7.5D position is only 2.5D apart from the Nibe A wind turbine and thus it is considered to be the near wake position of turbine Nibe A. As seen from Figure 7a,c, wake profiles are similar at positions x = 2.5D and x = 7.5D. However, at x = 7.5D, a smaller velocity deficit is observed, which is due to the wake mixing effect introduced by the Nibe B wind turbine. At x = 7.5D, it is also seen that the fluctuation of turbulence intensity is quite large which cannot be captured by any RANS model.

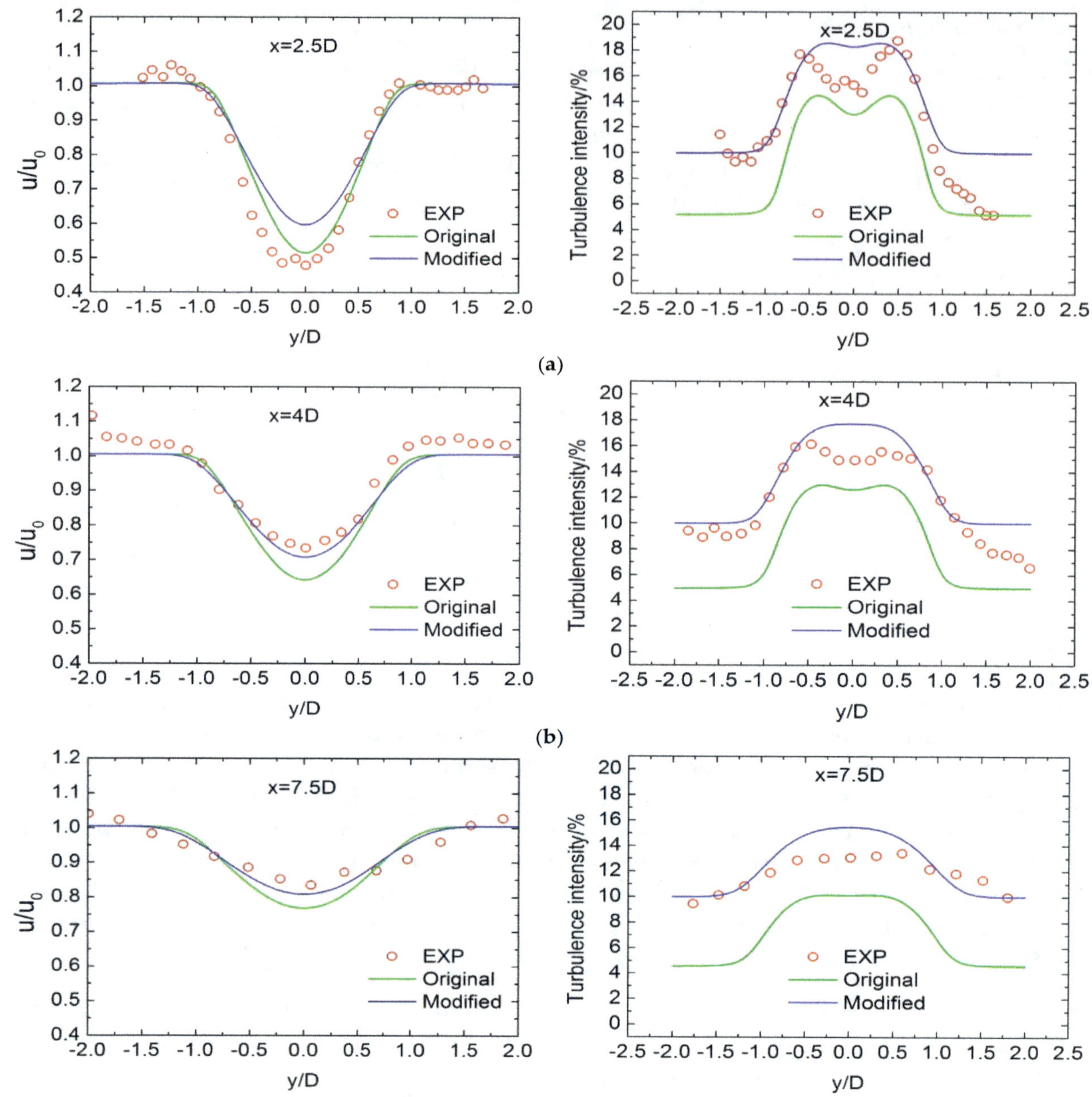

Figure 5. Comparisons between the measured and simulated wake data of a single wind turbine (Nibe B) at down-stream positions of 2.5D, 4.0D and 7.5D. (**a**) 2.5D; (**b**) 4D; (**c**) 7.5D.

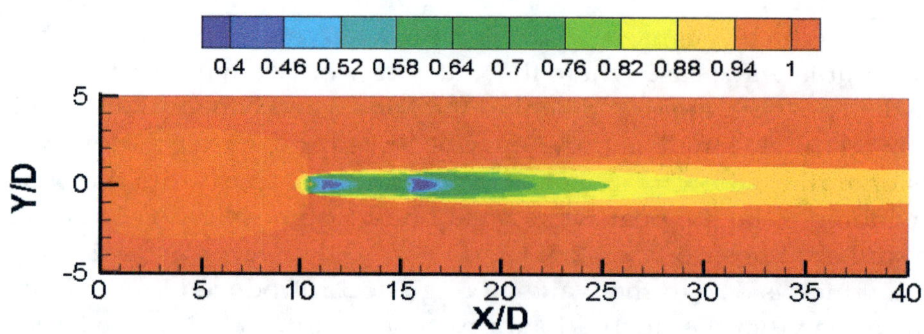

Figure 6. Contours of the time-averaged stream-wise velocity for two Nibe wind turbines.

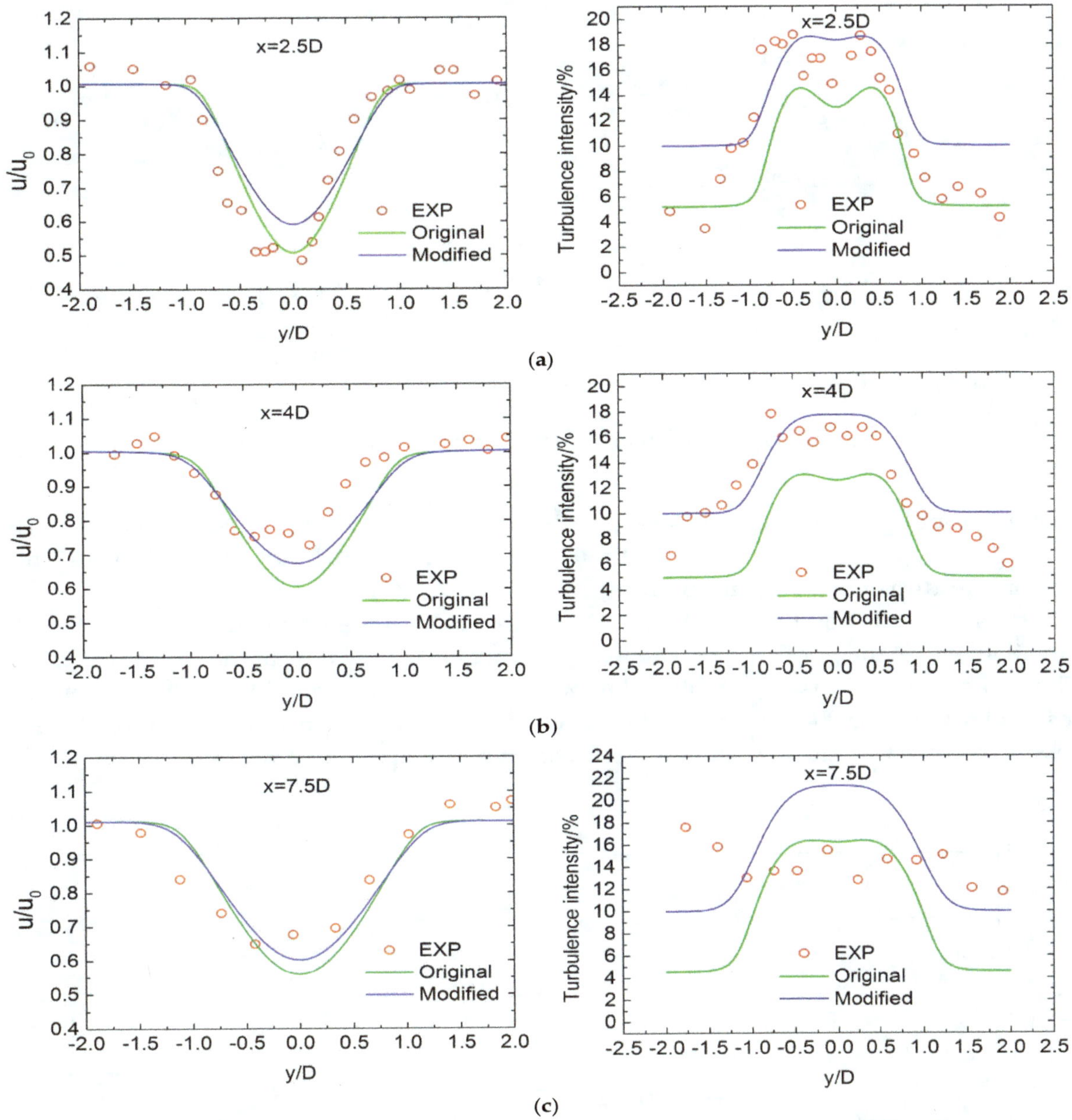

Figure 7. Comparisons between the measured and simulated wake data of the two wind turbines at down-stream positions of 2.5D, 4.0D and 7.5D. (**a**) 2.5D; (**b**) 4D; (**c**) 7.5D.

3.3. Horns Rev I Wind Farm Case

The Horns Rev I offshore wind farm is in the eastern North Sea near Denmark. Horns Rev wind farm consists of 80 Vestas V80 2 MW wind turbines within an area of about 20 km^2, as shown in Figure 8. The wind turbines are arranged in a regular array of 8 by 10 turbines which is very suitable to the study of multiple wake effects. The wind turbines are installed with an internal spacing along the main directions of 7D [53]. The wind turbines have a hub height of 70 m, a rotor diameter of D = 80 m, and a rated rotational speed of 18 rpm. The detailed pitch angle, rotational speed and blade geometry data are provided from the reference paper [54]. The 2 MW V80 wind turbine blade is composed of NACA63-series airfoils between the blade tip and the mid-span, and the FFA W3-series airfoils are used between the mid-span towards the blade inboard part. To validate the aerodynamic accuracy, the results from the present simulations and from the reference LES [26,54] simulations are compared. The inflow wind speed is 5.33 m/s, and turbulence intensity is 9.4% which are selected for fair comparison with some LES results.

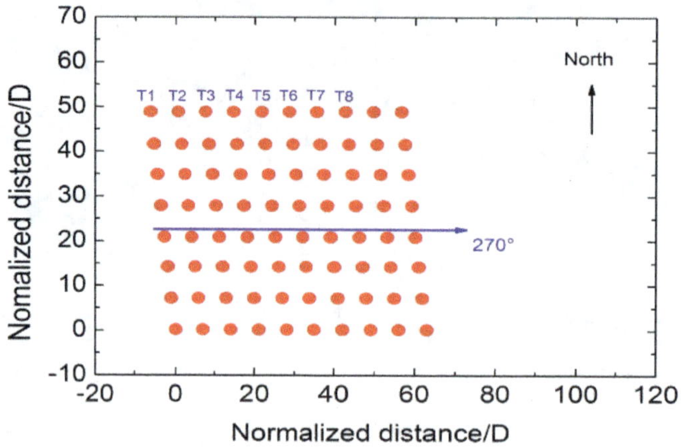

Figure 8. Layout of Horns Rev I wind farm.

The mesh topology used for this study is the same as in the previous simulations. The input airfoil data are now replaced to represent the Horns Rev wind turbines. The wind turbines under investigation are depicted in Figure 8 along the 270° directions. The wake profiles are shown in Figure 9 at downstream positions of 3D, 5D, 7D and 10D behind the first wind turbine where LES data are also available. General good agreements with LES are clearly seen from the results except very near wake. The wake expansion is well captured as well as the maximum velocity deficit. For the 7D and 10D cases, a slight over prediction of velocity deficit is seen from the current simulation. This indicates that the wake recovery at far downstream is slower using the present approach.

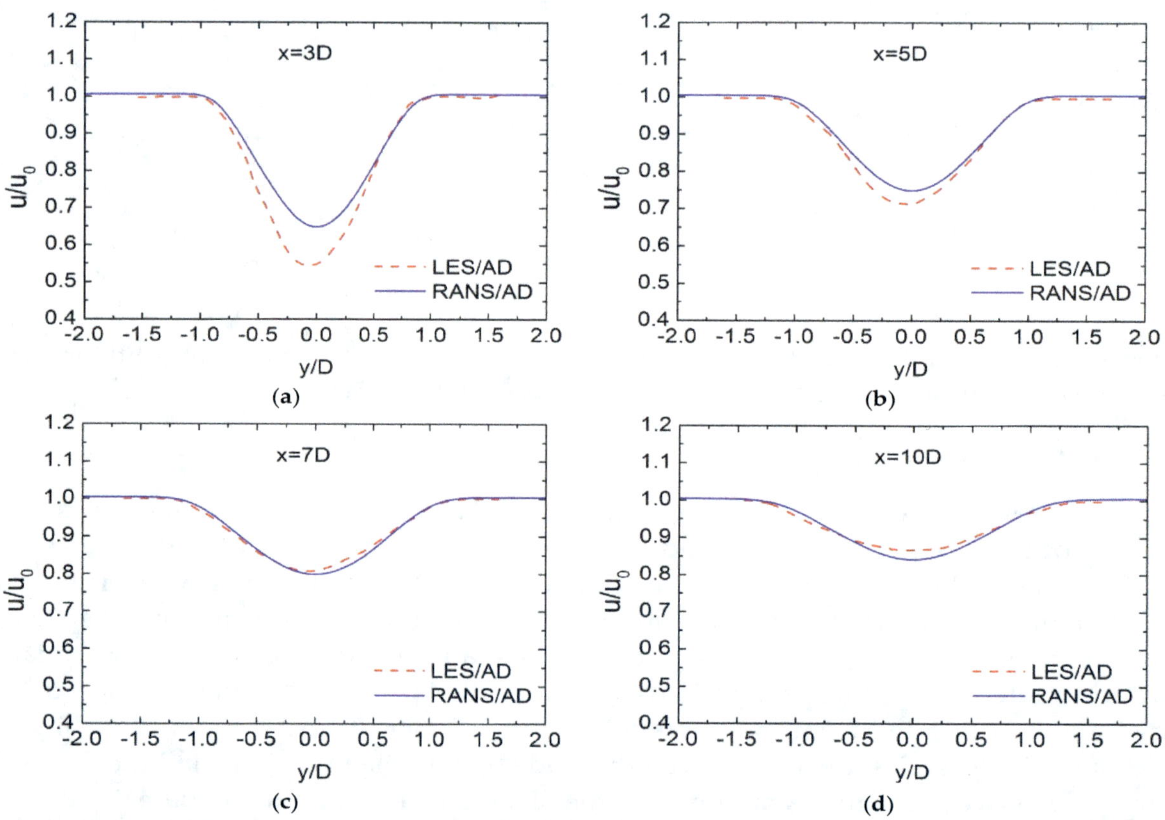

Figure 9. Comparisons between the LES and RANS lateral profiles normalized wake velocity for single wind turbine at down-stream positions of 3D, 5D, 7D and 10D. (**a**) 3D; (**b**) 5D; (**c**) 7D; (**d**) 10D.

The CFD simulation of the full column was carried out by extending the computational domain with a total number of 40 blocks. The resulted stream-wise velocity contours are shown in Figure 10 where the flow around wind turbine cluster can be observed.

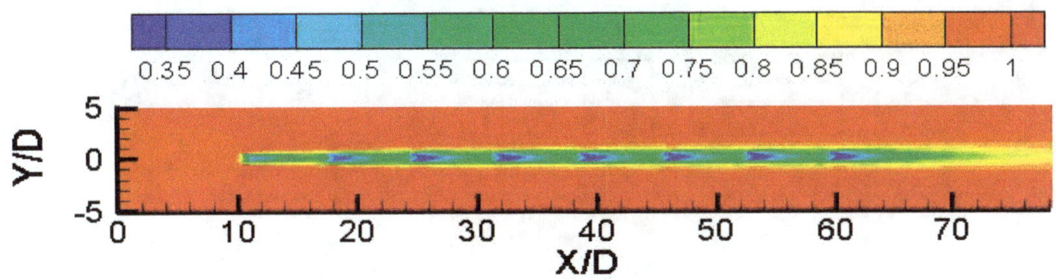

Figure 10. Contours of the stream-wise velocity for eight wind turbines.

In Figure 11, information about the power and thrust of the present wind turbines are also shown. Good agreements are seen from the power and thrust curves, which proves the model accuracy in a wide range of operational conditions.

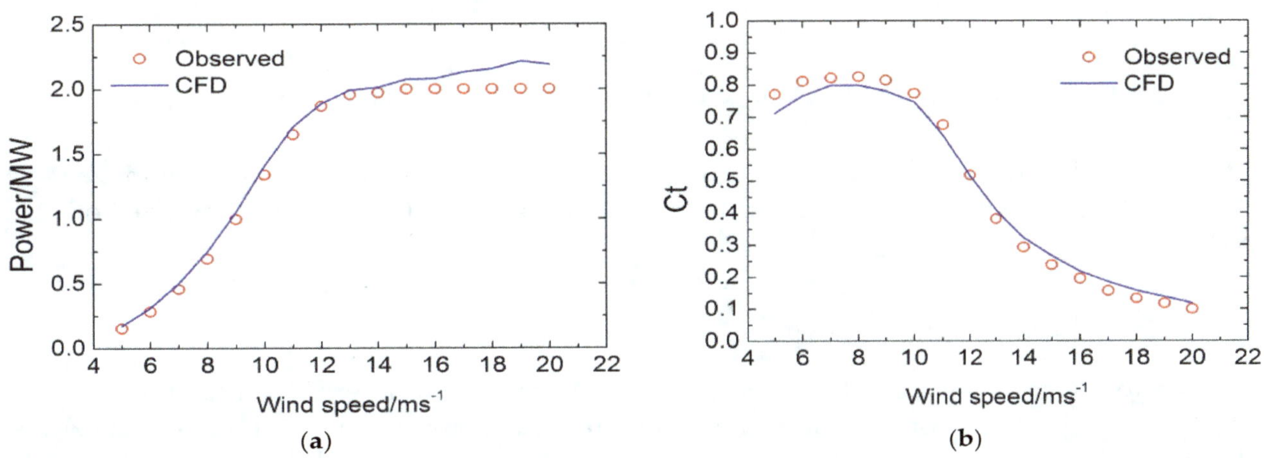

Figure 11. Comparisons between the observed data and RANS numerical for Cp, Ct, Power, and normalized power for 2 MW V80 wind turbines. (**a**) Power; (**b**) Ct.

3.4. Rotor Optimizations

3.4.1. Optimization Setup

This section describes the optimization problem used in present rotor aerodynamic design. With the optimization objectives of AEP and ATP, and the design variables of chord length, twist angle, relative thickness and rotational speed, the NSGA-II optimization method [55] is used to optimize the wind turbine rotor. In the NSGA-II optimization algorithm, the binary coding, crossover mutation and gene mutation are built to avoid premature phenomena during the optimization process. To better understand the optimization process, a sketch of the optimization diagram is given in Figure 12. The design variables are the blade geometrical data as well as the rotational speed. The "Bspline" curve is applied to reduce the number of control points. The CFD solver is embedded in the optimization loop which is a time-consuming part of the whole design process. The number of necessary iterations is determined by the NSGA-II optimization routine. Each evolution contains 30–50 populations and each population takes about 60 min.

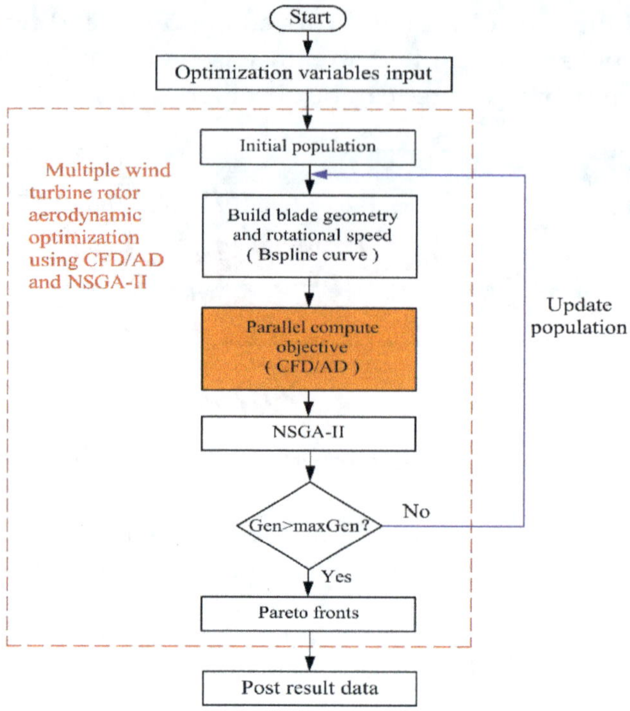

Figure 12. Wind turbine rotor aerodynamic optimization diagram.

The design objective is to maximize the annual energy production (AEP) and minimize the rotor thrust force (ATP, annual thrust production), thus the design objectives are to minimize both 1/AEP and ATP

$$\text{subject to } X \in R^n, \tag{20}$$

$$x_k^L \leq x_k \leq x_k^U,\ k = 1, \ldots, n \tag{21}$$

The vector X contains a total of n variables that are real numbers, R. The design variables, $X = [x_1, x_2, \ldots, x_n]$, are the control points that define the chord, twist angle and relative thickness as a function of blade span. The upper U and lower L notations, $x_k^L \leq x_k \leq x_k^U$, are the upper and lower limits of the control points. The 3rd-order Bsplines [56] are used to parameterize the chord, twist angle and relative thickness distributions, see Figure 13. The design is performed from a position at a radius of 0.337R to the tip of the blade. Thus, three control points for the chord, and two control points each for the twist angle and relative thickness distributions at the outboard of the blade are optimized. The 3rd-order Bspline curve is defined by

$$\begin{aligned}
q_i(u) &= \sum_{i=0}^{n} N_{i,p}(u) P_i \qquad u_a < u < u_b,\ i = 1, 2, \cdots, n \\
N_{i,0}(u) &= \begin{cases} 1 & if\ u_i \leq u \leq u_{i+1} \\ 0 & otherwise \end{cases} \\
N_{i,p}(u) &= \frac{u - u_i}{u_{i+p} - u_i} N_{i,p-1}(u) + \frac{u_{i+p+1} - u}{u_{i+p+1} - u_{i+1}} N_{i+1,p-1}(u) \\
u_0 &= u_1 = \cdots = u_p = 0 \\
u_{i+p} &= u_{i+p-1} + |p_i - p_{i-1}|,\ i = 1, 2, \cdots, n-1 \\
u_{n+p} &= u_{n+p+1} = \cdots = u_{n+p+3} = 1
\end{aligned} \tag{22}$$

where $q_i(u)$ is a vector-valued function of the independent variable, u. $\{P_i\}$ are the control points, and the $\{N_{i,p}(u)\}$ are the 3rd degree Bspline basis functions. The ith Bspline basis function of p-degree (order p+1). u_i is the knot vector.

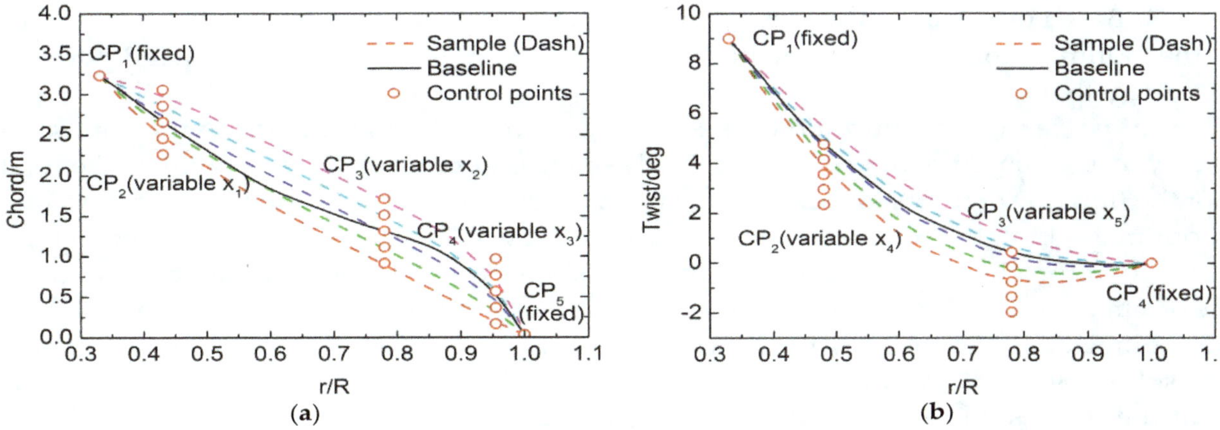

Figure 13. Distribution of the control points and the sample line of blade geometry. (**a**) Chord; (**b**) Twist angle.

To calculate the AEP and ATP, it is necessary to combine the power curve with the probability density of wind. The Weibull distribution function defining the probability density can be written in the following form

$$f(V_i < V < V_{i+1}) = \exp\left(-\left(\frac{V_i}{A}\right)^k\right) - \exp\left(-\left(\frac{V_{i+1}}{A}\right)^k\right) \quad (23)$$

where A is the scaling parameter, k is the shape factor and V_i is the wind speed distribution. If the wind turbine operates about 8760 h per year, its AEP and ATP can be evaluated as

$$AEP = \sum_{i=1}^{n-1} \frac{1}{2}(P(V_{i+1}) + P(V_i)) \times f(V_i < V < V_{i+1}) \times 8760 \quad (24)$$

$$ATP = \sum_{i=1}^{n-1} \frac{1}{2}(T(V_{i+1}) + T(V_i)) \times f(V_i < V < V_{i+1}) \times 8760 \quad (25)$$

where $P(V_i)$ and $T(V_i)$ are the power and thrust force at the wind speed of V_i.

3.4.2. Optimization Studies

The wind turbine rotor type of the Horns Rev I wind farm is chosen for the aerodynamic optimization. The selected wind turbines are located at the first and second rows (T1, T2 as depicted in Figure 8). It is worth noting that with the present method, it is possible to optimize the whole column of 8 wind turbines together. First, a larger number of wind turbines involve more design variables and thus a larger computational demand; second, it is the first two rows of wind turbines that have the major aerodynamic differences, as widely observed from measurements in several wind farms. According to the different optimization objectives, four optimization cases are carried out during the blade aerodynamic design.

Case 1: Baseline optimization for a single rotor. Using the optimization objectives of AEP and ATP, optimize the upstream wind turbine (the 1st row of rotors).

Case 2: Optimization of the 2nd row of rotors considering the wake effect from the 1st row of wind turbines. Using the optimization objectives of AEP and ATP, perform optimization for the 2nd wind turbine optimization.

Case 3: Further optimization of the rotational speed of the 1st row of wind turbines. The optimization objectives are the AEP and ATP for the two rows of wind turbines. The blade shape of the two types of rotors is obtained from Case 1 and Case 2, respectively, but the rotational speed of the 1st row of rotors will be optimized further.

Case 4: Based on Case 3, further optimization is performed for the two rows of wind turbines. Using the optimization objectives of both AEP and ATP, the two rows of wind turbines are optimized together.

To summarize these optimization cases: Case 1 follows a classical blade optimization routine; Case 2 only focuses on a downwind rotor, its blade shape is optimized; Case 3 optimizes the rotational speed of the first row of rotors to seek the global maximum energy production for two rows of rotors; Case 4 optimizes both the rotational speed of the first row of rotors and the blade geometry of the second row of rotors. By performing the 4 steps analysis, it expected that after the Case 4 study a more realistic optimization can be achieved for such a typical multiple wind turbine system.

The optimization Pareto fronts are shown in Figure 14 with the resulted Pareto front that measures how the solutions are distributed. The four subfigures (a–d) represents the above mentioned four cases. In the figures, the areas marked with *A*, *B* and *C* indicate the regions of feasible optimal solutions. The trend of area *A* suggests the solutions with high AEP, whereas low thrust is more focused in area *C*. As a multi-objective method, balance is then needed to weight between the high AEP and low ATP. In Figure 14a, Case 1 represents the standard optimization of a single rotor (the 1st row of rotors) without any wake effect. The high AEP and high ATP region is marked as area *A*, whereas area *B* finds good balance between low ATP and high AEP. In Figure 14b, the second row of wind turbine is optimized. Due to the wake effect from the first row of wind turbine, both the ATP and AEP are reduced comparing with Case 1 results. In Figure 14c, a high AEP area can hardly be achieved. The reason is that in Case 3 the AEP from the upstream rotor is reduced to increase the energy output for the downwind rotor. Such that, it is necessary to further optimize the downwind rotor. In Figure 14d, the optimization is continued such that the two rows of wind turbines are fully optimized with satisfactory results obtained.

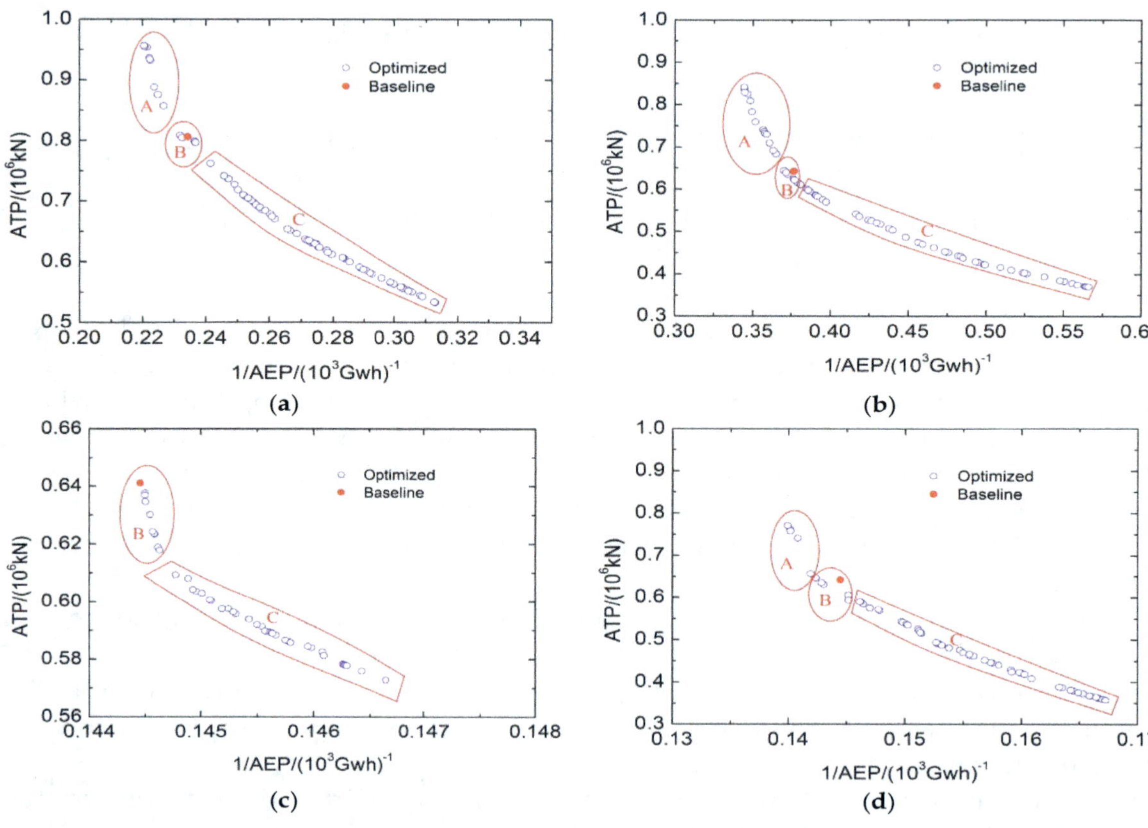

Figure 14. Pareto fronts of AEP and ATP in the 4 different cases. (**a**) Case 1; (**b**) Case 2; (**c**) Case 3; (**d**) Case 4.

Some key numbers are shown in Tables 2 and 3 after each optimization. For the Case 1 design, the upstream rotor is first optimized. As shown in Figure 14a, a trade-off between thrust and power can be found in area *B*. The final selected data is listed in Table 2 where the same ATP is maintained but the AEP is increased by 1.14%. For the Case 2 design, the downwind rotor is optimized. After optimization, the rotor in the wake has an increased power production of 1.78%, and there is also a small increase of thrust of 0.34%. Using the optimized solutions from Case 1 and Case 2, further investigation is carried out in Case 3. The rotational speed of the upstream wind turbine is considered to be a new design variable that may affect the total energy product. In Table 3, by comparing the AEP of each wind turbine with its reference value, it is seen that the 1st wind turbine has a negative change in AEP of −0.81% and the second wind turbine has a positive change in AEP of 1.2%. The overall gain of AEP is −0.04% because of the contribution of higher wind resource from the 1st wind turbine. The total thrust change of the two wind turbines is about 1.69%. It is seen that loads can be decreased with a certain amount by decreasing the front rotor speed, but it is not possible to gain more power from Case 3 study. A study in Case 4 is therefore performed to seek more energy from the 2nd rotor. The individual changes from each wind turbine and the overall changes are listed in Table 3. The reduction of the 1st rotor speed leads to a change of its AEP with −3.46%. The 2nd rotor is further optimized with a relatively large amount of AEP increase up to 9.71%. The total amount of AEP in the two-wind-turbine system is increased with 1.6%. The thrust is largely reduced for the 1st wind turbine under the new design condition at a relatively low TSR.

Table 2. Optimization results of Case 1 and Case 2.

Case	**Case 1 (1st Rotor)**		**Case 2 (2nd Rotor)**	
Objective	AEP	ATP	AEP	ATP
Reference	4.2665	0.8058	2.6559	0.6410
Optimized	4.3153	0.8058	2.7031	0.6432
Relative	1.14%	0	1.78%	0.34%

Case 1: Single wind turbine, consider both AEP&ATP; Case 2: Two wind turbines, consider both AEP and ATP of Downstream wind turbine.

Table 3. Optimization results of Case 3 and Case 4.

Case		**Case 3 (1st Rotor)**		**Case 4 (Two Rotors)**	
Objective		**AEP**	**ATP**	**AEP**	**ATP**
Reference	1st	4.2653	0.8058	4.2653	0.8058
	2rd	2.6559	0.6410	2.6559	0.6410
	Overall	6.9212	1.4468	6.9212	1.4468
Optimized	1st	4.2307	0.7919	4.1177	0.7522
	2rd	2.6875	0.6304	2.9139	0.6432
	Overall	6.9182	1.4224	7.0317	1.3955
Relative	1st	−0.81%	−1.72%	−3.46%	−6.65%
	2nd	1.2%	−1.65%	9.71%	0.34%
	Overall	−0.04%	−1.69%	1.6%	−3.55%

Case 3: Upstream WT rotation speed; consider AEP of overall WT, and ATP of Downstream WT; Case4: Upstream WT rotation speed and blade geometry; consider AEP of overall WT, and ATP of Downstream WT.

The final aerodynamic layouts of the optimized blades are gathered in Figure 15. The chord, twist and relative thickness distributions are given in Figure 15a–c, respectively. Based on the multi-objective optimizations, the chord length of the upstream wind turbine (Case 1) and downwind wind turbine (Case 2) deviates from the baseline chord distribution. The twist distribution of the two optimized blades is rather similar. The optimized rotor speed is shown in Figure 15d. With limited lower and upper bounds, the Case 4 optimization suggests a much lower rotational speed as compared with Case 3. As a result, the upwind rotor lost a certain amount of power which is gained by the downwind wind turbine. It is worth noting that the present study is only limited at aerodynamic optimization. Since the 1st wind turbine also has a decrease of thrust of 6.65%, there is a big potential to further reduce the structure weight.

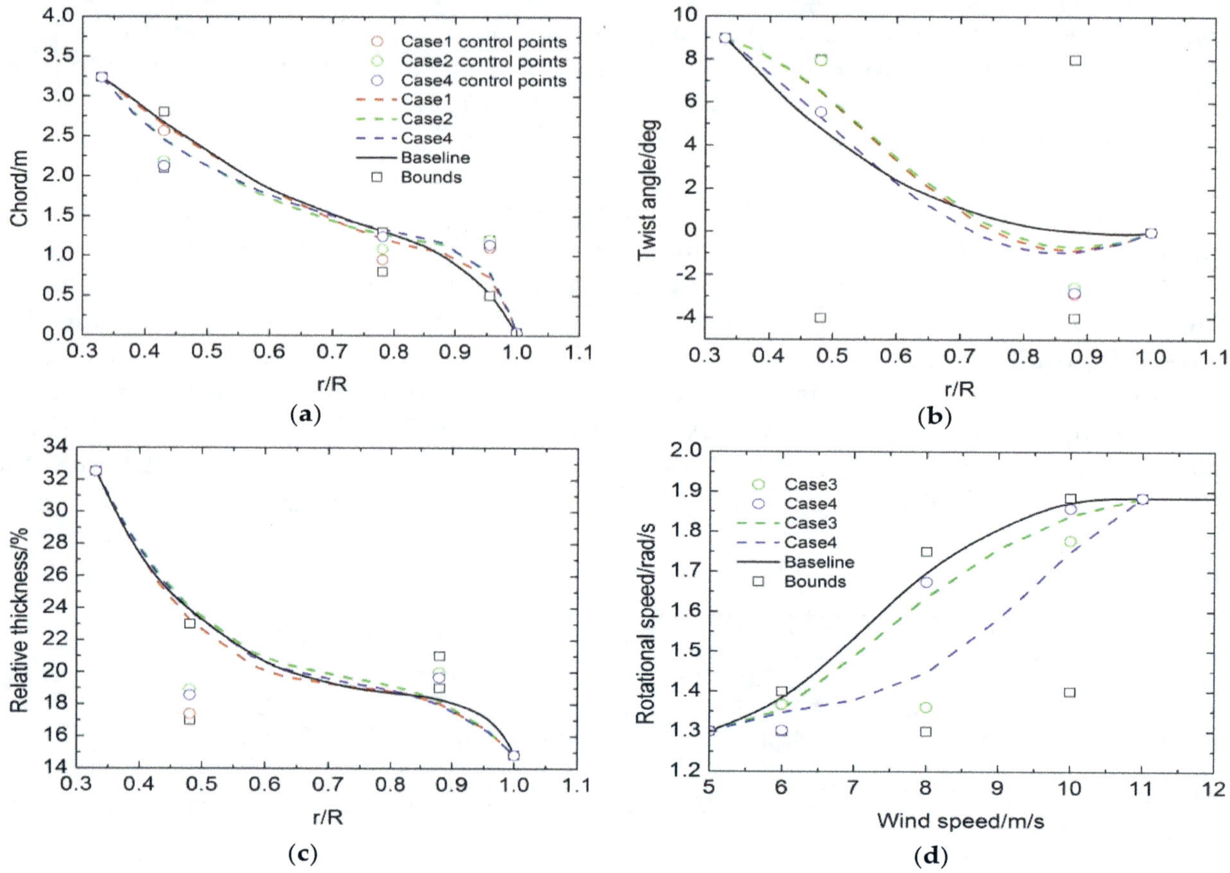

Figure 15. The comparisons of optimization results of wind turbine ((**a**–**c**) belong to Case 1, Case 2, Case 4; (**d**) belongs to Case 3 and Case 4). (**a**) Chord; (**b**) Twist angle; (**c**) Relative thickness; (**d**) Rotational speed.

4. Conclusions

This paper presented detailed numerical methodologies and convincing results for wind turbine rotor aerodynamic optimizations. To simulate the aerodynamic forces on the wind turbine rotor, a 2D actuator disc method is implemented in the incompressible flow solver favored by the axisymmetric boundary condition defined at the center line. The combined flow solver is more accurate and powerful than existing engineering models such as blade element momentum theory. On the other hand, it is order of magnitude faster than the 3D flow solver. However, the 2D and 3D solver implemented with same AD method should give same results. In addition to the AD flow simulations, two turbulent flow cases were investigated to proof the advantage of the modified k-ω turbulence model. With the fast CFD solver, the rotor optimization was carried out for single wind turbine with uniform flow and two wind turbines combined with wake interaction. The multi-objective optimization method NSGA-II is applied in the design loop and the two design objectives of high AEP and low ATP are required. The four optimization cases aimed at single rotor optimization without wake effect, single rotor optimization with wake effect, multi-wind turbine optimization with constant rotor speed and multi-wind turbine optimization with variable rotor speed. In general, single rotor optimizations are faster and they provide clear optimization trend. For the multi-wind turbine optimization cases, it was found that high aerodynamic performance is achieved when the upstream and downstream wind turbine both change their blade shape and rotational speed. These results indicate that traditional rotor optimization methods might not be sufficient due to the complex wake effect.

On a single PC with a developed parallel computation tool, the estimated time is 30 h for optimizing a single rotor, and for the coupled two-rotor optimizations, the simulation time is almost doubled. Future work will be carried out for more complicated design cases which may include the whole column of wind turbines such as the Horns Rev wind farm case. In such an optimization, the wind turbine spacing will be included as an important design parameter, to this end, the method can be applied for simple layout optimization.

Author Contributions: W.Z. conceived the rotor optimization method and wrote most part of the paper; J.C. carried out most of the programming part and the simulations; W.S., J.N.S. and T.W. put a lot of effort on guidance the overall structure of the paper as well as results and discussions.

Acknowledgments: The corresponding author wish to express special acknowledgement to the National Nature Science Foundation under grant number 11672261, the Jiangsu Province Natural Science Foundation (BK20160476).

References

1. Glauert, H. Airplane propellers. In *Aerodynamic Theory*; Springer: Berlin/Heidelberg, Germany, 1935; pp. 169–360.
2. Hansen, M.O.L. *Aerodynamics of Wind Turbines*, 2nd ed.; Earthscan, Routledge: Abingdon, UK, 2010; pp. 45–62.
3. Sørensen, J.N. *General Momentum Theory*; Springer: Berlin/Heidelberg, Germany, 2016; pp. 43–58.
4. Sørensen, J.N.; Mikkelsen, R. On the Validity of the Blade Element Momentum Method. In Proceedings of the European Wind Energy Conference, Copenhagen, Denmark, 2–6 July 2001.
5. Ghadirian, A.; Dehghan, M.; Torabi, F. Considering induction factor using BEM method in wind farm layout optimization. *J. Wind Eng. Ind. Aerodyn.* **2014**, *129*, 31–39. [CrossRef]
6. Vaza, P.; Pinho, T. An extension of BEM method applied to horizontal axis wind turbine design. *Renew. Energy* **2011**, *36*, 1734–1740. [CrossRef]
7. Leishman, J.G.; Bhagwat, M.J.; Bagai, A. Free-vortex filament methods for the analysis of helicopter wakes. *J. Aircr.* **2003**, *39*, 759–775. [CrossRef]
8. Bareiss, R.; Wagner, S. The Free Wake/Hybrid Wake Code RVOLM—A Tool for Aerodynamic Analysis of Wind Turbines. In Proceedings of the European Community Wind Energy Conference, Lubeck-Travemunde, Germany, 8–12 March 1993; pp. 424–427.
9. Voutsinas, S.G.; Belessis, M.; Huberson, S. Dynamic Inflow Effects and Vortex Particle Methods. In Proceedings of the European Community Wind Energy Conference, Lubeck-Travemunde, Germany, 8–12 March 1993; pp. 428–431.
10. Ramos, G.N.; Hejlesen, M.M.; Sørensen, J.N.; Walther, J.H. Hybird vortex simulations using a three-dimensional viscous-inviscid panel method. *Wind Energy* **2017**, 1871–1889. [CrossRef]
11. Shen, X.; Chen, J.G.; Zhu, X.C.; Liu, P.Y.; Du, Z.H. Multi-objective optimization of wind turbine blades using lifting surface method. *Energy* **2015**, *90*, 1111–1121. [CrossRef]
12. Cao, J.F.; Wang, T.G.; Long, H.; Ke, S.T.; Xu, B.F. Dynamic loads and wake prediction for large wind turbines based on free wake method. *Trans. Nanjing Univ. Aeronaut. Astronaut.* **2015**, *32*, 240–249.
13. Zhu, W.J.; Shen, W.Z.; Sørensen, J.N. Integrated airfoil and blade design method for large wind turbines. *Renew. Energy* **2014**, *70*, 172–183. [CrossRef]
14. Sørensen, J.N.; Shen, W.Z.; Munduate, X. Analysis of the Wake States by a Full-Field Actuator Disc Model. *Wind Energy* **1998**, *1*, 73–88. [CrossRef]
15. Shen, W.Z.; Zhu, W.J.; Sørensen, J.N. Validation of the actuator line/Navier Stokes technique using mexico easurements. In *The Science of Making Torque from Wind*; Wiley: Hoboken, NJ, USA, 2010; pp. 227–234.
16. Shen, W.Z.; Zhang, J.H.; Sørensen, J.N. The Actuator Surface Model: A New Navier-Stokes Based Model for Rotor Computations. *J. Sol. Energy Eng.* **2009**, *13*, 011002. [CrossRef]
17. Bak, C. *Research in Aeroelasticity EFP-2006: Key Parameters in Aerodynamic Rotor Design*; RISØ-R-1611(EN); RISØ National Laboratory: Roskilde, Denmark, 2007.

18. Wang, X.D.; Shen, W.Z.; Zhu, W.J.; Sørensen, J.N.; Chen, J. Shape Optimization of Wind Turbine Blades. *Wind Energy* **2009**, *12*, 781–803.
19. Gupta, S.; Leishman, J.G. Comparison of momentum and vortex methods for the aerodynamic analysis of wind turbines. In Proceedings of the 43rd AIAA Aerospace Sciences Meeting and Exhibit, Reno, NV, USA, 10–13 January 2005; p. 594.
20. Vermeer, L.J.; Sørensen, J.N.; Crespo, A. Wind turbine wake aerodynamics. *Prog. Aerosp. Sci.* **2003**, *39*, 467–510. [CrossRef]
21. Sanderse, B.; Koren, B. Review of computational fluid dynamics for wind turbine wake aerodynamics. *Wind Energy* **2011**, *14*, 799–819. [CrossRef]
22. Göçmen, T.; Lann, P.; Réthoré, P.E.; Diaz, A.P.; Larsen, G.C.; Ott, S. Wind turbine wake models developed at the technical university of Denmark: A review. *Renew. Sustain. Energy Rev.* **2016**, *60*, 752–769. [CrossRef]
23. Jensen, N.O. *A Note on Wind Generator Interaction*; Report M-2411; Risø National Laboratory: Roskilde, Denmark, 1983.
24. Larsen, G.C. *A Simple Stationary Semi-Analytical Wake Model*; Technical Report Risø; Technical University of Denmark: Roskilde, Denmark, 2009.
25. Larsen, G.C.; Madsen, H.A.; Bingöl, F.; Mann, J. *Dynamic Wake Meandering Modeling*; Report R-1607(EN); Risø National Laboratory: Roskilde, Denmark, 2007.
26. Tian, L.L.; Zhu, W.J.; Shen, W.Z.; Zhao, N.; Shen, Z. Development and validation of a new two-dimensional wake model for wind turbine wakes. *J. Wind Eng. Ind. Aerodyn.* **2015**, *137*, 90–99. [CrossRef]
27. Tian, L.L.; Zhu, W.J.; Shen, W.Z.; Song, Y.L.; Zhao, N. Prediction of multi-wake problems using an improved Jensen wake model. *Renew. Energy* **2017**, *102*, 457–469. [CrossRef]
28. Wang, T.G. Unsteady Aerodynamic Modelling of Horizontal Axis Wind Turbine Performance. Ph.D. Thesis, University of Glasgow, Scotland, UK, 1999.
29. Ott, S.; Berg, J.; Nielsen, M. *Linearized CFD Models for Wakes*; Report R-1772(EN); Risø National Laboratory: Roskilde, Denmark, 2011.
30. Ainslie, J.F. Calculating the Flowfield in the Wake of Turbines. *J. Wind Eng. Ind. Aerodyn.* **1988**, *27*, 213–224. [CrossRef]
31. Sørensen, N.N.; Bechmann, A.; Pierre, E.R.; Frederik, Z. Near wake Reynolds-averaged Navier–Stokes predictions of the wake behind the MEXICO rotor in axial and yawed flow conditions. *Wind Energy* **2014**, *17*, 75–86. [CrossRef]
32. The, J.; Yu, H. A critical review on the simulations of wind turbine aerodynamics focusing on hybrid RANS-LES methods. *Energy* **2017**, *138*, 257–289. [CrossRef]
33. Allah, V.A.; Mayam, M.H.S. Large eddy simulation of flow around a single and two in-line horizontal-axis wind turbines. *Energy* **2017**, *121*, 533–544. [CrossRef]
34. Abdulqadir, S.A.; Iacovides, H.; Nasser, A. The physical modelling and aerodynamics of turbulent flows around horizontal axis wind turbines. *Energy* **2017**, *119*, 767–799. [CrossRef]
35. Troldborg, N.; Sørensen, J.N.; Mikkelsen, R.F.; Sørensen, N.N. A simple atmospheric boundary layer model applied to large eddy simulations of wind turbine wakes. *Wind Energy* **2014**, *17*, 657–669. [CrossRef]
36. Sørensen, J.N.; Shen, W.Z. Numerical modeling of wind turbine wakes. *J. Fluids Eng.* **2002**, *124*, 393–399. [CrossRef]
37. Shen, W.Z.; Zhu, W.J.; Sørensen, J.N. Actuator line/Navier-Stokes computations for the MEXICO rotor: comparison with detailed measurements. *Wind Energy* **2012**, *15*, 811–825. [CrossRef]
38. Deiterding, R.; Wood, S.L. A dynamical adaptive lattic Boltzmann method for predicting wake phenomena in fully coupled wind engineering problems. In Proceedings of the VI International Conference on Computational Methods for Coupled Problems in Science and Engineering Coupled Problems, Venice, Italy, 18–20 May 2015.
39. Angelidis, D.; Sotiropoulos, F. An immersed boundary-adaptive mesh refinement solver (IB-AMR) for high fidelity fully resolved wind turbine simulations. In Proceedings of the 68th Annual Meeting of the APS Division of Fluid Dynamics, Boston, MA, USA, 22–24 November 2015.
40. Zhu, W.J.; Behrens, T.; Shen, W.Z.; Sørensen, J.N. Hybrid immersed boundary method for airfoils with a trailing-edge flap. *AIAA J.* **2013**, *51*, 30–41. [CrossRef]
41. Menter, F.R. Two-Equation Eddy-Viscosity Turbulence Models for Engineering Applications. *AIAA J.* **1994**, *32*, 1598–1605. [CrossRef]

42. Wilcox, D.C. *Turbulence Modelling for CFD*, 1st ed.; DCW Industries: La Cañada Flintridge, CA, USA, 1993.

43. Prospathopoulos, J.M.; Politis, E.S. Evaluation of the effects of turbulence model enhancements on wind turbine wake predictions. *Wind Energy* **2011**, *14*, 285–300. [CrossRef]

44. Richards, P.; Hoxey, R. Appropriate boundary conditions for computational wind engineering models using the k-ϵ turbulence model. *J. Wind Eng. Ind. Aerodyn.* **1993**, *46*, 145–153. [CrossRef]

45. Spalart, P.R.; Rumsey, C.L. Effective In-flow Conditions for Turbulence Models in Aerodynamic Calculations. *AIAA J.* **2007**, *45*, 2544–2553. [CrossRef]

46. Tian, L.L.; Zhu, W.J.; Shen, W.Z.; Sørensen, J.N.; Zhao, N. Investigation of modified AD/RANS models for wind turbine wake predictions in large wind farm. *J. Phys. Conf. Ser.* **2014**, *524*, 012151. [CrossRef]

47. Shen, W.Z.; Mikkelsen, R.; Sørensen, J.N. Tip loss correction for Actuator/Navier-Stokes computations. *J. Sol. Energy Eng.* **2005**, *127*, 209–213. [CrossRef]

48. Michelsen, J.A. *Basis3D—A Platform for Development of Multiblock PDE Solvers*; Technical Report AFM; Technical University of Denmark: Kongens Lyngby, Denmark, 1992.

49. Sørensen, N.N. *General Purpose Flow Solver Applied Over Hills*; RISØ-R-827-(EN); Risø National Laboratory: Roskilde, Denmark, 1995.

50. Shen, W.Z.; Michelsen, J.A.; Sørensen, J.N. An improved Rhie-Chow interpolation for unsteady flow computations. *AIAA J.* **2001**, *39*, 2406–2409. [CrossRef]

51. Shen, W.Z.; Michelsen, J.A.; Sørensen, N.N.; Sørensen, J.N. An improved SIMPLEC method on Collocated grids for steady and unsteady flow computations. *Numer. Heat Transf.* **2003**, *43*, 221–239. [CrossRef]

52. Taylor, G.L. *Wake Measurements on the Nibe Wind Turbines in Denmark*; Contractor Report ETSU WN 5020; National Power-Technology and Environment Center: Leatherhead, UK, 1990.

53. Hansen, K.S.; Barthelmie, R.J.; Jensen, L.E. The impact of turbulence intensity and atmospheric stability on power deficits due to wind turbine wakes at Horns Rev wind farm. *Wind Energy* **2011**, *15*, 183–196. [CrossRef]

54. Wu, Y.T.; Fernando, P.A. Modeling turbine wakes and power losses within a wind farm using LES: An application to the Horns Rev offshore wind farm. *Renew. Energy* **2015**, *75*, 945–955. [CrossRef]

55. Deb, K.; Agrawal, S.; Pratap, A. A fast elitist nondominated sorting genetic algorithm for multi-objective optimization: NSGA-II. In Proceedings of the Parallel Problem Solving from Nature VI Conference, Paris, France, 18–20 September 2000; pp. 849–858.

56. Piegl, L.; Tiller, W. *The NURBS Book*, 2nd ed.; Springer: New York, NY, USA, 1997; ISBN 3-540-61545-8.

2

Comparative Study of Dynamic Stall under Pitch Oscillation and Oscillating Freestream on Wind Turbine Airfoil and Blade

Chengyong Zhu and Tongguang Wang *

Jiangsu Key Laboratory of Hi-Tech Research for Wind Turbine Design, Nanjing University of Aeronautics and Astronautics, Nanjing 210016, China; rejoicezcy@nuaa.edu.cn
* Correspondence: tgwang@nuaa.edu.cn

Abstract: This study aims to assess the dynamic stall of the wind turbine blade undergoing pitch oscillation (PO) and oscillating freestream (OF), respectively. Firstly, a thin-airfoil theoretical analysis was performed to differentiate between these two dynamic effects. During upstroke, PO results in a positive effective airfoil camber, while OF has an additional negative effective airfoil camber, and yet in contrast during downstroke, PO decreases the effective camber, while OF increases the effective camber. Secondly, the equivalence relation between PO and OF is investigated by numerically solving the unsteady Reynolds-averaged Navier-Stokes equations. The difference between PO and OF mainly exists in the linear part of the aerodynamic loads. Because the difference is great at high reduced frequencies or angle of attack (AOA) amplitudes, PO and OF should be treated separately for dynamic stall from different aerodynamic sources. Thirdly, the Beddoes-Leishman dynamic model coupled with Bak's rotational stall delay model was used to predict the yawed responses of the blade section. The obtained results show different aerodynamic responses between PO and OF, although consideration of rotational augmentation can greatly improve the accuracy of the lift and drag coefficients. To improve the understanding and coupling modeling of rotational augmentation and dynamic stall, an extended analysis of the coupled effect was performed as well.

Keywords: dynamic stall; pitch oscillation; oscillating freestream; rotational augmentation; wind turbine

---

1. Introductions

Horizontal axis wind turbines (HAWTs) often experience unsteady air loads, to which wind turbine failures, reduced machine life, and increased operating maintenance are all directly linked. However, deep understanding and accurate prediction of the unsteady blade air loads and rotor performance face many challenges [1], among which 'dynamic stall' is of particular significance. Dynamic stall is characterized by the shedding and passage of a strong vortical disturbance over the suction surface, inducing a highly nonlinear fluctuating pressure field [2]. The unsteady effects on HAWTs can be classified into two categories according to different aerodynamic sources. One is blade motion, such as blade pitching, elastic bending, and flapping. The other is varying flow field structure induced by atmospheric turbulence, wind shear, wind gust, yawed inflow, tower shadow, interactions among rotors in a wind farm, and so on. These sources ultimately act on the local angle of attack (AOA) and velocity field at the blade-section level, hence producing unsteady loads on the blades.

In the 1980s, McCroskey [2] and Carr [3] conducted a comprehensive review on the main physical features of dynamic stall of helicopter rotors. Then, Butterfield [4] verified the great effect of dynamic stall on HAWT performance. In order to consider dynamic stall in engineering, several empirical and semi-empirical models have been developed for rotor air load prediction and rotor design analysis. These models often contain simplified representation of the essential physics using sets of linear and non-linear equations for the lift, drag, and pitching moment [5–7]. Leishman and Beddoes [6] developed the well-known Beddoes-Leishman (B-L) model based on proper considerations of flow mechanisms. An apparent advantage of the B-L model is that it yields good results with relatively few empirical coefficients, most of which can be derived from the static airfoil data. Despite the original application to helicopters (compressible flow), the B-L model has been introduced into wind turbines with several modifications [8–11]. Another promising method is the computational fluid dynamics (CFD) simulation achieved by numerically solving the Navier-Stokes equations with a suitable turbulence model. CFD methods have a very good capability of predicting both two-dimensional (2D) and three-dimensional (3D) dynamic stall events [12,13]. Ekaterinaris and Platzer [14] provided a comprehensive review on the computational prediction of dynamic stall. For Reynolds-averaged Navier-Stokes simulations (RANS), they found that the SST k-ω eddy viscosity model [15] performs better than any other two-equation models. However, their results are not yet satisfactory in the stalled regime and during flow reattachment. In this regard, they further found that the accurate prediction of transition from a laminar-to-turbulent boundary layer is a key element of improved hysteresis predictions.

As is well known, most of the information concerning dynamic stall is obtained by sinusoidal pitch oscillation about the quarter-chord axis, because this motion is considered to be adequate to represent the characteristics of dynamic stall. However, other types of motion, such as plunging oscillation, fore-aft motion, and oscillating freestream, can make a big difference in the dynamic-stall process [3,16]. In terms of relative motion, the blade motion and flow field variation may be further subdivided into the following three pairs (Figure 1):

- Firstly, the blade sections experience a time-varying tangential velocity (Figure 1a) under yawed inflow or in-plane gusts; as a counterpart, the blade sections undergo a fore-aft motion with edgewise vibration (Figure 1b).
- Secondly, unsteady inflow will cause a time-varying velocity component perpendicular to the rotor plane (Figure 1c); as a counterpart, the blade sections undergo a plunge motion during flapwise vibration (Figure 1d).
- Thirdly, a periodic AOA change (i.e., oscillating freestream, OF) results from the superposition of rotational velocity and the in-plane freestream velocity component (Figure 1e) under yawed inflow; as a counterpart, pitch oscillation (PO) occurs and affects the effective AOA under blade pitching or elastic bending (Figure 1f).

Van der Wall and Leishman [16] clarified the difference in the nature of the first pair, time-varying tangential velocity and fore-aft motion; the latter leads to a uniform perturbation velocity across the airfoil chord, while the former produces a perturbation velocity gradient across the chord. When the reduced frequency is low, both of them may be considered almost identical, which, however, is invalid for high reduced frequencies. Similarly, it can be speculated that time-varying freestream velocity will bring about a different effect on perturbation velocity from that of the plunging motion. For the third pair, Karbasian et al. [17] stated that OF can be referred to as PO. However, the equivalence relation between PO and OF still remains open from the point of view of Van der Wall and Leishman [16]. Practically, PO is generally used to conduct experimental studies and establish the engineering models. In order to improve the wind turbine unsteady aerodynamic prediction, it is necessary to clarify the relation between PO and OF.

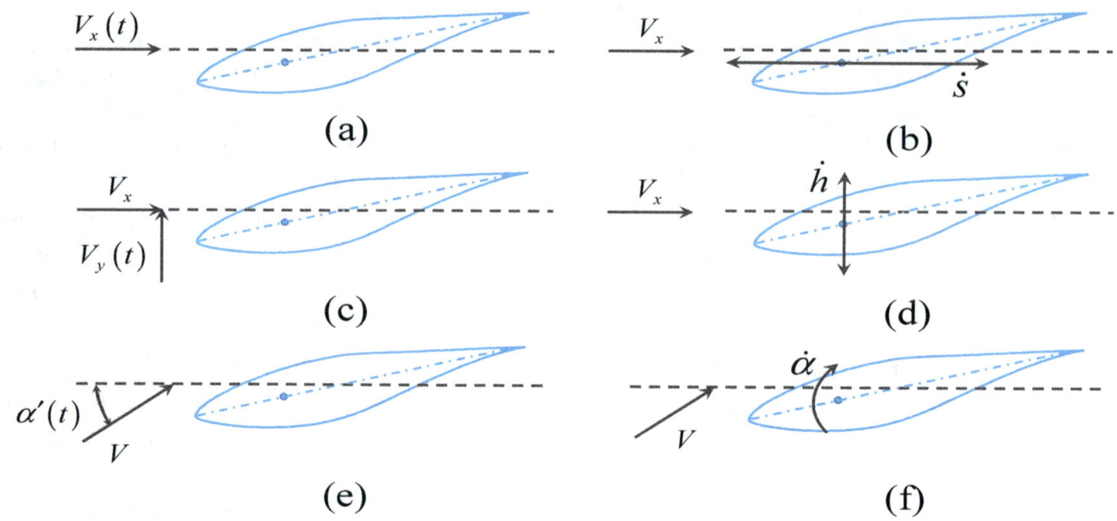

Figure 1. Schematic of different unsteady aerodynamic sources from a blade-section view. The left comes from varying flow field structures, and the right comes from the blade motion. The dashed line denotes the rotor plane. (**a**) Time-varying tangential velocity; (**b**) Fore-aft motion; (**c**) Time-varying normal velocity; (**d**) Plunge motion; (**e**) Oscillating freestream; (**f**) Pitch oscillation.

To investigate the dynamic stall of the blade sections under PO and OF, the problem of dynamic stall must be considered as fully 3D due to the rotational effect [1]. During the last few decades, both dynamic stall and rotational effects have captured significant attention but have been studied independently. The rotational effects play an important role in the boundary layer development due to the centrifugal and Coriolis forces. The centrifugal force results in radial flow, thinning the boundary layer of the blade. On the other hand, the Coriolis force adds a chordwise positive pressure gradient, flattening the separated zone and delaying the flow separation. Because the gross output power and sectional lift are clearly elevated with flow separation delayed, the rotational effect is also called a rotational augmentation. Several stall delay models have been developed to correct the 2D airfoil data by Snel et al. [18], Chaviaropoulos and Hansen [19], Du and Selig [20], Raj [21], Bak et al. [22], Corrigan and Schillings [23], and Lindenburg [24]. Breton et al. [25] evaluated these existing models on the Phase VI rotor with a prescribed wake vortex code and discussed their deficiencies and strengths. The 3D rotational effect is relatively strong on small-scaled stall-controlled HAWTs, and almost all of the correction models are established on the studies of stall-controlled HAWT aerodynamics. However, Zahle et al. [26] and Bangga et al. [27] noted that rotational augmentation is still prominent in the root region of a 10 MW modern pitch-controlled HAWT. Bangga et al. [28] also observed the unsteady effects on rotational augmented blade loads under a turbulent inflow.

A comparative analysis of the dynamic stall under PO and OF on the NREL S809 airfoil [29] and the Phase VI rotor [30] is presented in this paper. Firstly, the quasi-steady thin-airfoil theory is used to clarify the difference between PO and OF in the induced perturbation velocity across the chord. Secondly, the unsteady RANS method is used to assess the equivalence relation between PO and OF. The turbulence is modeled by the SST k-ω eddy viscosity model [15] incorporated with the γ-Re_θ transition model [31]. Thirdly, the classical B-L model is used to predict the yawed response of the Phase VI rotor under PO and OF. Bak's rotational stall delay model is incorporated into the dynamic stall model to consider rotational augmentation. The inverse BEM method is used to obtain the sectional AOAs and relative velocities from the experimental data. These extracted AOAs and velocities are then used to isolate the 3D rotational effect from the 3D unsteady aerodynamic data. All the predicted results agree well with the corresponding experimental data. It is found that PO and OF have an opposite effect on the effective airfoil camber. Therefore, it is necessary to separately consider the effects of PO and OF.

2. Methodology

The research objects are the Phase VI rotor and its basic airfoil, the NREL S809 airfoil. Here, PO denotes the sinusoidal pitch of the airfoil about the quarter-chord at a constant freestream velocity (Figure 1f), while OF denotes the sinusoidal change in the direction of inflow velocity at a constant magnitude to the stationary airfoil (Figure 1e). What is satisfied for both PO and OF is the same instantaneous angle between the freestream velocity and the chord as follows:

$$\alpha(\mathrm{t}) = \alpha_m + A\sin(2\pi f t) \tag{1}$$

where α_m, A, and f are the mean AOA, the AOA amplitude, and the frequency of oscillation, respectively. The reduced frequency is often defined as $k = \pi fc/V$ with V being the freestream velocity and c the chord length.

2.1. Thin-Airfoil Theoretical Analysis

Allowing for a simple analysis of the flow mechanism, the airfoil is simplified as a flat plate with its camber and thickness neglected here, and then, a quantitative comparison can be made with the quasi-steady thin-airfoil theory. The governing equation for thin-airfoil theory is Laplace's equation. The idea is to place a vortex sheet singularity of unknown strength on the airfoil [32]. The strength of this vortex sheet is determined by satisfying the flow tangency on the camber line in conjunction with the Kutta condition at the trailing edge. Formulated in Fourier series, the general form of pressure distribution on the airfoil is as follows:

$$\Delta C_p(\alpha,\theta) = 4\left[A_0\left(\frac{1+\cos\theta}{\sin\theta}\right) + \sum_{n=1}^{\infty} A_n \sin n\theta\right] \tag{2}$$

where α is the AOA, and $\theta = \cos^{-1}(1 - 2x/c)$ (x is the chordwise location with the leading edge denoted by $x = 0$). The coefficients A_0 and A_n are given by following equations:

$$\begin{aligned} A_0 &= \alpha - \tfrac{1}{\pi}\int_0^{\pi} \tfrac{\mathrm{d}y}{\mathrm{d}x}\mathrm{d}\theta \\ A_n &= \tfrac{2}{\pi}\int_0^{\pi} \tfrac{\mathrm{d}y}{\mathrm{d}x}\cos n\theta\,\mathrm{d}\theta \end{aligned} \tag{3}$$

where $y(x)$ describes the camber line of the airfoil. Integrating Equation (2) along the chord, we can obtain the lift coefficient, $C_l = 2\pi(A_0 + 0.5A_1)$, and the pitching moment coefficient about the quarter-chord axis, $C_{m0.25} = -0.25\pi(A_1 - A_2)$. For PO, the pitch-rate term produces a linear variation in normal perturbation velocity [5], $w(x) = -\dot{\alpha}(x - 0.25c)$ (where $\dot{\alpha}$ is the pitch rate and positive during upstroke) so that the induced camber is a parabolic arc (Figure 2). For OF, the normal perturbation velocity can be formulated as $w(x) = V\alpha(\mathrm{t}) - V\alpha(t-x/V)$. Because the quasi-steady condition means a low frequency, this formula can be linearized as $w(x) = \dot{\alpha}x$. Then, the term of $\mathrm{d}y/\mathrm{d}x$ in Equation (3) can be replaced with $w(x)/V$. Hence, the lift coefficients are given by $C_l = 2\pi(\alpha + \dot{\alpha}c/2V)$ under PO and $C_l = 2\pi(\alpha - 3\dot{\alpha}c/4V)$ under OF, indicating an additional effective AOA, $\alpha_{\mathrm{eq}} = \dot{\alpha}c/2V$ and $\alpha_{\mathrm{eq}} = -3\dot{\alpha}c/4V$, respectively.

In short, PO and OF have an opposite effect on the effective airfoil camber. During upstroke, PO increases the effective camber, while OF decreases the effective camber. During downstroke, PO decreases the effective camber, while OF increases the effective camber. Therefore, the lift under PO is higher during upstroke and lower during downstroke compared with the lift under OF. Notice that this effect is directly related to the pitch rate $\dot{\alpha} = 2\pi fA\cos(2\pi ft)$. Consequently, the difference between PO and OF is negligible at low variation frequencies and substantial at high ones.

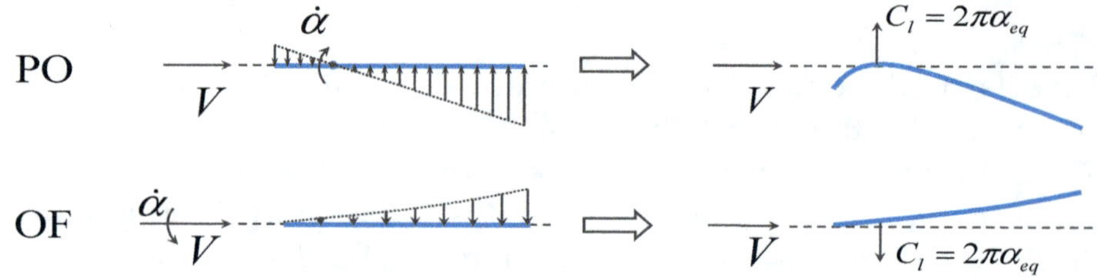

Figure 2. Induced effective camber for pitch oscillation (PO, upper) and oscillating freestream (OF, lower).

2.2. Numerical Modeling

For predictions of the dynamic stall, the commercial program ANSYS/FLUENT 16.0 (ANSYS, Inc., Canonsburg, PA, USA) [33] was used to solve the unsteady Reynolds-averaged Navier-Stokes equations. The mesh around airfoil geometry is generated in a structured O-type configuration to assure the wall orthogonality (Figure 3). The computational domain is exactly square. The first layer spacing off the wall is 10^{-5} c; this setting can ensure that the viscous sublayer is directly resolved with y^+ less than 1. There are 48 normal layers in the boundary layer mesh, with the growth rate being 1.08. After a proper grid independence study, the total size of the computational mesh is set to be 51,090 with 245 wrap-around points and 206 normal layers. A far-field distance of 20 c away from the airfoil is sufficient. The left, upper, and lower sides are all inlet boundary with velocity imposed. The right side is outlet boundary with pressure imposed (Figure 3). To simulate the dynamic stall under PO, the sliding-mesh method [33] was used. Therefore, the computational zone was divided into two subdomains, a rotating region (inner) and a stationary region (outer). These two regions interact by a circular sliding interface; this control circle of the sliding interface is 1.2 c in radius (Figure 3). To simulate the dynamic stall under OF, a user-defined function was implemented on the velocity inlet boundaries. The typical Reynolds number of 1×10^6 was selected in numerical simulations.

In order to obtain a good resolution, the third-order Monotone Upstream-Centered Schemes for Conservation Laws (MUSCL) convection scheme [33] is used for spatial discretization of the whole set of RANS and turbulence equations. This scheme was conceived of from the original MUSCL [34] by blending a central differencing scheme and a second-order upwind scheme. The second-order implicit scheme [33] is used for time differencing. Although it is theoretically valid that implicit methods are unconditionally stable with respect to the time step, nonlinearity effects would become prominent and oscillatory solutions may occur when the time step is increased. An operational cell convective courant number between 5 and 10 for viscous turbomachinery flows with an implicit scheme provides the best error damping properties [35]. On these bases, the number of inner iterations per time step is chosen to be 40 in this study to ensure the cell convective courant number is less than 10.

The pressure-based Coupled algorithm [33] is employed to handle the pressure-velocity coupling. Originally, the PISO algorithm [33] was preferable for transient problems, because it was derived from the SIMPLE algorithm [33] specifically for unsteady calculations. The PISO algorithm can yield accurate results with lower computational costs than the SIMPLE algorithm when time steps are sufficiently small. Both the SIMPLE and PISO algorithms, however, are known to converge slowly because the momentum equation and pressure-correction equation are solved separately. In the Coupled algorithm, the Navier-Stokes equations are directly solved through an implicit discretization of pressure in the momentum equations, with benefits in terms of robustness and convergence, especially with large time steps or with a poor-quality mesh. Balduzzi et al. [35] found that the obtained results with the Coupled algorithm are much closer to the experimental data than the SIMPLE and PISO algorithms. The lower sensitivity to the temporal discretization leads to the choice of the Coupled algorithm as the preferable and more robust formulation for the pressure-velocity coupling.

Because the incorporation of the transitional flow effects is a key element for an improved prediction of dynamic stall [14], the turbulence in the boundary layer is modeled by the four-equation SST k-ω eddy viscosity model [15] incorporated with the γ-Re$_\theta$ transition model [31].

2. Methodology

The research objects are the Phase VI rotor and its basic airfoil, the NREL S809 airfoil. Here, PO denotes the sinusoidal pitch of the airfoil about the quarter-chord at a constant freestream velocity (Figure 1f), while OF denotes the sinusoidal change in the direction of inflow velocity at a constant magnitude to the stationary airfoil (Figure 1e). What is satisfied for both PO and OF is the same instantaneous angle between the freestream velocity and the chord as follows:

$$\alpha(t) = \alpha_m + A\sin(2\pi f t) \tag{1}$$

where α_m, A, and f are the mean AOA, the AOA amplitude, and the frequency of oscillation, respectively. The reduced frequency is often defined as $k = \pi fc/V$ with V being the freestream velocity and c the chord length.

2.1. Thin-Airfoil Theoretical Analysis

Allowing for a simple analysis of the flow mechanism, the airfoil is simplified as a flat plate with its camber and thickness neglected here, and then, a quantitative comparison can be made with the quasi-steady thin-airfoil theory. The governing equation for thin-airfoil theory is Laplace's equation. The idea is to place a vortex sheet singularity of unknown strength on the airfoil [32]. The strength of this vortex sheet is determined by satisfying the flow tangency on the camber line in conjunction with the Kutta condition at the trailing edge. Formulated in Fourier series, the general form of pressure distribution on the airfoil is as follows:

$$\Delta C_p(\alpha,\theta) = 4\left[A_0\left(\frac{1+\cos\theta}{\sin\theta}\right) + \sum_{n=1}^{\infty} A_n \sin n\theta\right] \tag{2}$$

where α is the AOA, and $\theta = \cos^{-1}(1 - 2x/c)$ (x is the chordwise location with the leading edge denoted by $x = 0$). The coefficients A_0 and A_n are given by following equations:

$$\begin{aligned} A_0 &= \alpha - \tfrac{1}{\pi}\textstyle\int_0^{\pi} \frac{dy}{dx} d\theta \\ A_n &= \tfrac{2}{\pi}\textstyle\int_0^{\pi} \frac{dy}{dx} \cos n\theta \, d\theta \end{aligned} \tag{3}$$

where $y(x)$ describes the camber line of the airfoil. Integrating Equation (2) along the chord, we can obtain the lift coefficient, $C_l = 2\pi(A_0 + 0.5A_1)$, and the pitching moment coefficient about the quarter-chord axis, $C_{m0.25} = -0.25\pi(A_1 - A_2)$. For PO, the pitch-rate term produces a linear variation in normal perturbation velocity [5], $w(x) = -\dot{\alpha}(x - 0.25c)$ (where $\dot{\alpha}$ is the pitch rate and positive during upstroke) so that the induced camber is a parabolic arc (Figure 2). For OF, the normal perturbation velocity can be formulated as $w(x) = V\alpha(t) - V\alpha(t - x/V)$. Because the quasi-steady condition means a low frequency, this formula can be linearized as $w(x) = \dot{\alpha}x$. Then, the term of dy/dx in Equation (3) can be replaced with $w(x)/V$. Hence, the lift coefficients are given by $C_l = 2\pi(\alpha + \dot{\alpha}c/2V)$ under PO and $C_l = 2\pi(\alpha - 3\dot{\alpha}c/4V)$ under OF, indicating an additional effective AOA, $\alpha_{eq} = \dot{\alpha}c/2V$ and $\alpha_{eq} = -3\dot{\alpha}c/4V$, respectively.

In short, PO and OF have an opposite effect on the effective airfoil camber. During upstroke, PO increases the effective camber, while OF decreases the effective camber. During downstroke, PO decreases the effective camber, while OF increases the effective camber. Therefore, the lift under PO is higher during upstroke and lower during downstroke compared with the lift under OF. Notice that this effect is directly related to the pitch rate $\dot{\alpha} = 2\pi fA\cos(2\pi ft)$. Consequently, the difference between PO and OF is negligible at low variation frequencies and substantial at high ones.

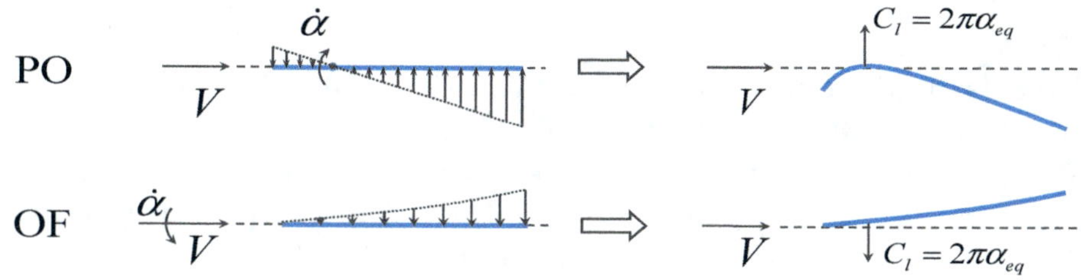

Figure 2. Induced effective camber for pitch oscillation (PO, upper) and oscillating freestream (OF, lower).

2.2. Numerical Modeling

For predictions of the dynamic stall, the commercial program ANSYS/FLUENT 16.0 (ANSYS, Inc., Canonsburg, PA, USA) [33] was used to solve the unsteady Reynolds-averaged Navier-Stokes equations. The mesh around airfoil geometry is generated in a structured O-type configuration to assure the wall orthogonality (Figure 3). The computational domain is exactly square. The first layer spacing off the wall is 10^{-5} c; this setting can ensure that the viscous sublayer is directly resolved with y^+ less than 1. There are 48 normal layers in the boundary layer mesh, with the growth rate being 1.08. After a proper grid independence study, the total size of the computational mesh is set to be 51,090 with 245 wrap-around points and 206 normal layers. A far-field distance of 20 c away from the airfoil is sufficient. The left, upper, and lower sides are all inlet boundary with velocity imposed. The right side is outlet boundary with pressure imposed (Figure 3). To simulate the dynamic stall under PO, the sliding-mesh method [33] was used. Therefore, the computational zone was divided into two subdomains, a rotating region (inner) and a stationary region (outer). These two regions interact by a circular sliding interface; this control circle of the sliding interface is 1.2 c in radius (Figure 3). To simulate the dynamic stall under OF, a user-defined function was implemented on the velocity inlet boundaries. The typical Reynolds number of 1×10^6 was selected in numerical simulations.

In order to obtain a good resolution, the third-order Monotone Upstream-Centered Schemes for Conservation Laws (MUSCL) convection scheme [33] is used for spatial discretization of the whole set of RANS and turbulence equations. This scheme was conceived of from the original MUSCL [34] by blending a central differencing scheme and a second-order upwind scheme. The second-order implicit scheme [33] is used for time differencing. Although it is theoretically valid that implicit methods are unconditionally stable with respect to the time step, nonlinearity effects would become prominent and oscillatory solutions may occur when the time step is increased. An operational cell convective courant number between 5 and 10 for viscous turbomachinery flows with an implicit scheme provides the best error damping properties [35]. On these bases, the number of inner iterations per time step is chosen to be 40 in this study to ensure the cell convective courant number is less than 10.

The pressure-based Coupled algorithm [33] is employed to handle the pressure-velocity coupling. Originally, the PISO algorithm [33] was preferable for transient problems, because it was derived from the SIMPLE algorithm [33] specifically for unsteady calculations. The PISO algorithm can yield accurate results with lower computational costs than the SIMPLE algorithm when time steps are sufficiently small. Both the SIMPLE and PISO algorithms, however, are known to converge slowly because the momentum equation and pressure-correction equation are solved separately. In the Coupled algorithm, the Navier-Stokes equations are directly solved through an implicit discretization of pressure in the momentum equations, with benefits in terms of robustness and convergence, especially with large time steps or with a poor-quality mesh. Balduzzi et al. [35] found that the obtained results with the Coupled algorithm are much closer to the experimental data than the SIMPLE and PISO algorithms. The lower sensitivity to the temporal discretization leads to the choice of the Coupled algorithm as the preferable and more robust formulation for the pressure-velocity coupling.

Because the incorporation of the transitional flow effects is a key element for an improved prediction of dynamic stall [14], the turbulence in the boundary layer is modeled by the four-equation SST k-ω eddy viscosity model [15] incorporated with the γ-Re_θ transition model [31].

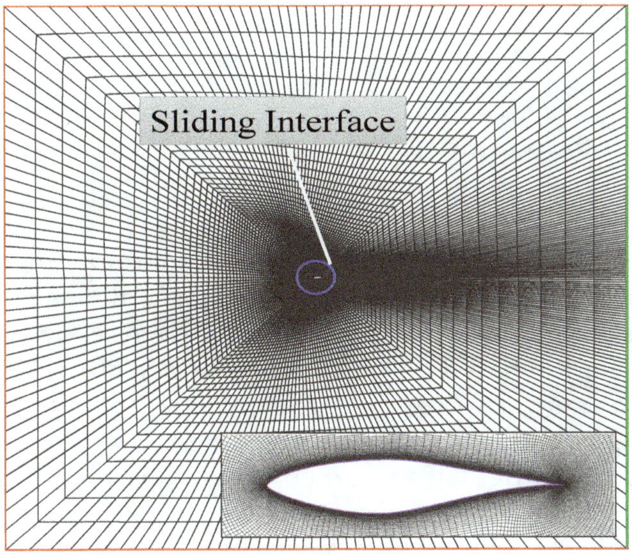

Figure 3. Computational mesh and boundary layer grids. The red line denotes the velocity inlet boundary, and the green line denotes the pressure outlet boundary.

2.3. *Beddoes-Leishman Semi-Empirical Model*

The classical B-L dynamic stall model [6] is capable of effectively assessing the unsteady loads with the flow physical features considered, and therefore, is fully implemented in the present work, including the leading-edge separation effect. Pereira et al. [10] have proven that it is unrealistic to disregard the leading-edge separation, because the wind turbine airfoils are thick. The main features and formulae of the B-L dynamic stall model may be summarized as follows.

2.3.1. Unsteady Attached Flow

Unsteady effects in attached flow conditions are simulated by the superposition of indicial aerodynamic responses for time step changes in the AOA. For a step change in the AOA, the normal force and pitching moment coefficients can be written as follows:

$$\begin{aligned} \Delta C_n &= C_n^c + C_n^{nc} \\ \Delta C_m &= C_m^c + C_m^{nc} \end{aligned} \tag{4}$$

where the superscripts c and nc denote the circulatory part and non-circulatory part of the aerodynamic loads, respectively. The indicial response can be approximated empirically in terms of the exponential function. Hence, the circulatory loading is as follows:

$$C_n^c = C_{n\alpha}\alpha_e^c = C_{n\alpha}(\alpha(s) - X(s) - Y(s)) \tag{5}$$

where $C_{n\alpha}$ is the normal force slope, and the non-dimensional time s represents the distance travelled at the relative velocity by the airfoil in semi-chords; that is, $s = 2Vt/c$. $X(s)$ and $Y(s)$ are deficiency functions written in a special finite difference approximation to Duhamel's integral as given by the following:

$$\begin{aligned} X(s) &= X(s - \Delta s)\exp(-b_1\Delta s) + A_1\Delta\alpha\exp(-b_1\Delta s/2) \\ Y(s) &= Y(s - \Delta s)\exp(-b_2\Delta s) + A_2\Delta\alpha\exp(-b_2\Delta s/2) \end{aligned} \tag{6}$$

Here, A_1, A_2, b_1, and b_2 are the coefficients of the indicial functions and are a function of the airfoil and the Mach number. Then, the circulatory pitching moment coefficient is given by the following:

$$C_m^c = (0.25 - x_{ac}/c)C_n^c \tag{7}$$

where x_{ac} is the position of the airfoil aerodynamic center measured from the leading edge. A similar recurrence relation is used for the non-circulatory part of the air loads, C_n^{nc} and C_m^{nc} (see [6] for details).

2.3.2. Unsteady Separated Flow

The non-linear effects of the trailing-edge separation are implemented using a Kirchhoff flow model. The movement of the unsteady flow separation point position is related to the static separation position via a deficiency function. The onset of the leading-edge separation, and hence the dynamic stall, is identified using a criterion based on the attainment of a critical local leading-edge pressure, which is further related to the normal force.

Under unsteady conditions, there is a lag in the leading-edge pressure with the increasing AOA, which can be expressed as a first-order lag as given by the following:

$$C'_n = C_n^p - D_p \tag{8}$$

where

$$\begin{gathered} C_n^p = C_n^c + C_n^{nc} \\ D_p(s) = D_p(s - \Delta s)\exp(-\Delta s/T_p) + \left[C_n^p(s) - C_n^p(s - \Delta s)\right]\exp(-\Delta s/2T_p) \end{gathered} \tag{9}$$

The constant T_p can be determined empirically from unsteady airfoil data. The effective AOA $\alpha_f = C'_n/C_{n\alpha} + \alpha_0$ after incorporating the unsteady pressure response is used to obtain the effective flow separation point f'. In this study, f' is determined by look-up-table interpolation from the static airfoil data. The additional effects of a boundary layer response are considered as a first-order lag as given by the following:

$$f'' = f' - D_f \tag{10}$$

where

$$D_f(s) = D_f(s - \Delta s)\exp\left(-\Delta s/T_f\right) + [f'(s) - f'(s - \Delta s)]\exp\left(-\Delta s/2T_f\right). \tag{11}$$

The constant T_f can be determined empirically from unsteady airfoil data. Then, the non-linear normal force coefficient is obtained by the Kirchhoff flow theory

$$C_n^f = C_{n\alpha}\left(\frac{1 + \sqrt{f''}}{2}\right)^2(\alpha - \alpha_0) \tag{12}$$

where α_0 is the zero-lift AOA. Afterwards, the chordwise force coefficient is obtained by Gupta and Leishman's approximation [11] as given by the following:

$$C_t^f = \begin{cases} C_{n\alpha}\sqrt{f''}\alpha\sin\alpha & C'_n \le C_{n1} \\ K_1 + C_{n\alpha}\sqrt{f''}f''^{\Phi}\alpha\sin\alpha & C'_n > C_{n1} \end{cases} \tag{13}$$

where C_{n1} is the critical normal force value and the leading-edge separation is initiated when $C'_n > C_{n1}$. The parameter K_1 and Φ are obtained from the static test data [11]. The pitching moment coefficient is obtained by fitting the ratio C_m/C_n in a least-square method from the static airfoil data as given by the following:

$$C_m^f = C_{m0} + [k_0 + k_1(1 - f'') + k_2\sin(\pi f''^{\mu})]C_n^f \tag{14}$$

where C_{m0} is the zero-lift pitching moment coefficient. The values of k_0, k_1, k_2, and μ can be adjusted for different airfoils.

2.3.3. Vortex Force

The induced vortex force and the associated pitching moment are represented empirically in a time-dependent manner during dynamic stall. The total accumulated vortex normal force coefficient under unsteady conditions is given by the following:

$$C_n^v = C_n^v(s-\Delta s)\exp(-\Delta s/T_v) + [C_v(s) - C_v(s-\Delta s)]\exp(-\Delta s/2T_v) \tag{15}$$

where

$$C_v(s) = C_n^c(s)\left[1 + \left(\frac{1+\sqrt{f''}}{2}\right)^2\right]. \tag{16}$$

T_v is the vortex time decay constant. The center of pressure on the airfoil also varies with the chordwise position of the shedding vortex and will achieve a maximum value when the vortex reaches the trailing edge after a non-dimensional time period T_{vl}. A general representation of the center of pressure behavior (aft of the quarter-chord) is empirically formulated as follows:

$$(CP)_v = 0.25\left(1 - \cos\frac{\pi T_v}{T_{vl}}\right) \tag{17}$$

where T_v denotes the non-dimensional vortex time, $0 \le T_v \le 2T_{vl}$ (i.e., $T_v = 0$ at the onset of dynamic stall and $T_v = T_{vl}$ when the vortex reaches the trailing edge). Thus, the increment in the pitching moment about the quarter-chord due to dynamic stall is given by the following:

$$C_m^v = -(CP)_v \cdot C_n^v. \tag{18}$$

The total unsteady air loads on the airfoil is then obtained by superposition as follows:

$$\begin{aligned} C_n &= C_n^f + C_n^{nc} + C_n^v \\ C_t &= C_t^f + (C_n^{nc} + C_n^v)\sin\alpha_e^c \\ C_m &= C_m^c + C_m^{nc} + C_m^f + C_m^v \end{aligned} \tag{19}$$

Apart from dynamic stall, rotational augmentation is of crucial importance for wind turbine aerodynamic performance prediction. The rotational stall delay model of Bak et al. [22] is implemented. This model was developed on the difference in chordwise pressure distribution between the rotating blade section and the static airfoil. Although the obtained lift and drag coefficients were generally good, the modeled pitching moment coefficient indicated a poor agreement with the measurements, because the center of pressure was often underpredicted at large AOAs. Thus, a simple modification that the integrated location is changed from 0.25 to 0.1 was made to the pitching moment coefficient when fully separated flow occurs. Figure 4 implies an acceptable performance of this modification. It is generally acknowledged that the rotational augmentation is relatively apparent in separated flow; therefore, rotational augmentation mainly affects the nonlinear air loads.

To incorporate the rotational stall delay model into the dynamic stall model, the static airfoil data are corrected by Bak's model firstly. Then, the static separation point is obtained from the Kirchhoff formulation (Equation (12)). The dynamic separation point is calculated by a first-order lag of the static separation point (Equation (10)), while the nonlinear normal force is reconstructed by Equation (12). Afterwards, the chordal force coefficient is obtained by Equation (13), with an increment added to consider the rotational augmentation. Finally, the nonlinear pitching moment coefficient is obtained by Equation (14), and a rotational increment is added as well. According to the recommendations of Pereira et al. [10] and Gupa et al. [11], the set of corresponding empirical time constants, T_p, T_f, T_v, T_{vl}, is as follows: 1.7, 8.0, 1.0, and 7.0, where T_p and T_f are related to the first-order lag of the leading-edge separation and trailing-edge separation, respectively, and both T_v and T_{vl} are the non-dimensional time constants related to the movement of the dynamic vortex.

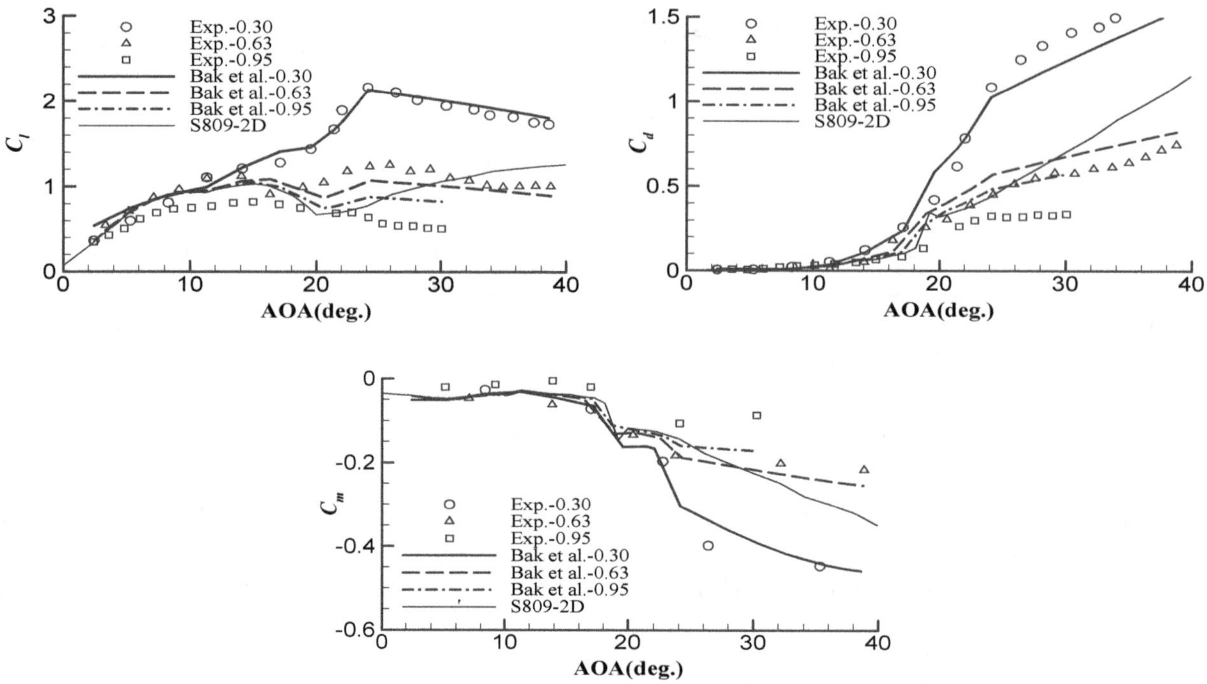

Figure 4. Three-dimensional (3D) corrected airfoil characteristics of different radial locations compared to two-dimensional (2D) measurements and measured 3D data.

The difference between PO and OF is addressed in the circulatory part of unsteady linear air loads, based on the thin-airfoil theory. To obtain input conditions, the sectional velocity components are extracted from the yawed measurements with the inverse BEM method [22,24,36–38]. This method has widely been used to determine the local AOA with the pre-determined local forces used to calculate the local induction factors and hence the induced velocities. In the inverse BEM method, Prandtl's tip correction is used to consider the tip loss effect.

3. Results and Discussion

3.1. Time-Varying Sectional Incident Velocity under Yawed Inflow

Schepers [39] provided a detailed discussion on two aerodynamic effects under yawed inflow. One is the skewed wake effect, inducing an unbalanced axial velocity, producing asymmetric air loads, and then resulting in yawing stability. The other is the advancing and retreating blade effect, making the in-plane tangential velocity vary with the azimuth. Figure 5 provides the definitions of yaw angle and azimuth angle. Here, the blade azimuth is defined as zero at the 6-o'clock position, where the rotor is assumed to rotate clockwise. Figure 6 indicates the variations of the local AOA and relative velocity at different blade spanwise locations. These variations can be approximately formulated in trigonometric functions due to the rotational periodicity. It can be found that the spanwise location strongly influences the mean AOA and the AOA amplitude; the highest unsteadiness can be observed on the inboard blade despite a low velocity magnitude. Therefore, dynamic stall is often severe at inboard sections, where the rotational augmentation has manifested.

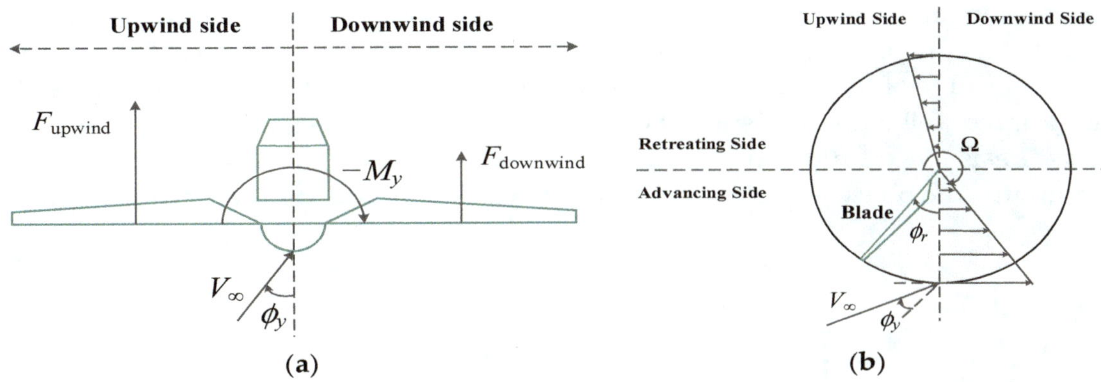

Figure 5. Schematic of a wind turbine in the yawed condition. (**a**) Top view; (**b**) Front view.

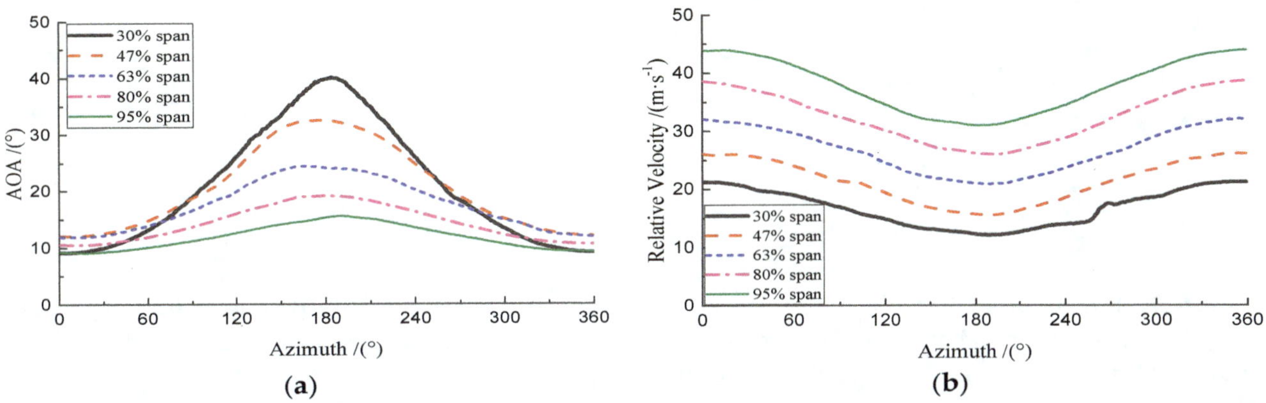

Figure 6. The variation of the angle of attack (AOA) and relative velocity with the azimuth at different spans. Wind speed is 13 m/s, and yaw angle is 30°. (**a**) AOA; (**b**) Relative velocity.

3.2. Validations of Numerical Modeling Methods

The pitch-oscillation case in the deep stall regime is selected for analyses of the grid independence and time step independence, where $\alpha_m = 14°$, $A = 10°$, and $k = 0.078$. Different grid resolutions are obtained by changing the numbers of wrap-around points on the airfoil and normal layers; then, three types of grids are generated with total numbers of 37,850, 51,090, and 70,525, respectively. Three different time steps of 0.002 s, 0.001 s, and 0.0005 s are used with the steps per cycle being 273, 546, and 1092, respectively. However, there is no clear difference in the lift coefficient among the test cases. The streamwise velocity profile is used to estimate the capability of capturing flow structures; this velocity profile is at mid-chord at the phase angle of 90° (i.e., $\phi = 90°$, where $\phi = 2\pi ft$). The coarsen grid and large time step can yield clear deviation (Figure 7). Finally, the grid in the size of 51,090 and the time step of 0.001 s are chosen as the computational settings.

In light stall, the numerical results agree well with the experimental data [40] as shown in Figure 8. An improvement over Karbasian, Esfahani, and Barati's CFD results [17] can be observed due to the consideration of the transition effect. The B-L model shows a reasonable representation of unsteady aerodynamics. Because of a lack of measurements under OF, the CFD results are compared with those of Gharali et al. in the same conditions [41]. First and foremost, it is necessary to clarify the determination of the AOA under OF. In steady flow, the AOA is defined as the angle between the chord and freestream velocity, and this instantaneous AOA (denoted as 'inflow angle') is assured at same value for both motions in the computations. Actually, the true AOA (denoted as 'local angle') is different, because the relative velocity direction to the airfoil can be influenced by the freestream time lag and flow disturbance of the airfoil. To obtain the local angle and compare it with the inflow angle, we extract the unsteady relative velocity vector at the location of 3 c before the airfoil stagnation point (Figure 9). An apparent lag between the two types of AOAs can be observed during downstroke. The variation amplitude of the local angle becomes slightly larger. Both the trend and the shape of the hysteresis loop agree well with the results of Gharali et al. [41], as shown in Figure 10. A notable feature is that the hysteresis loop is more apparent with respect to the inflow angle than to the local angle. To avoid ambiguity, the phase angle will replace the AOA as an independent variable afterwards.

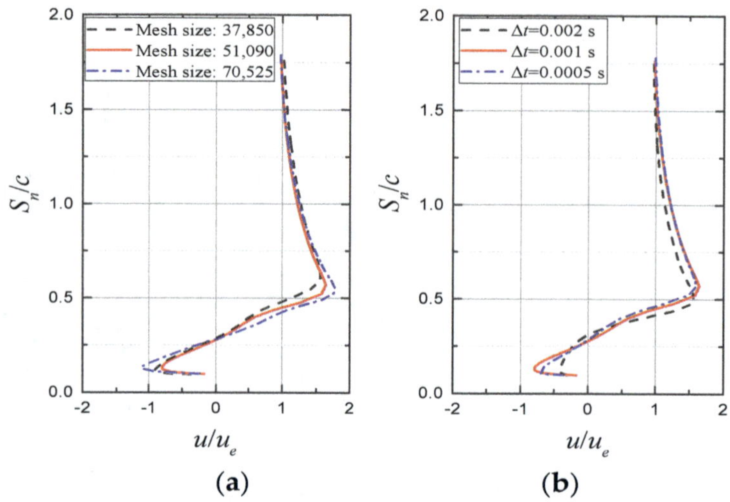

Figure 7. Effects of grid resolution and time steps on streamwise velocity profile at mid-chord at $\phi = 90°$. S_n is a normal distance away from the mid-chord, and u_e is the streamwise velocity of external flow. (**a**) Effect of grid resolution; (**b**) Effect of time steps.

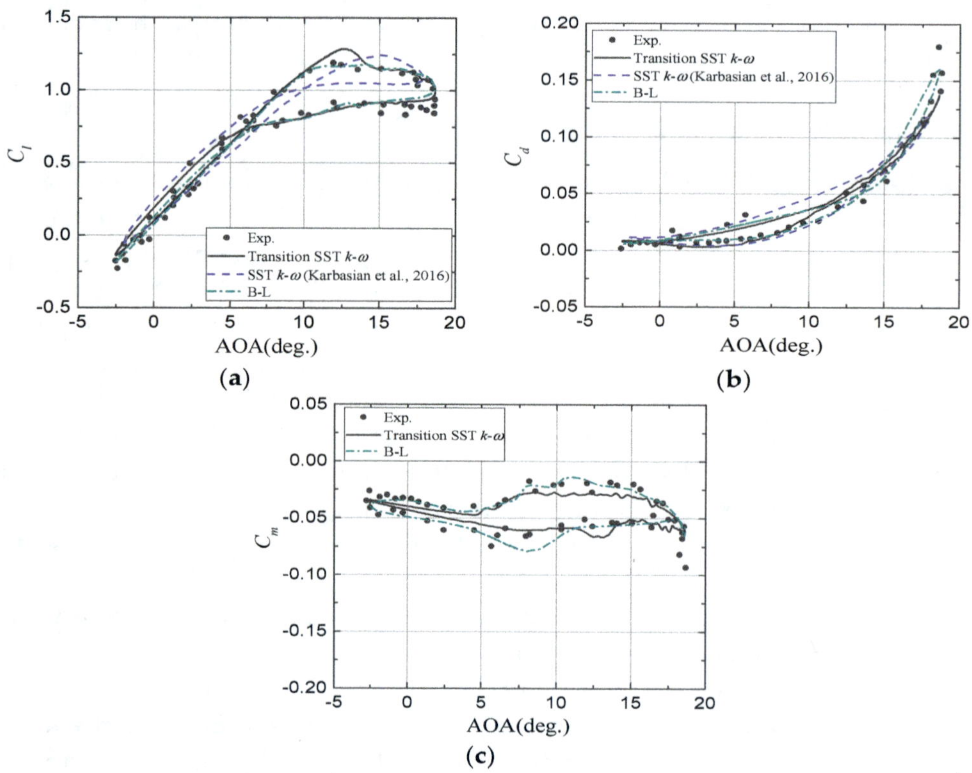

Figure 8. Comparison of aerodynamic coefficients under pitch oscillation. $\alpha_m = 8°$, $A = 10°$, and $k = 0.026$. (**a**) The lift coefficient; (**b**) The drag coefficient; (**c**) The pitching moment coefficient.

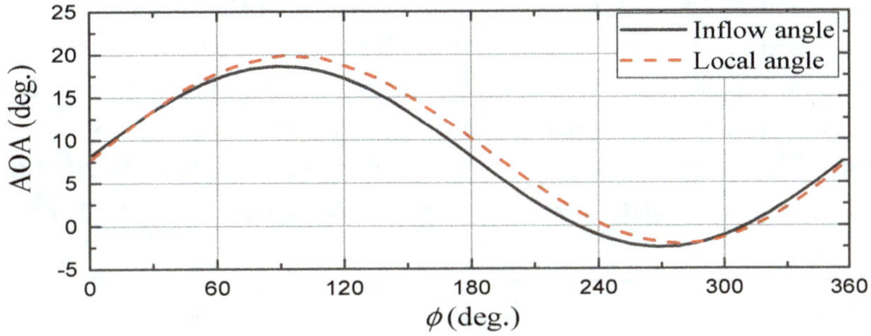

Figure 9. Comparison between the inflow angle and the local angle.

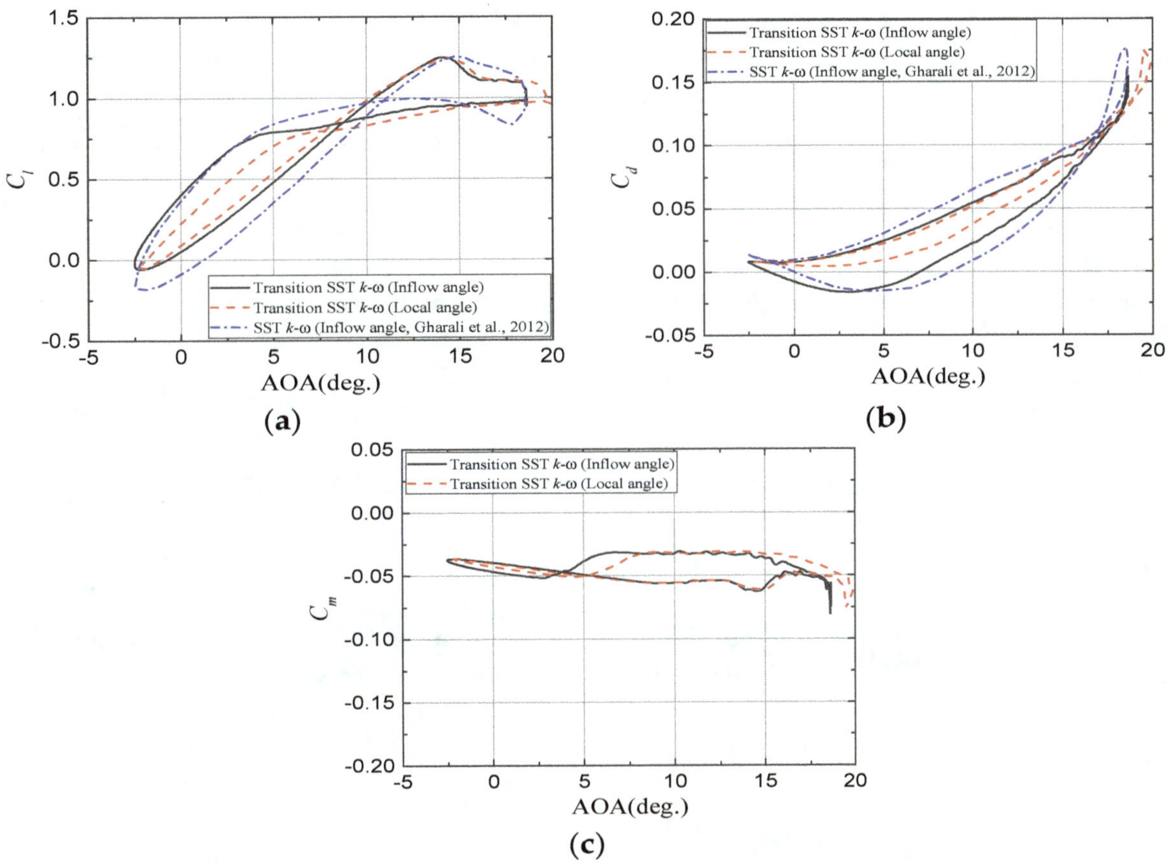

Figure 10. Comparison of aerodynamic coefficients under oscillating freestream. $\alpha_m = 8°$, $A = 10°$ and $k = 0.026$. (**a**) The lift coefficient; (**b**) The drag coefficient; (**c**) The pitching moment coefficient.

3.3. Equivalence Analysis between Pitch Oscillation and Oscillating Freestream

The aforementioned theoretical analysis indicates an opposite effect of PO and OF on the effective airfoil camber. Figure 11 shows the variations of the lift and the drag coefficients along one cycle. Notice that the difference in the nonlinear air loads (where ϕ is from 30° to 180°) is relatively negligible. A clear difference in the linear lift coefficient can be observed with a higher value under PO during upstroke and a higher value under OF during downstroke; the CFD results agree well with the thin-airfoil theory. From the vorticity contours in Figure 12, flow is attached to the airfoil and there is no obvious difference between the two motions from $\phi = 270°$ to $\phi = 360°$. When the AOA reaches the maximum at $\phi = 90°$, the flow separation is apparent under both PO and OF. The separated flow under PO is more persistent. In contrast, the vortex under OF rolls up and then sheds into wake. Despite a great difference in separated flow structure between the two motions, the associated aerodynamic loads do not show great differences. The possible reason is that the pressure distribution under PO possesses a sharp leading-edge suction peak, while the pressure distribution under OF experiences an elevated pressure on the suction side due to the separated vortex. These two factors may lead to an almost equal value in calculation. Figure 13 shows the unsteady pressure variations at different chordwise locations at $\alpha_m = 2°$, $A = 8°$, and $k = 0.078$. It is obvious that there is a stronger suction peak under PO than under OF. However, the pressure coefficient under OF decreases with a time lag.

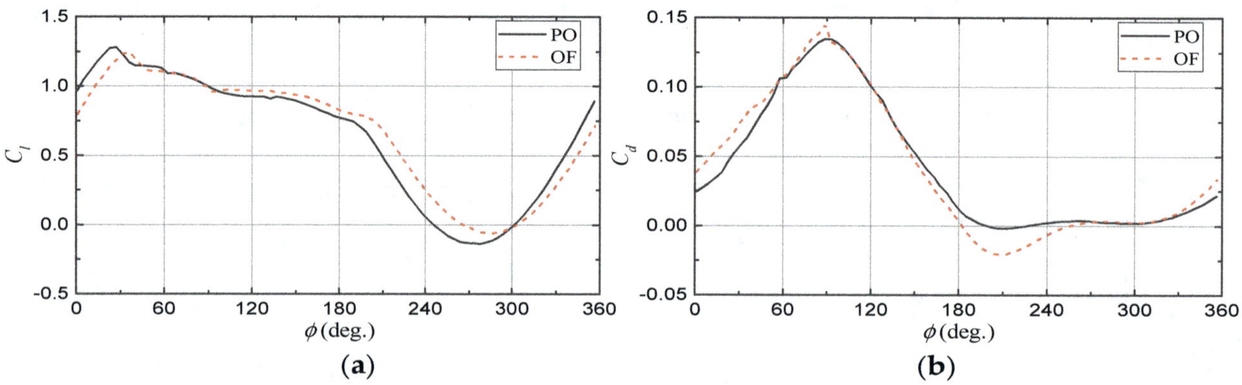

Figure 11. Comparison of lift coefficient and drag coefficient between pitch oscillation and oscillating freestream. $\alpha_m = 8°$, $A = 10°$, and $k = 0.026$. (**a**) The lift coefficient; (**b**) The drag coefficient.

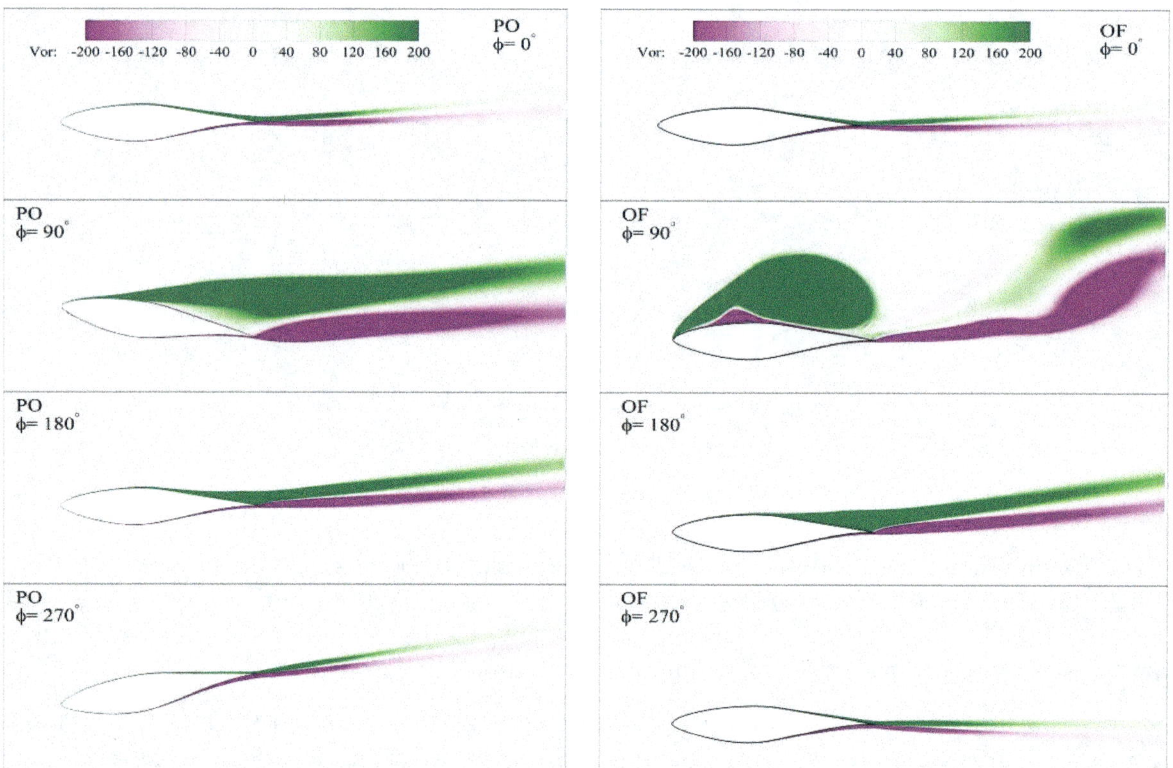

Figure 12. Vorticity contours of pitch oscillation and oscillating freestream at different phase angles. $\alpha_m = 8°$, $A = 10°$ and $k = 0.026$.

Because the difference between the two motions focuses on the linear part of the aerodynamic loads, the unsteady aerodynamic effect only at the low AOA is assessed. The mean AOA, AOA amplitude, and oscillation frequency are chosen as parameters to evaluate the difference. The deviation of the normal coefficient along one cycle between the two motions is defined as follows:

$$Dev. = \frac{max(|C_{n,PO}(\varnothing) - C_{n,OF}(\varnothing)|)}{C_{n,qs}} \tag{20}$$

where $C_{n,qs}$ is the quasi-steady normal coefficient at the mean AOA. Figure 14 indicates that the oscillation frequency f and the AOA amplitude A affect the deviation more strongly than the mean AOA. From the thin-airfoil theory, the key is the pitch rate $\dot{\alpha} = 2\pi f A\cos(2\pi f t)$. Because f and A are factors of the pitch rate, the deviation becomes great when f and A are high. Under yawed inflow, the oscillation frequency is directly determined by the rotor speed, and the AOA amplitude is seriously affected by the yaw angle. Consequently, it is necessary to consider the effects of PO and OF separately.

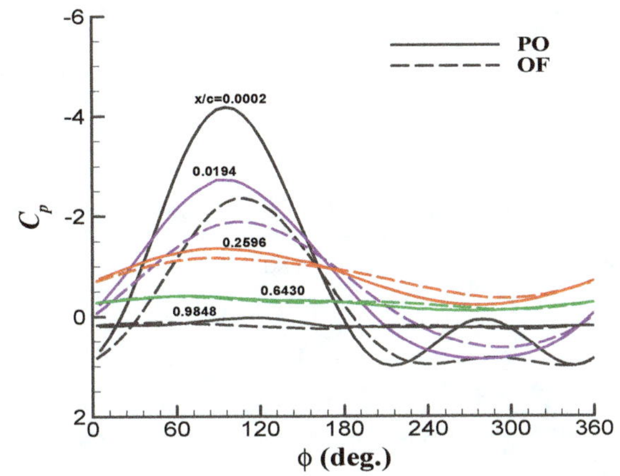

Figure 13. Comparison of unsteady pressures at different chordwise locations on the upper surface between pitch oscillation and oscillating freestream. $\alpha_m = 2°$, $A = 8°$, and $k = 0.078$. The same locations are in a same color.

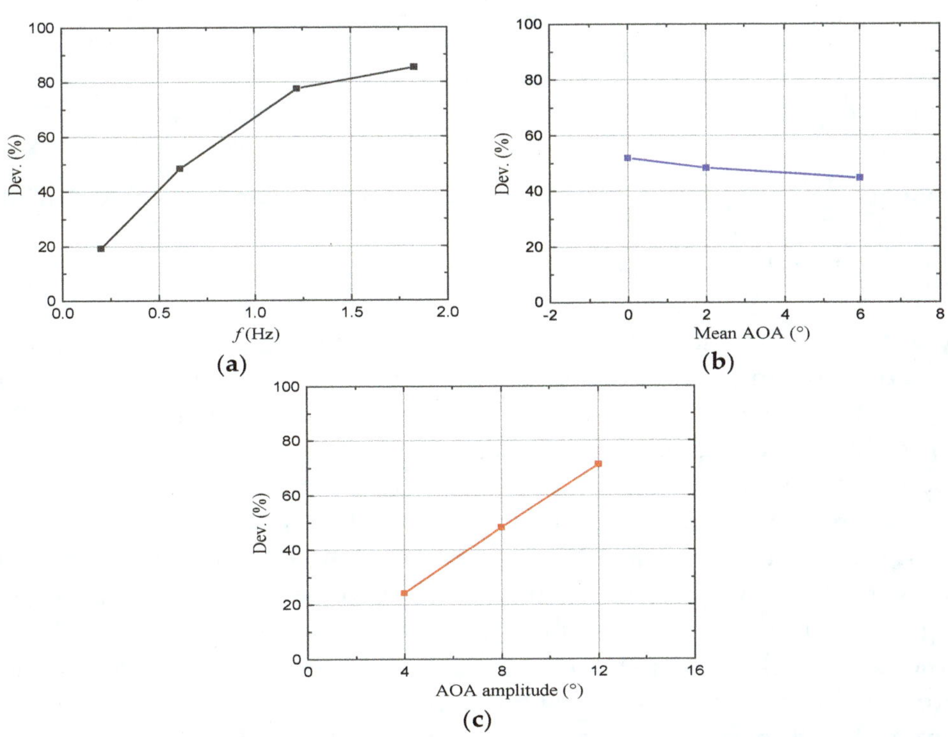

Figure 14. Effects of the oscillation frequency, the mean AOA and the AOA amplitude on the deviation between pitch oscillation and oscillating freestream. (**a**) Effect of the oscillation frequency; (**b**) Effect of the mean AOA; (**c**) Effect of the AOA amplitude.

3.4. Comparison of Dynamic Stall in Yawed Condition between the Two Motions

The 30% span of the Phase VI blade is investigated due to the manifest dynamic stall on the inboard blade. The wind speed is 10 m/s, and the yaw angle is 10° and 30°, respectively (the cases of S1000100 and S1000300 in NREL UAE Phase VI [30]). The B-L dynamic stall model incorporated with Bak's rotational stall delay model shows a good agreement with the experimental data for the yaw angle of 10°, as shown in Figure 15. A noticeable difference in both the lift coefficient and drag coefficient between the 2D and 3D results is observed, manifesting the 3D rotational effect.

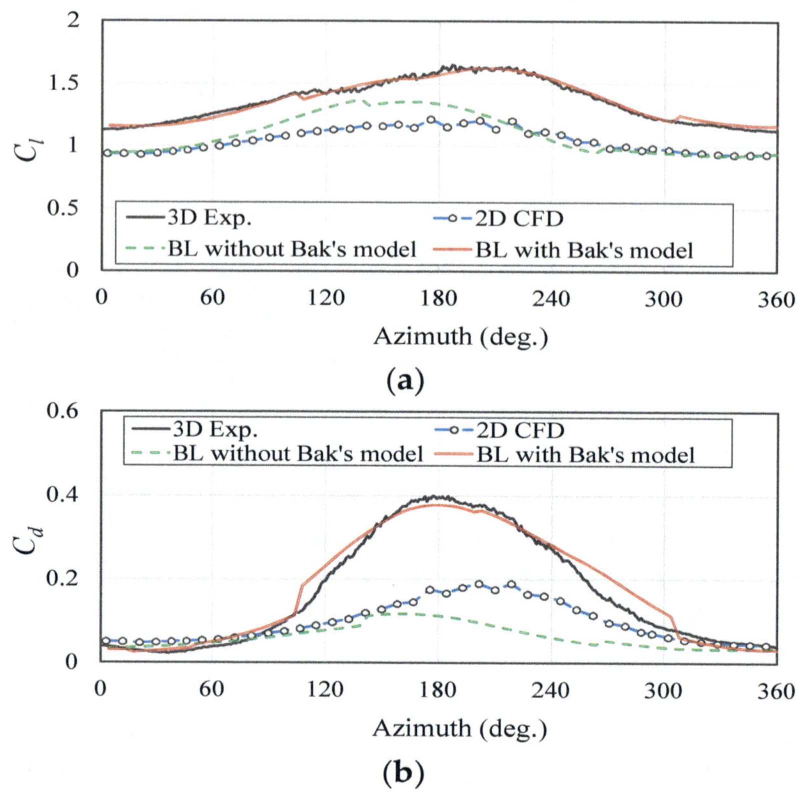

Figure 15. Validation of the corrected Beddoes-Leishman (B-L) dynamic stall model in lift coefficient and drag coefficient. The wind speed is 10 m/s, the yaw angle is 10°, and radial location is 30% span. (**a**) The lift coefficient; (**b**) The drag coefficient.

Figure 16 shows the aerodynamic force coefficients and pitching moment coefficient varying with azimuthal angle at the yaw angles of 10° and 30°, respectively. The calculated results are generally in a good agreement with the experimental data. When the yaw angle increases, the 3D dynamic stall becomes more intense with the high lift coefficient, but trends of air load variations are consistent. The difference between PO and OF is moderate. As aforementioned, in 2D cases, the aerodynamic coefficients under PO are higher during upstroke and lower during downstroke. A slight improvement can be observed in the prediction under OF; the possible reason is that dynamic stall under yawed inflow is essentially a velocity-varying problem without pitch motion and fore–aft motion. Notice that prediction of the pitching moment coefficients is quite poor in both amplitudes and trends. Given that the pitching moment depends not merely on the integrated normal force but also on the location of the center of pressure, it is somewhat not surprising that the predicted pitching moment is unsatisfactory under the dynamic condition, because 3D effects have an influence on the normal force and on the movement of the center of pressure, which depends on appropriate coupling of the dynamic stall and the 3D rotational stall delay models. This problem may be solved only if deep understanding and accurate modeling of both the dynamic stall and the 3D stall delay are achieved. Nevertheless, the results from the B-L dynamic stall model show a different effect between PO and OF when their different additional camber effects are included.

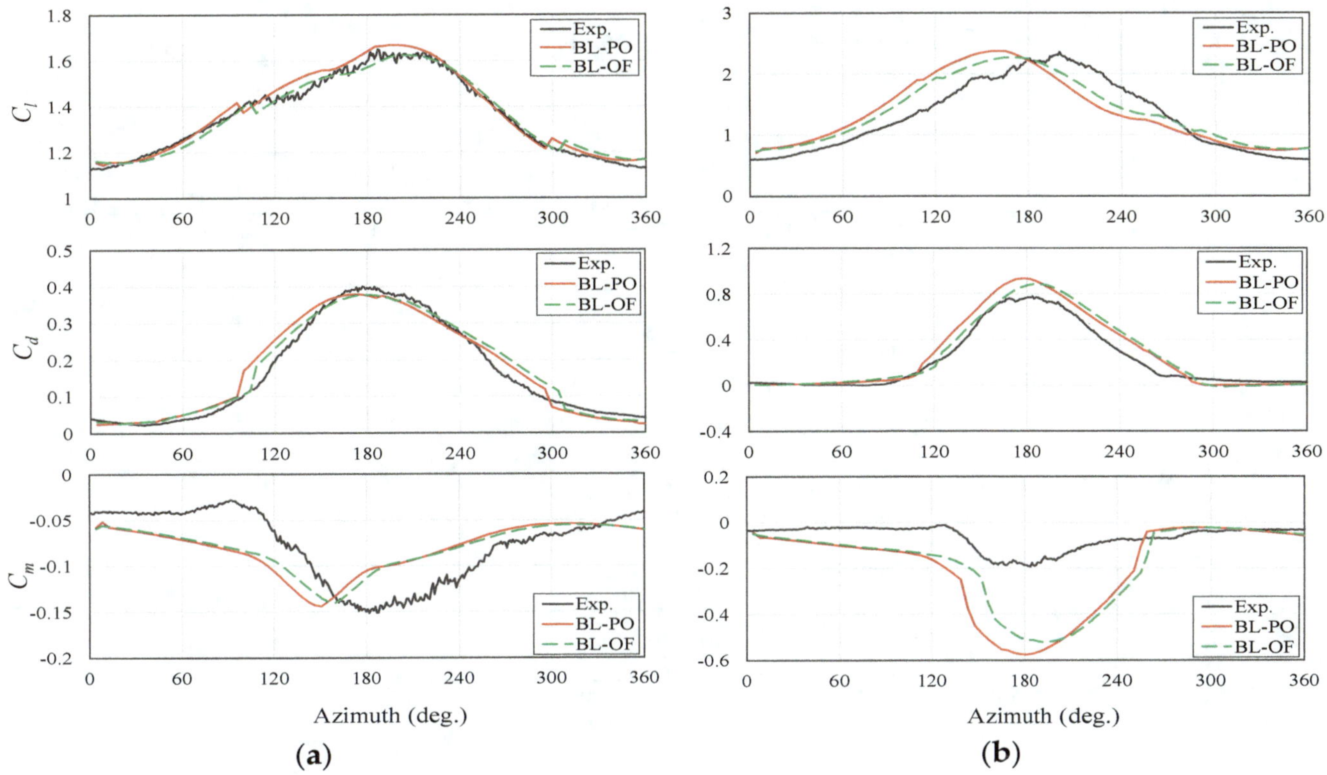

Figure 16. Effect of the yaw angle on the aerodynamic coefficients. The wind speed is 10 m/s, and the radial location is 30% span. (**a**) 10° yaw angle, and (**b**) 30° yaw angle.

3.5. Effect of Rotational Augmentation on the Dynamic Stall of the Inboard Blade

Because the rotational effect greatly suppresses the flow separation, the lift and nose-down pitching moment coefficients are increased. However, it is unclear whether the rotational stall delay effect under steady conditions is consistent with its performance under unsteady conditions. To provide an insight into the coupled effect of the rotational augmentation and dynamic stall, an extended analysis is conducted. As rotational augmentation is often isolated by making comparisons between 2D flow and 3D rotational flow [42], our strategy is to isolate the rotational augmentation by comparing the CFD results under an equivalent planer motion with yawed experimental data; the former contains only dynamic stall and the latter is coupled with rotational augmentation. The equivalent planer motion of a rotating blade section is obtained by the 2D time-varying incident velocity (2D TVIV, Figure 1a,c), including variations of the sectional AOA and the relative velocity magnitude.

Figure 17 shows the hysteresis loops of the aerodynamic coefficients at the 30% span. Because both the mean AOA and the AOA amplitude are low, the aerodynamic forces are generally attached to the unsteady linear part with almost no stall. Rotational augmentation apparently elevates the lift and drag coefficients, as under steady conditions. The value is even higher than the steady experimental data under axial inflow, meaning that the separation is more delayed in 3D rotational unsteady flow than in 3D rotational steady flow. Hence, both rotational augmentation and unsteady effects lead to a delayed flow separation and the elevated lift coefficient at a low yaw angle.

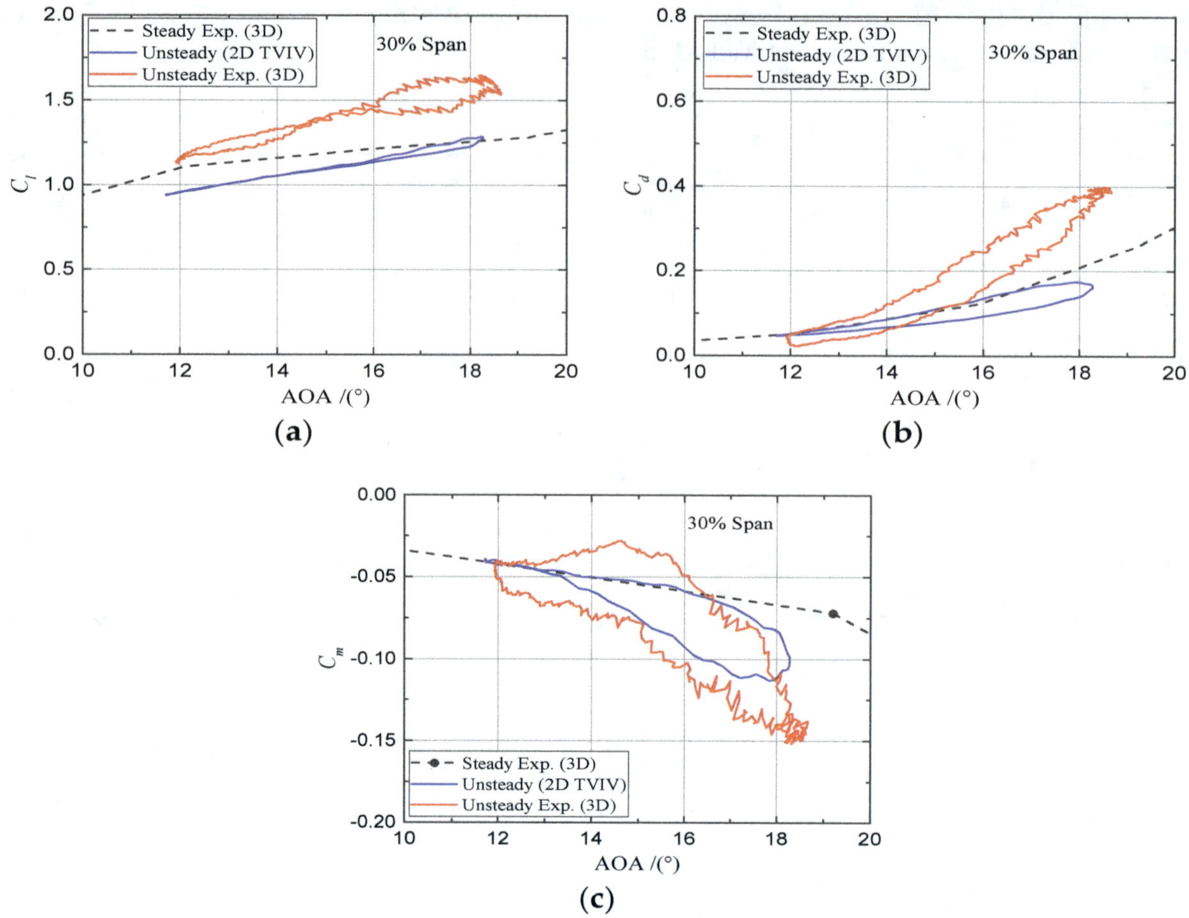

Figure 17. Hysteresis loops of the lift, drag and pitching moment coefficients at 30% spanwise location. The wind speed is 10 m/s, and the yaw angle is 10°. Steady Exp. (i.e., experimental) (3D) denotes the experimental data under axial inflow, while Unsteady Exp. (3D) denotes the experimental data under yawed inflow. (**a**) The lift coefficient; (**b**) The drag coefficient; (**c**) The pitching moment coefficient.

When the yaw angle is high, the air load variation is more severe with the large AOA amplitude. In comparison with Figure 17, Figure 18 shows hysteresis loops of the aerodynamic coefficients at 30° yaw angle at 30% span. It is rather remarkable that the 2D and 3D hysteresis loops of the lift coefficient are in the opposite direction. The 3D lift coefficient is even lower than the 2D value with the AOA increasing, meaning no rotational augmentation effect in the lift force. However, when the AOA decreases, the rotational augmentation dramatically advances the reattachment and results in a flattened hysteresis loop in an anticlockwise direction. In other words, the rotational augmentation effectively pushes the reattachment point rearward and reduces the hysteresis of the air loads on the downwind side (where the AOA is decreasing) (Figure 5b). Because of the unsteady effects, the unsteady 3D experimental data shows an apparent increase in the lift coefficient over the steady 3D experimental data. However, the 3D drag and pitching moment coefficients slightly vary around the steady data. With the AOA decreasing, the hysteresis loop under 2D TVIV indicates a notable additional fluctuation from secondary vortices in the pitching moment coefficient. In short, at a high yaw angle, the rotational augmentation is minor on the upwind side and effectively reduces the hysteresis on the downwind side, while unsteady effects greatly delay the flow separation and elevate the lift coefficient along the whole cycle.

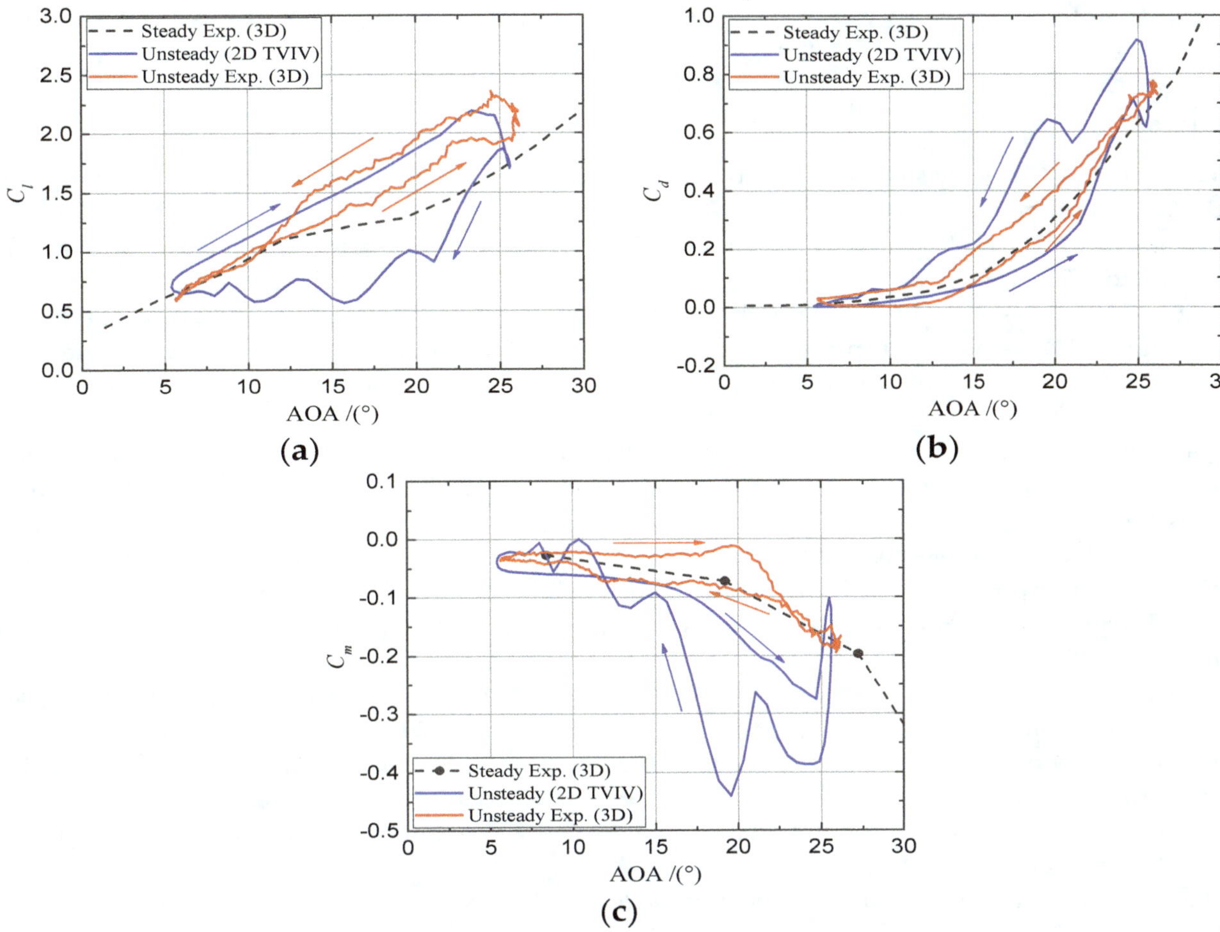

Figure 18. Hysteresis loops of the lift, drag, and pitching moment coefficients at 30° yaw angle. The wind speed is 10 m/s, and the spanwise location is 30%. (**a**) The lift coefficient; (**b**) The drag coefficient; (**c**) The pitching moment coefficient.

4. Conclusions

Dynamic stall is very common on HAWTs, resulting from many aerodynamic sources. Dynamic stall plays an important role in unsteady air loads. The accurate prediction of dynamic stall is extremely important for structural analysis. This study provides a comparative analysis of dynamic stall under pitch oscillation (PO) and oscillating freestream (OF) on wind turbine airfoil and blade, using the quasi-steady thin-airfoil theory, CFD method, and Beddoes-Leishman dynamic stall engineering model.

From the quasi-steady thin-airfoil theory, PO and OF have an opposite effect on the effective airfoil camber and the associated lift coefficient. The CFD results show that the difference between PO and OF mainly exists in the linear part of the aerodynamic loads. The deviation between these two motions is greatly influenced by the reduced frequency and the AOA amplitude. For the yawed response predictions, the Beddoes-Leishman dynamic stall model coupled with a rotational stall delay model also show a different effect between PO and OF. In addition, the effect of the rotational augmentation on the unsteady air loads is assessed. At a low yaw angle, the rotational augmentation apparently elevated the lift and drag coefficients along the whole cycle, as under steady conditions. At a high yaw angle, however, the rotational augmentation apparently flattens the hysteresis loop and its effect during which the AOA dynamically decreasing is more obvious than when the AOA is dynamically increasing.

It may be concluded from this work that dynamic stall behaviors under PO and OF are different at certain conditions and should separately be modeled in engineering.

Author Contributions: C.Z. and T.W. conceived of and designed the research; C.Z. conducted the data collection and wrote the paper; T.W. revised the paper.

References

1. Leishman, J.G. Challenges in modelling the unsteady aerodynamics of wind turbines. *Wind Energy* **2002**, *5*, 85–132. [CrossRef]
2. McCroskey, W.J. *The Phenomenon of Dynamic Stall*; National Aeronautics and Space Administration: Washington, DC, USA, 1981.
3. Carr, L.W. Progress in analysis and prediction of dynamic stall. *J. Aircr.* **1988**, *25*, 6–17. [CrossRef]
4. Butterfield, C.P. Aerodynamic pressure and flow-visualization measurement from a rotating wind turbine blade. In Proceedings of the Eighth ASME Wind Energy Symposium, Houston, TX, USA, 22–25 January 1989; pp. 245–256.
5. Leishman, J.G. *Principles of Helicopter Aerodynamics*, 2nd ed.; Cambridge University Press: Cambridge, UK, 2006.
6. Leishman, J.G.; Beddoes, T.S. A Semi-Empirical Model for Dynamic Stall. *J. Am. Helicopter Soc.* **1989**, *34*, 3–17. [CrossRef]
7. Tran, C.T.; Petot, D. Semi-empirical model for the dynamic stall of airfoils in view of the application to the calculation of responses of a helicopter blade in forward flight. *Vertica* **1981**, *5*, 35–53.
8. Øye, S. *Dynamic Stall Simulated as Time Lag of Separation*; Technical University of Denmark: Lyngby, Denmark, 1991.
9. Hansen, M.H.; Gaunaa, M.; Madsen, H.A. *A Beddoes-Leishman Type Dynamic Stall Model in State-Space and Indicial Formulation*; Risø National laboratory: Roskilde, Denmark, 2004.
10. Pereira, R.; Schepers, G.; Pavel, M.D. Validation of the Beddoes-Leishman dynamic stall model for horizontal axis wind turbines using MEXICO data. *Wind Energy* **2013**, *16*, 207–219. [CrossRef]
11. Gupta, S.; Leishman, J.G. Dynamic stall modelling of the S809 aerofoil and comparison with experiments. *Wind Energy* **2006**, *9*, 521–547. [CrossRef]
12. Johansen, J. *Unsteady Airfoil Flows with Application to Aeroelastic Stability*; Risø National laboratory: Roskilde, Denmark, 1999.
13. Visbal, M.R. Numerical Investigation of Deep Dynamic Stall of a Plunging Airfoil. *AIAA J.* **2011**, *49*, 2152–2170. [CrossRef]
14. Ekaterinaris, J.A.; Platzer, M.F. Computational prediction of airfoil dynamic stall. *Prog. Aeosp. Sci.* **1998**, *33*, 759–846. [CrossRef]
15. Menter, F.R. Two-Equation Eddy-Viscosity Transport Turbulence Model for Engineering Applications. *AIAA J.* **1994**, *32*, 1598–1605. [CrossRef]
16. Van der Wall, B.G.; Leishman, J.G. On the Influence of Time-Varying Flow Velocity on Unsteady Aerodynamics. *J. Am. Helicopter Soc.* **1994**, *39*, 25–36. [CrossRef]
17. Karbasian, H.R.; Esfahani, J.A.; Barati, E. Effect of acceleration on dynamic stall of airfoil in unsteady operating conditions. *Wind Energy* **2016**, *19*, 17–33. [CrossRef]
18. Snel, H.; Houwink, R.; Bosschers, J. *Sectional Prediction of Lift Coefficients on Rotating Wind Turbine Blades in Stall*; Energy Research Center of the Netherlands: Petten, The Netherlands, 1993.
19. Chaviaropoulos, P.K.; Hansen, M.O.L. Investigating three-dimensional and rotational effects on wind turbine blades by means of a quasi-3D Navier-Stokes solver. *J. Fluids Eng. Trans. ASME* **2000**, *122*, 330–336. [CrossRef]
20. Du, Z.; Selig, M. A 3-D stall-delay model for horizontal axis wind turbine performance prediction. In Proceedings of the 36th AIAA Aerospace Sciences Meeting and Exhibit, 1998 ASME Wind Energy Symposium, Reno, NV, USA, 12–15 January 1998.
21. Raj, N.V. *An Improved Semi-Empirical Model for 3-D Post-Stall Effects in Horizontal Axis Wind Turbines*; University of Illinois: Urbana-Champaign, IL, USA, 2000.
22. Bak, C.; Johansen, J.; Andersen, P.B. Three-Dimensional Corrections of Airfoil Characteristics Based on Pressure Distributions. In Proceedings of the European Wind Energy Conference and Exhibition (EWEC), Athens, Greece, 27 February–2 March 2006.
23. Corrigan, J.J.; Schillings, J.J. Emprical Model for Stall Delay Due to Rotation. In Proceedings of the American Helicopter Society Aeromechanics Specialists Conference, San Francisco, CA, USA, 19–21 January 1994.

24. Lindenburg, C. *Investigation into Rotor Blade Aerodynamics*; Energy research Centre of the Netherlands Petten: Petten, The Netherlands, 2003.
25. Breton, S.P.; Coton, F.N.; Moe, G. A Study on Rotational Effects and Different Stall Delay Models Using i a Prescribed Wake Vortex Scheme and NREL Phase VI Experiment Data. *Wind Energy* **2008**, *11*, 459–482. [CrossRef]
26. Zahle, F.; Bak, C.; Guntur, S.; Soslashrensen, N.N.; Troldborg, N. Comprehensive aerodynamic analysis of a 10 MW wind turbine rotor using 3D CFD. In Proceedings of the 32nd ASME Wind Energy Symposium, National Harbor, MD, USA, 13–17 January 2014; p. 15.
27. Bangga, G.; Lutz, T.; Jost, E.; Kramer, E. CFD studies on rotational augmentation at the inboard sections of a 10 MW wind turbine rotor. *J. Renew. Sustain. Energy* **2017**, *9*, 023304. [CrossRef]
28. Bangga, G.; Yusik, K.; Lutz, T.; Weihing, P.; Kramer, E. Investigations of the inflow turbulence effect on rotational augmentation by means of CFD. *J. Phys. Conf. Ser.* **2016**, *753*, 022026. [CrossRef]
29. Somers, D.M. *Design and Experimental Results for the S809 Airfoil*; National Renewable Energy Laboratory: Golden, CO, USA, 1997.
30. Hand, M.M.; Simms, D.A.; Fingersh, L.J.; Jager, D.W.; Cotrell, J.R.; Schreck, S.J.; Larwood, S.M. *Unsteady Aerodynamics Experiment Phase VI: Wind Tunnel Test Configurations and Available Data Campaigns*; National Renewable Energy Laboratory: Golden, CO, USA, 2001.
31. Menter, F.R.; Langtry, R.B.; Likki, S.R.; Suzen, Y.B.; Huang, P.G.; Volker, S. A correlation-based transition model using local variables—Part I: Model formulation. *J. Turbomach. Trans. ASME* **2006**, *128*, 413–422. [CrossRef]
32. Glauert, H. *The Elements of Airfoil and Airscrew Theory*; Cambridge University Press: Cambridge, UK, 1947.
33. ANSYS Inc. *FLUENT Theory Guide, Release 16.0*; ANSYS Inc.: Canonsburg, PA, USA, 2015.
34. Van Leer, B. Towards the ultimate conservative difference scheme. V. A second-order sequel to Godunov's method. *J. Comput. Phys.* **1979**, *32*, 101–136. [CrossRef]
35. Balduzzi, F.; Bianchini, A.; Maleci, R.; Ferrara, G.; Ferrari, L. Critical issues in the CFD simulation of Darrieus wind turbines. *Renew. Energy* **2016**, *85*, 419–435. [CrossRef]
36. Guntur, S.; Sørensen, N.N. An evaluation of several methods of determining the local angle of attack on wind turbine blades. In *Science of Making Torque from Wind 2012*; Iop Publishing Ltd.: Bristol, UK, 2014; Volume 555.
37. Guntur, S.; Sørensen, N.N.; Schreck, S.; Bergami, L. Modeling dynamic stall on wind turbine blades under rotationally augmented flow fields. *Wind Energy* **2016**, *19*, 383–397. [CrossRef]
38. Guntur, S.; Bak, C.; Sørensen, N.N. Analysis of 3D Stall Models for Wind Turbine Blades Using Data from the MEXICO Experiment. In Proceedings of the 13th International Conference on Wind Engineering, Amsterdam, The Netherlands, 10–15 July 2012.
39. Schepers, J.G. *IEA Annex XX: Comparison between Calculations and Measurements on a Wind Turbine in Yaw in the NASA-Ames Wind Tunnel*; Energy Research Center of the Netherlands: Petten, The Netherlands, 2007.
40. Ramsay, R.F.; Hoffman, M.J.; Gregorek, G.M. *Effects of Grit Roughness and Pitch Oscillation on the S809 Airfoil*; National Renewable Energy Laboratory: Golden, CO, USA, 1995.
41. Gharali, K.; Johnson, D.A. Numerical modeling of an S809 airfoil under dynamic stall, erosion and high reduced frequencies. *Appl. Energy* **2012**, *93*, 45–52. [CrossRef]
42. Guntur, S.; Sorensen, N.N. A study on rotational augmentation using CFD analysis of flow in the inboard region of the MEXICO rotor blades. *Wind Energy* **2015**, *18*, 745–756.

3

Development of an Efficient Numerical Method for Wind Turbine Flow, Sound Generation and Propagation under Multi-Wake Conditions

Zhenye Sun [1], Wei Jun Zhu [1,2,*], Wen Zhong Shen [2], Emre Barlas [3], Jens Nørkær Sørensen [2], Jiufa Cao [1] and Hua Yang [1]

1 School of Hydraulic Energy and Power Engineering, Yangzhou University, Yangzhou 225127, China; zhenye_sun@yzu.edu.cn (Z.S.); jfcao@yzu.edu.cn (J.C.); yanghua@yzu.edu.cn (H.Y.)
2 Technical University of Denmark, 2800 Lyngby, Denmark; wzsh@dtu.dk (W.Z.S.); jnso@dtu.dk (J.N.S.)
3 Goldwind Energy ApS, 8382 Hinnerup, Denmark; eba@goldwinddenmark.com
* Correspondence: wjzhu@yzu.edu.cn

Abstract: The propagation of aerodynamic noise from multi-wind turbines is studied. An efficient hybrid method is developed to jointly predict the aerodynamic and aeroacoustics performances of wind turbines, such as blade loading, rotor power, rotor aerodynamic noise sources, and propagation of noise. This numerical method combined the simulations of wind turbine flow, noise source and its propagation which is solved for long propagation path and under complex flow environment. The results from computational fluid dynamics (CFD) calculations not only provide wind turbine power and thrust information, but also provide detailed wake flow. The wake flow is computed with a 2D actuator disc (AD) method that is based on the axisymmetric flow assumption. The relative inflow velocity and angle of attack (AOA) of each blade element form input data to the noise source model. The noise source is also the initial condition for the wave equation that solves long distance noise propagation in frequency domain. Simulations were conducted under different atmospheric conditions which showed that wake flow is an important part that has to be included in wind turbine noise propagation.

Keywords: wind turbine noise source; wind turbine noise propagation; wind turbine wake

---

1. Introduction

The generation and propagation of wind turbine aerodynamic noise exhibit several special characteristics as compared to some other industrial noise problems. Considering the generation of wind turbine aerodynamic noise, the rotor aerodynamic noise level depends on airfoil profiles at cross sections, blade shape (the spanwise distribution of chord, twist, and airfoils), rotor size, rotational speed, yaw angle, tilt angle, pitch setting, the angular position of blades, wind speed profile, inflow turbulence level, density, and viscosity of air. From the perspective of aerodynamic noise propagation (ANPropagation), aerodynamic noise source (ANSource) of large wind turbines is often located at high altitude which leads to more significant problem for ANPropagation over long range. Intrinsically, the study of wind turbine noise is a multi-disciplinary subject which should consider wind farm aerodynamic noise generation (ANGeneration), ANPropagation, and energy production. However, ANGeneration from a single airfoil has a similar basis in the nature of aerodynamics and aeroacoustics as that from a large wind farm. Sophisticated numerical methods are needed to understand the physical mechanisms of flow induced noise. Computational aero-acoustics (CAA) methods were widely applied favored by the high-performance computing technology. However, advanced large eddy simulation (LES)/CAA method still requires heavy computational effort.

Consequently, conventional CAA methods are neither efficient to model wind farm noise source, nor to predict long range ANPropagation. In the following paragraphs, feasible aerodynamic and aero-acoustic modeling techniques are briefly reviewed and the numerical methods employed in the current study are introduced.

As a prerequisite for wind turbine noise source and propagation modeling, a proper aerodynamic tool should be considered. There are many methods developed for wind turbine aerodynamic simulations. The classic blade element momentum (BEM) method [1,2] is a commonly used semi-engineering technique that simulates the aerodynamics of a single wind turbine. With a BEM model, one can apply engineering approaches to compute the noise source level of a wind turbine. When the prediction accuracy increases, the model complexity as well as the computational cost also increases. It is known that the CFD based methods predict the rotor aerodynamic forces and the wake flow behind the rotor more accurately. Considering long-distance ANPropagation of wind turbines, CFD methods become time consuming because the wake flow should be simulated in the large range whose size is the same as that of the ANPropagation. In the present study, the velocity field is the input data for the ANPropagation solver. Instead of conventional CFD methods, other sophisticated methods are developed for wind turbine applications, such as the AD/AL/AS (actuator disc /actuator line /actuator surface) methods, these methods have good trade-off between model accuracy and computational efficiency. The actuator disc method [3,4] is applied in this study which has relatively high computational efficiency. The rotor is assumed as a permeable disc. The aerodynamic loadings of the rotor are treated as external volume force terms of the momentum equations of Navier–Stokes (NS) equations. The rotor disc of a horizontal axis wind turbine (HAWT) has an inherent axisymmetric characteristic. In this study, to reduce computational time the 3D-AD method is simplified to 2D-AD. The boundary condition of rotor center is axisymmetric, such that the obtained flowfield with 2D-NS equations will be exactly identical to that of the 3D case. Since the axisymmetric assumption is made, it has to be mentioned that the solution only holds correct for axisymmetric uniform inflow conditions and rotor under no yaw and tilt manipulations.

Dealing with the modeling of ANSource, the CAA methods, based on high fidelity techniques [5–10], can model the physical details of flow induced noise mechanisms. Nevertheless, the atmospheric acoustic scientists mainly focus on sound propagation problems with larger scales or sizes [11–13]. It is true that the governing equations of most CAA based methods can solve both wind turbine ANGeneration and ANPropagation problems, but the heavy simulation cost often limits the computational domain into small size. It is clear that more efficient methods with less computational effort are needed. To satisfy this requirement, the flow passing wind turbine, the ANSource and the ANPropagation through long path are naturally combined together in this study. The engineering ANSource model for wind turbine [14–16] is built into the above mentioned AD flow solver. Such that the relative inflow velocity and AOA of each blade section are obtained from the AD-CFD simulation and then fed into the noise source model of wind turbine.

Afterwards, the relative sound pressure level (RSPL) from wind turbines is simulated over large distances from the turbine positions to the desired receivers. The wind turbine ANSource are the key inputs to the propagation solver [11,12]. The wind turbine wake flow, computed by the AD-CFD method, is considered as the medium of ANPropagation. The complex flow environment is found as an important factor for long-path wind turbine ANPropagation. In the previous study of Lee [17] and Heimann [18], the influences of atmospheric conditions on ANPropagation were investigated. Both studies showed that there is large noise variation under different wind profile and atmospheric stability conditions. More recently, this parabolic equation (PE) method is used by Barlas et al. [19,20] for wind turbine noise applications. It was clearly shown that the wind turbine ANPropagation has different refraction characteristics under various wake conditions. The present work focuses on an integrated method for modeling of flow, ANSource and ANPropagation, and the detailed phenomenon of ANPropagation through multi-wind turbine wakes are studied.

This paper is organized following the sequences: In Section 2, the numerical method for wind turbine flow simulation, ANGeneration and ANPropagation modeling are described; Section 3 provides results of various validation cases first and more results from the integrated flow-noise model afterwards; conclusions are provided in the last section.

2. Numerical Methods

The related numerical methods applied in the study are described separately in this section: (1) The RANS/AD model for wind turbine flow simulation; (2) The BPM (named after Brooks TF, Pope DS, Marcolini MA [14]) wind turbine ANGeneration model; (3) The wind turbine ANPropagation model. The relations between the solvers are shown in the schematic diagram below in Figure 1. It is seen that the flowfield data are the input to both the wind turbine noise source and propagation models.

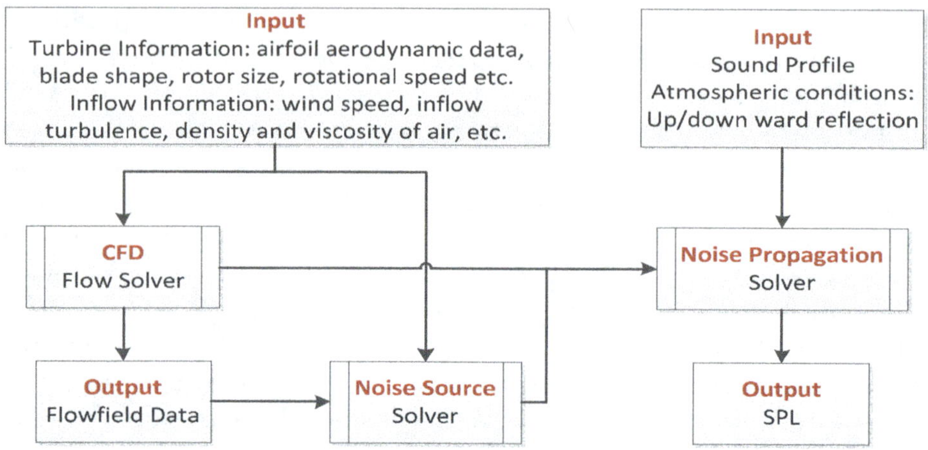

Figure 1. Schematic diagram of relations between the developed solvers.

2.1. RANS/AD Method

To simulate the flow passes wind turbine rotor, the general equations are modified by adding the aerodynamic loadings of the rotor as external volume force terms in the momentum equations of the NS equations

$$\frac{\partial u_i}{\partial x_i} = 0 \quad (i = 1\,,\,2\,) \tag{1}$$

$$\frac{\partial u_i}{\partial t} + u_j \frac{\partial u_i}{\partial x_j} = -\frac{1}{\rho}\frac{\partial P}{\partial x_i} + \frac{\partial}{\partial x_j}(\nu \frac{\partial u_i}{\partial x_j} - \overline{u'_i u'_j}) + f_{ext} \tag{2}$$

$$-\rho\overline{u'_i u'_j} = \mu_t \left(\frac{\partial u_i}{\partial x_j} + \frac{\partial u_j}{\partial x_i}\right) - \frac{2}{3}\rho k \delta_{ij} \tag{3}$$

where x_i is the displacement in the inertial coordinate system and u_i is the velocity components. t, ρ, and P represent time, air density and pressure respectively. ν is the kinematic viscosity term, f_{ext} is the added source term. With the assumption based on the axisymmetric property of the HAWT, 2D-RANS/AD method is utilized such that the computational effort can be greatly decreased.

The rotor is assumed as permeable disc such that the mass equation remains unchanged. To avoid numerical instability the volume force in Equation (2) should be smeared along the flow direction using 1D Gaussian function. For the elements which are away from the disc with a normal distance of d, the smeared force f' is computed by convolution computation

$$f\prime = f_{AD} \otimes \eta^{1D} \tag{4}$$

$$\eta^{1D}(d) = 1/(\varepsilon\sqrt{\pi})\exp\left[-(d/\varepsilon)^2\right] \tag{5}$$

where the free parameter $\varepsilon = 3\Delta z$ is three times larger than the axial size of reference grid Δz. What is more, the body forces of rotor can be computed using blade element theory with the aerodynamic data of airfoil and the simulated information of flowfield.

The present 2D-AD method solves the ideal axisymmetric flowfield. The axial and tangential induced velocities are naturally combined with the NS equations, such that the inflow angle and the relative inflow velocity are determined from

$$\phi = \tan^{-1}\left(\frac{V_0 - W_z}{\Omega r + W_\theta}\right),\ V_{rel}^2 = (V_0 - W_z)^2 + (\Omega r + W_\theta)^2 \tag{6}$$

The local AOA is easily computed by $\alpha = \phi - \gamma$, where γ is the real twist angle (added by pitch) at local blade element. The lift and drag forces per unit span are computed using Equation (7) with the aerodynamic data of airfoil.

$$L = \frac{1}{2}\rho V_{rel}^2 cBC_l,\ D = \frac{1}{2}\rho V_{rel}^2 cBC_d \tag{7}$$

The blade forces are updated after each iteration using newly computed velocity field, then the external volume force terms of the NS equations apply renewed blade forces in the next iteration.

2.2. BPM Model for Noise Source or Generation

The original BPM noise model was developed by Brooks et al. [14] which is able to predict airfoil self-noise. The model was based on scaling the wind tunnel experimental data using the NACA 0012 airfoil with chord lengths varies from 2.5 cm to 61 cm. Here, the BPM model was integrated with the BEM method [15] in order to deal with the complicated flow for wind turbine. The required boundary layer thickness data in the BPM model is obtained from the Xfoil computation so that this data varies with the airfoils rather than fixed with NACA 0012. A general expression of the BPM model is shown in Equation (8)

$$\mathrm{SPL} = 10log_{10}\left(\frac{\delta M^n l D}{r^2}\right) + G_1(St) + G_2(Re) + G_3(\delta) + C \tag{8}$$

In the equation, it is shown that rotor aerodynamic noise is mainly a function of parameter δ which represents boundary layer thickness. The boundary layer thickness depends on airfoil shape, Reynolds number, AOA, turbulence level etc. l is the spanwise length of an airfoil segment. G_1, G_2, and G_3 are related to the Strouhal number St, Mach number M, Reynolds number Re and boundary layer thickness parameter δ. Detailed descriptions of noise mechanisms can be found in references [14,15]. ANSource level of rotor generally depends on the blade shape and operational conditions, such as rotational speed, wind speed, yaw, tilt, etc. Instead of combining the engineering BEM model, the BPM noise model is directly integrated into the NS equations with the AD model. The BPM model is a passive subroutine that can be activated anytime with the inflow velocity and AOA provided from AD simulations. The trailing edge boundary layer thickness is interpolated from the existing database prepared with Xfoil through a wide range of Reynolds numbers and AOAs. To take clean and rough blade surface conditions into account, the boundary layer thickness database is created under different surface roughness cases, such that wind turbine noise under free transition and fully turbulent flows can be modelled. A fully-turbulent boundary layer thickness database is utilized to calculate the noise of rotor whose blade surface is rough, and free transition database is applied for a rotor with a clean surface.

2.3. Noise Propagation Method

Sound speed in an ambient flow is written as $c_{eff} = c + v_x$ where c is the constant sound speed at a given air temperature, v_x is the wind speed component through the propagation path. The wind speed components may come from field measurements, empirical expressions or numerical simulations.

For an incompressible wind flow, the air density is assumed as a constant, the wave equation in a moving atmosphere reads [12]

$$\left[\nabla^2 + k^2(1+\epsilon) - \frac{2i}{\omega}\frac{\partial v_i}{\partial x_j}\frac{\partial^2}{\partial x_i \partial x_j} + \frac{2ik}{c_0}v \cdot \nabla\right] P'(r) = 0 \tag{9}$$

where $k = \omega/c_0$, ω is the radian frequency of sound wave, c_0 is the reference sound speed, $P'(r)$ denotes the monochromatic field of sound pressure and $\varepsilon = (c_0/c)^2 - 1$. If the effect from ambient flow is neglected, this equation can be further reduced to Helmholtz equation. The ANSource of a wind turbine is placed on the left boundary of the computational domain depicted in Figure 2. A classical approach to represent a monopole ANSource is to distribute the source along the vertical direction which is also called starting field. In this study, the rotor noise source is located at the hub height as sketched at the left side of the boundary. Based on the initial source strength, the equation or the sound pressure amplitude is computed along the x-direction, such that solution is renewed from $P(x)$ to $P(x + \Delta x)$. The boundary condition applied on the ground surface is calculated from the impedance of the ground. The empirical impedance model for absorbing materials developed by Delany and Bazely [21] is used in this study,

$$Z = 1 + 0.0511\left(\frac{\sigma}{f}\right)^{0.75} + i\,0.0768\left(\frac{\sigma}{f}\right)^{0.73} \tag{10}$$

where Z is the normalized acoustic impedance, σ represents the flow resistivity and f is the frequency. The similar approach can be applied on the top surface. A constant of the normalized acoustic impedance $Z = 1$ is implemented on the top grid, such that the vertical waves through the top surface is vanished without any reflection. However, to sufficiently eliminate wave reflections, an absorbing layer is defined on top of the grid. As shown in the figure, the grey region sketched on top of the domain is the absorbing layer. In the study, the width of such an absorbing layer is defined as 50λ. This implies that for lower frequency sound propagation, a larger absorbing layer is needed. The sound pressure along the vertical line is updated from step n to $n + 1$ until the receiver position is reached. The solution procedure starts with the separation of the forward and backward propagation waves. Here, only the forward wave propagates from source towards receiver is considered (in the positive x-axis), which is in the downwind direction. Detailed mathematical manipulations of the equation are found in [12,13].

According to the methodologies described in Sections 2.1 and 2.2, the noise spectrum of a given wind turbine is the logarithmic sum of sound pressure level (SPL) from all the blade elements such that

$$L_P(f) = 10log_{10}\left(\sum_i^n 10^{0.1SPL^i_{total}}\right) \tag{11}$$

where n is the number of the blade elements and SPL^i_{total} is the SPL of the ith blade element, which accounts for all the noise mechanisms. The SPL of wind turbine ANSource $L_p(f)$ is the initial condition for calculating long-path ANPropagation. The sound power level $L_w(f)$, which is a representation of wind turbine total noise strength, it does not change with propagation range. The SPL at the receiver is

$$L_P(f) = L_w(f) - 10log_{10}4\pi D^2 - \alpha D + \Delta L \tag{12}$$

where the first term on the right side is the sound power level, the second term $-10\log_{10} 4\pi D^2$ is the geometric attenuation term that corresponds to the sound intensity loss when the sound wave sphere is expanding. The third term shows the attenuation due to air absorption, which is proportional to the absorption coefficient α and the propagation length D. ΔL is the relative SPL differences caused by factors such as ground reflection/absorption (flat/irregular terrain), atmospheric refraction, wind and turbulence and sound barriers, etc. It does not include the effects of geometric attenuation and air absorption. The ΔL varies along the propagation path with these factors, which is beneficial to clearly manifest the effects of propagation. Most of the discussions and results in Section 3 are dealing with this relative pressure level which is abbreviated as RSPL. The values of α depend on sound frequency, air temperature, pressure and humidity. In the following studies, unless specified separately, a relative humidity of 70% and temperature of 10 °C are assumed.

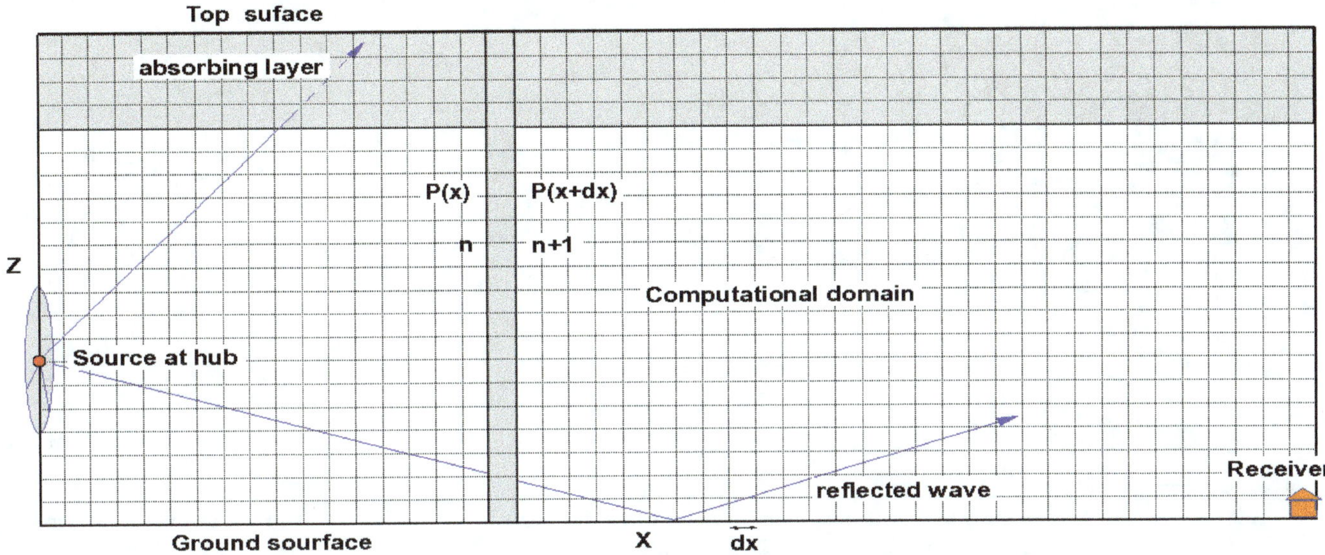

Figure 2. Demonstration of grids in the plane of propagation utilized in the simulations.

The wave equation is solved based on each frequency. The ANSource and wake information of given wind turbines from BPM/AD computations are combined into the ANPropagation solver. For each frequency, the corresponding ANSource, known as the starting field, is individually applied as the initial condition to the computational domain. Detailed descriptions about starting field formulation can be found in publication from Salomons [11]. The horizontal size of the simulation range is determined by the location of the receiver. However, the size of the vertical domain is flexible such that the height of domain can vary with the simulated frequency. The maximum vertical size of the grid should be less than $\Delta x = \Delta z = \lambda/8$ so that a sound wave can be resolved with at least eight grid points. As a result, the domain height is inversely proportional to frequencies for a fixed number of vertical grids. Besides, a larger CFD domain should be used in the simulation such that the flow field can fully cover the noise propagation domain.

3. Results and Discussions

A few validations about ANPropagation cases are shown first. Under various flow conditions, the flow passing wind turbine rotor, ANGeneration and ANPropagation problems are discussed in detail. The RSPL ΔL over a long distance is computed for each frequency. After the validations, the wind turbine ANPropagation across multi-wakes is shown. In all the cases, the temperature at the ground level is 15 °C, such that the equivalent sound speed is 340 m/s.

3.1. Validation of Long-Range Sound Propagations without Ambient Flow

This validation case considers the idealized long-range sound propagation without any moving medium or temperature gradient effects. The terrain contains a flat and acoustically soft surface with a total length of 200 m. The flow resistivity of a grassland are usually in range from 100 kPa $\cdot$ s $\cdot$ m^{-2} to 300 kPa $\cdot$ s $\cdot$ m^{-2}. The absorbing ground surfaces in the current simulation has a flow resistivity of 250 kPa $\cdot$ s $\cdot$ m^{-2}. The ANSource locates at $x = 0$ m, $z = 0.5$ m and the receiver at $x = 200$ m, $z = 1.5$ m. The current results are compared with the data presented in reference [22] where either measured data or data obtained from other models are available. The sound transmission losses over 200 m are shown in Figure 3 where the selected frequencies in the 1/3-octave band are gathered. At very low frequencies, i.e., f = 20 Hz and f = 40 Hz, the RSPL ΔL is increased with the propagation distance. At higher frequencies, ΔL gradually decreases along the propagation distance. The ΔL values at 200 m are compared in Figure 4 which shows that there is a large propagation loss between 400 Hz and 1000 Hz. The present simulation agrees with the reference data and the simulation obtained from the Nord2000 commercial model [23].

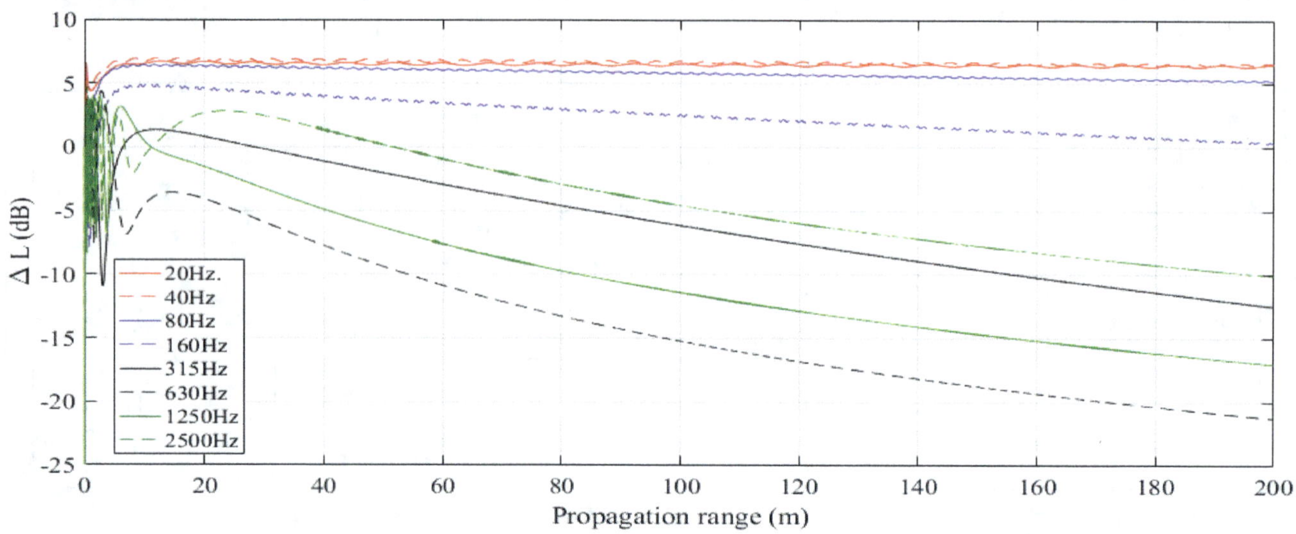

Figure 3. Sound transmission loss over 200 m at various frequencies (soft ground).

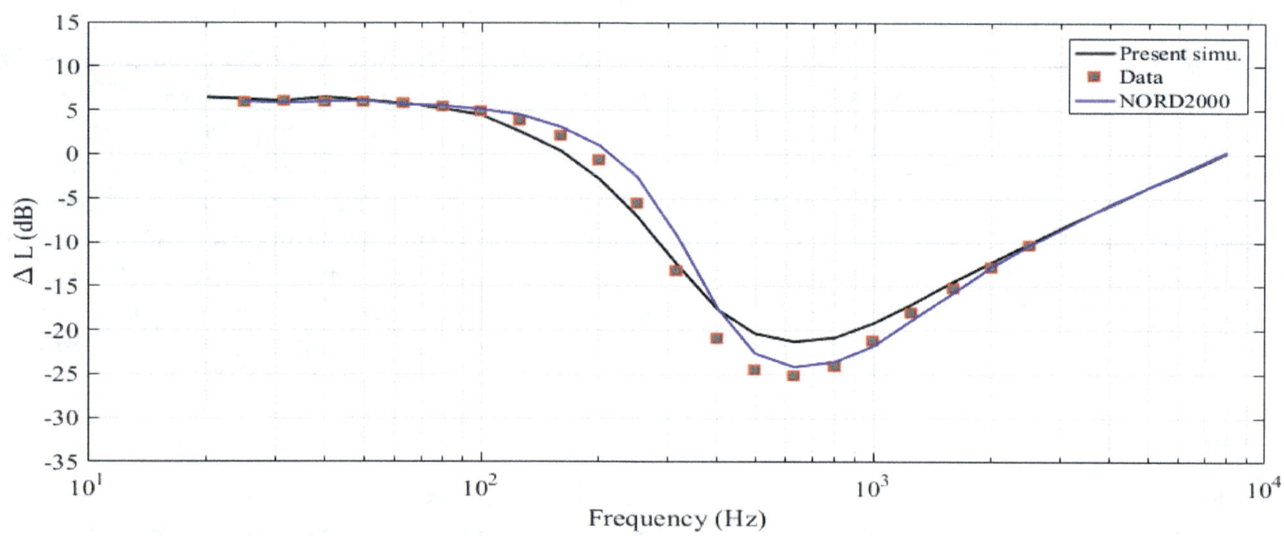

Figure 4. Comparisons of the RSPL in 1/3-octave bands.

3.2. Validation of Long-Range Sound Propagations with a Positive Temperature Gradient

In this case, the terrain distance, source, and receiver locations remain the same, but the ground is an acoustically hard surface, meaning that the flow resistivity is close to infinity. The temperature gradient was measured as 0.0846 K/m. If a linear temperature variation is assumed, the temperature and sound speed profiles are easily computed using the following relations

$$T = \frac{\partial T}{\partial z} \cdot z + T_0 \tag{13}$$

$$c = c_0 \cdot \sqrt{T/T_0} \tag{14}$$

Such a temperature gradient leads to a downward sound propagation. As a result, the RSPL along the propagation direction will probably increase. In Figure 5, it is seen that there is a general increase of ΔL along the propagation path except near the source locations (x < 40 m). The ΔL values are gathered at x = 200 m distance and are depicted in Figure 6. It is found that the ΔL values are positive between 20 Hz and 5 kHz. The current prediction fits well with the reference data.

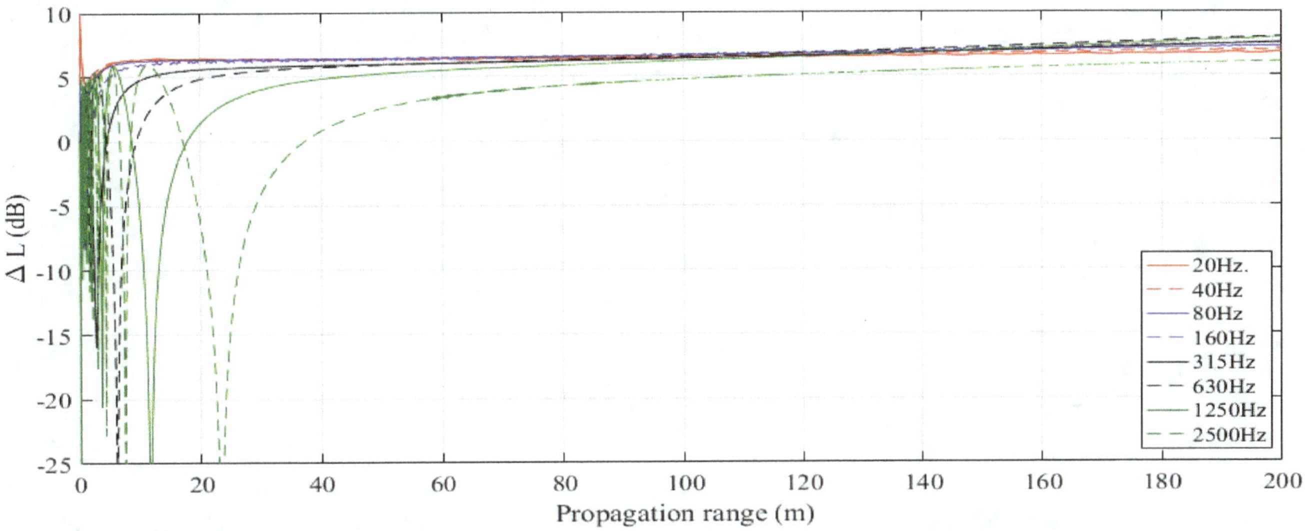

Figure 5. Sound transmission loss over 200 m at various frequencies (rigid ground and downward refraction).

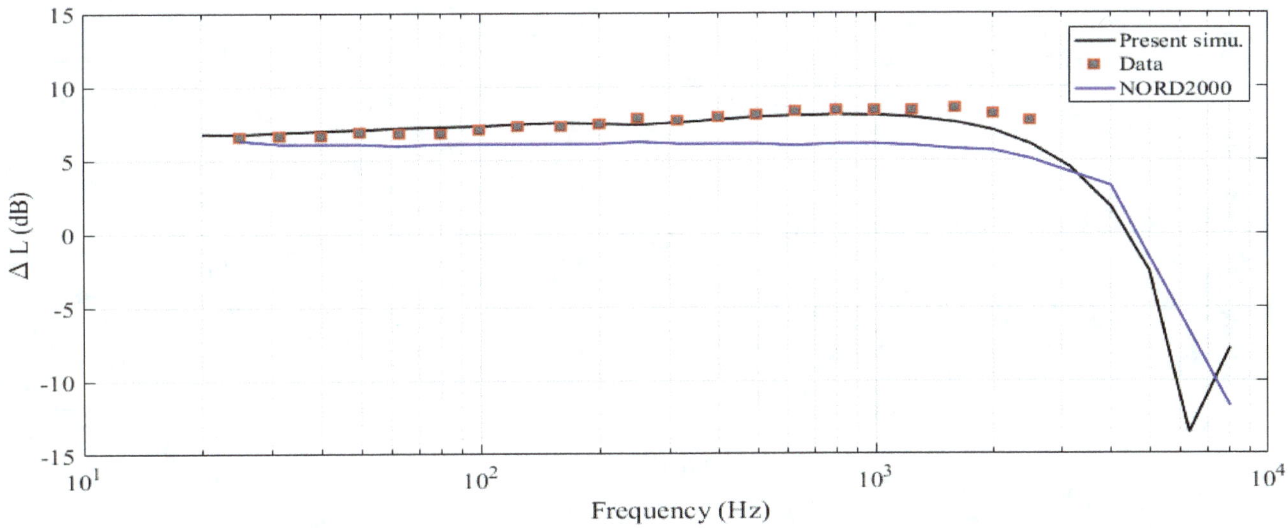

Figure 6. Comparisons of the RSPL in 1/3-octave bands at a distance of 200 m.

3.3. Validation of Long-Range Sound Propagations with Upward Refraction under Wind Condition

Similar as shown in the previous test case, the ground acoustic impedance is also infinity. However, an upward refraction case is considered which is caused by a moving atmosphere. The wind speed recorded at 10 m height above the ground is −4.165 m/s, and the surface roughness length is 0.1 m. In a moving medium, the sound speed can be computed using the equation below

$$c_{eff}(z) = c_0 + b \ln\left(\frac{z}{z_0} + 1\right) \tag{15}$$

where c_{eff} is the profile of effective sound speed along the vertical direction, c_0 = 340 m/s is the sound speed and z_0 = 0.1 m is the roughness height. Knowing the wind speed at 10 m height and the roughness length, it is easy to show that $b \approx -1$. Such a sound speed profile represents a slightly upward atmospheric refraction. In Figure 7, except at very low frequencies, the RSPL decays very fast along the propagation path. Figure 8 shows a very good agreement with the reference data which indicates the capability of the current method to handle noise propagations through moving media.

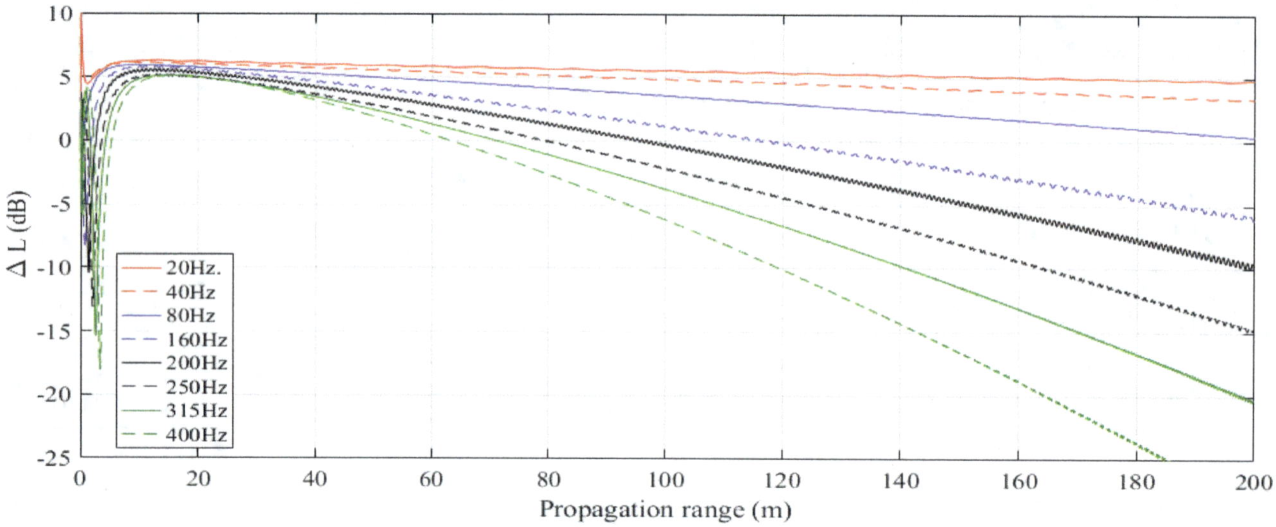

Figure 7. Sound transmission loss over 200 m at various frequencies (rigid ground and upward refraction).

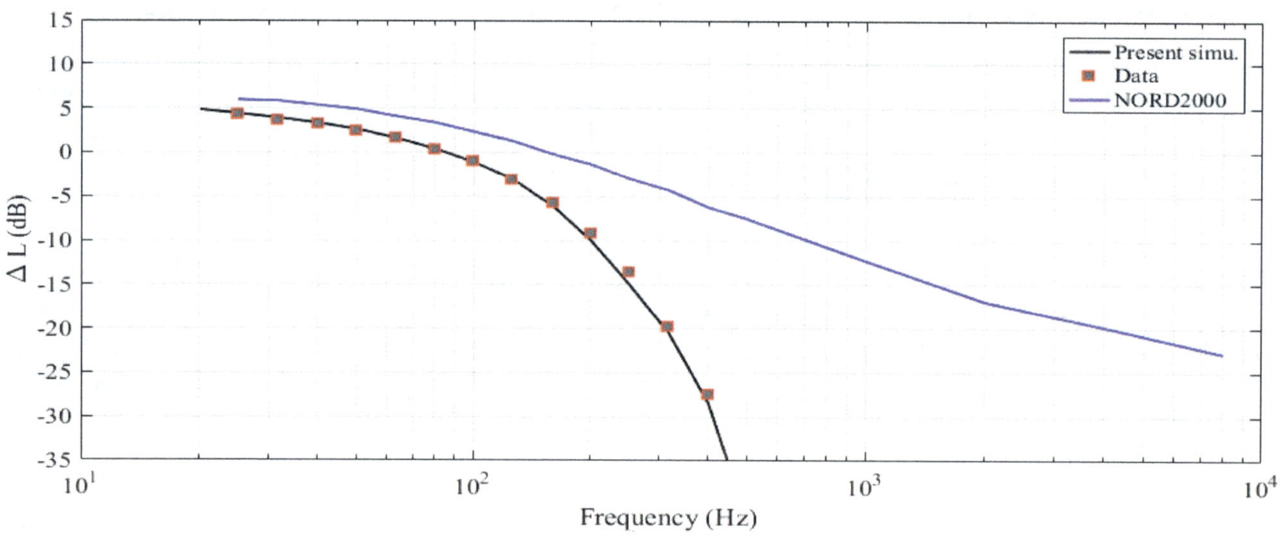

Figure 8. Comparisons of the RSPL in 1/3-octave bands.

3.4. Coupled Flow and Noise Simulations for a Wind Turbine Array

3.4.1. Wake Modeling of a Wind Turbine Cluster

Wind turbine flow are simulated for a Vestas wind turbine called NM80. The AOA and the relative velocity at all blade elements from RANS/AD simulations are inputted to the BPM model. The AD solver is much more efficient because it represents the rotor by volume force term rather than the dense body-fitted-mesh around blade surface. The flow computations are carried out with the in-house developed flow solver [24–26] which is an incompressible flow solver.

The grid for AD simulation and a typical flowfield with multi-wake structures are depicted in Figure 9. In Figure 9a, two wind turbines, depicted with thick white lines, are used for demonstration. The inflow boundary is applied on the left and top domain edge (numbered as 1 in Figure 9), the axisymmetric boundary on bottom edge (numbered as 2) and outflow boundary on the right edge (numbered as 3). The computational domain consists of the several blocks whose number is the same as that of wind turbines to be investigated. Each block contains grid points of 64 × 64. Using this approach, multi-wake flowfield can be efficiently modelled. A typical horizontal velocity contour plot is shown in Figure 9b.

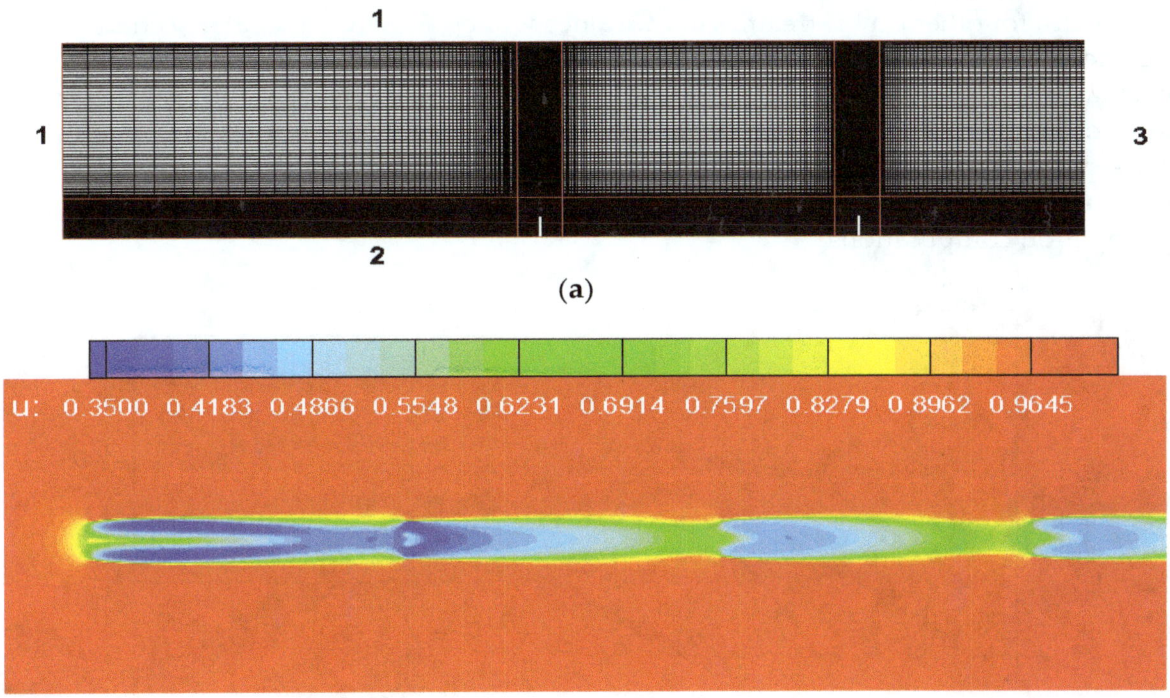

(b)

Figure 9. (**a**) Mesh topology; (**b**) stream-wise flow with multi-wakes.

To show the model accuracy, in Figure 10, some results from the current simulations using RANS turbulence model are compared with the LES results under same flow conditions [27,28]. On the modeling of wind turbine wake, many studies can be found [29–31]. In the present study, it is not of our primary interest to go into more details on this topic. Some more details of the AD modeling approach and cross comparisons can be found from our previous works [32–34].

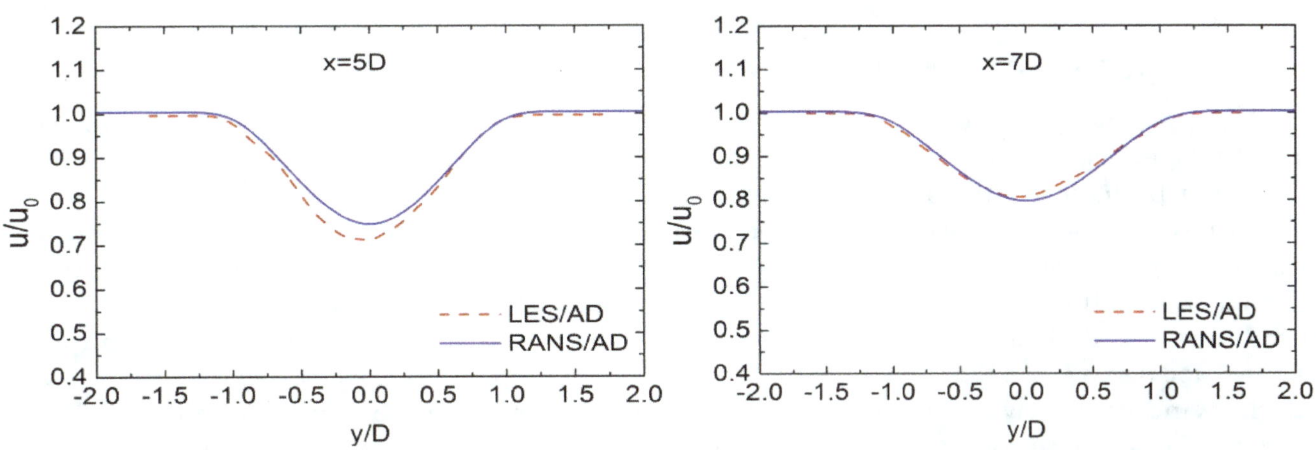

Figure 10. Comparisons of the normalized wake velocity from LES and RANS simulations, for down-stream locations of 5D and 7D after a single wind turbine.

3.4.2. Modeling of the Wind Turbine ANSource

As is shown in the above AD simulation, every wind turbine has different inflow conditions because some wind turbines are in the wake of another, which leads to different ANSources. The simulation were conducted for a free-stream wind velocity of 8 m/s. For a receiver at 120 m far away, the detailed noise spectra solely from the first upstream wind turbine are shown in Figure 11. The noise spectra for other downstream wind turbines will be shown in the later sections. As shown in the figure, the total SPL spectrum is name as SPL-TOTAL which is logarithmically summed from several noise mechanisms. These noise mechanisms include: noise from blade tip (SPL-TIP); from trailing edge bluntness (SPL-TEBLUNT); from the shedding of laminar boundary layer vortex (SPL-LBLVS); from flow separation (SPL-SEPARATION); from turbulent boundary layer trailing edge (SPL-TBLTE); from turbulent inflow (SPL-INFLOW) [14–16]. Later, the total noise spectrum will be applied into the noise propagation computations.

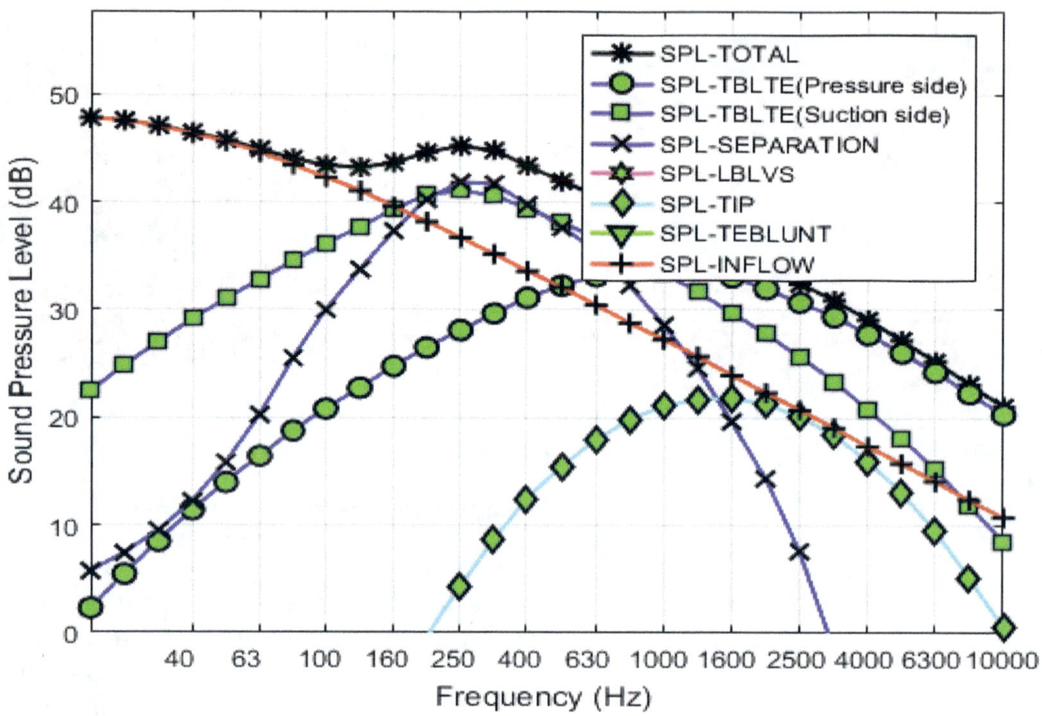

Figure 11. Aerodynamic noise source spectra of a single wind turbine.

3.4.3. Sound Profiles Inside wind Turbine Wakes

Although the flow and the noise source are modelled without wind and temperature gradients, in reality the effective sound speed profile will be affected by wind and temperature profiles. It was shown that the combinations of wind speed and temperature gradient forms different types of sound speed profiles [35]. Three idealized initial sound profiles are considered in the following sound propagation simulations. In Figure 12a, a uniform sound speed at temperature of 15 °C is assumed at the left boundary. The wind turbines are separated with a normalized distance of seven rotor diameters. For example, the second wind turbine is located at x = 560 m and the third wind turbine is located at x = 1120 m. Inside the computational domain, the sound speed is superimposed with the wind turbine wake field. Similarly, Figure 12b,c represent downward and upward sound speed contours. These sound profiles merge with the wind turbine multi-wakes and show distinct characteristics. The sound profiles have significant effects on wind turbine noise propagation as shown in the next subsection.

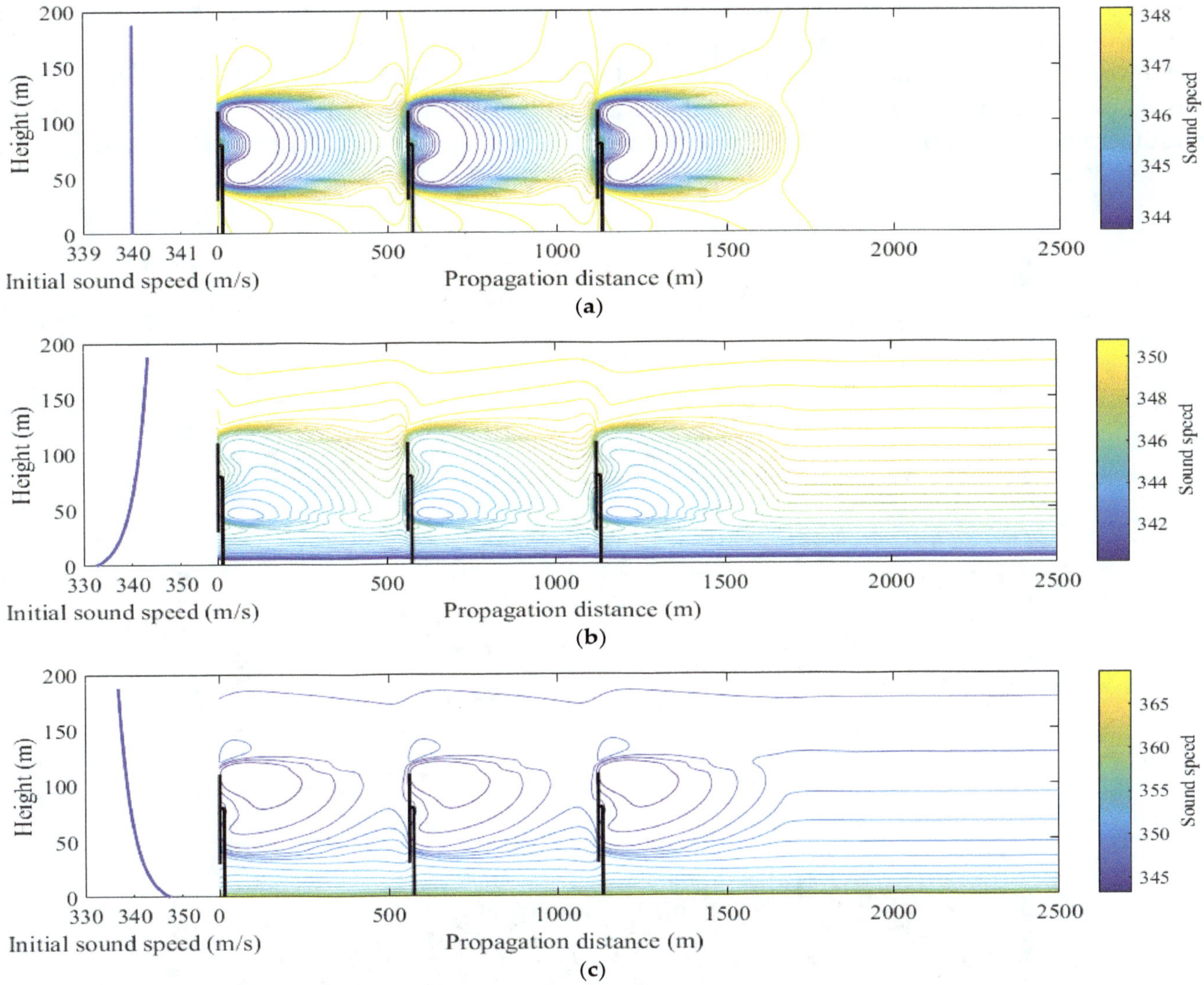

Figure 12. Initial sound speed profiles at upstream and effective sound speed contours in the wake flow. (**a**) Uniform initial sound profile transmitted into wake; (**b**) upward refraction sound profile transmitted into wake; (**c**) downward refraction sound profile transmitted into wake.

3.4.4. Wind Turbine ANPropagation across Multi-Wake

As shown in Figure 13, the RSPL or ΔL is simulated along 1500 m propagation distance, at $f = 300$ Hz and with the noise spectrum in Figure 11. The first wind turbine locates at $x = 0$ m, the second at 560 m and the third at 1120 m. The distance between each wind turbine are relative large. The wind turbine ANSource are placed at a height of 70 m. The ANSources from the second and third wind turbines, depicted as white lines, are not considered so as to focus on the propagation phenomenon through the wakes. In Figure 13, RSPL results are shown under several conditions: (1) Case 1, ANPropagation under homogeneous atmospheric condition, no wake effect; (2) Case 2, ANPropagation under homogeneous atmospheric condition, considering multi-wakes; (3) Case 3, ANPropagation under atmospheric condition of down-ward refraction, considering multi-wakes; (4) Case 4, ANPropagation under atmospheric condition of upward refraction, considering multi-wakes. It is shown that the wind turbine ANPropagation is clearly affected by the complicated flow environment. For the case without any wake effect shown in Figure 13a, the ANPropagation is mainly influenced by the reflection and absorption of ground. As is shown in Figure 13b, the contours of RSPL is more complicated than that of Figure 13a which distinctly shows the influences of wake. For atmospheric condition of downward refracting depicted in Figure 13c, the 'noise pollution' band also shows a downward reflecting behavior. Additionally, a low-noise zone, shown in blue, can be found near $x = 1000$ m. In Figure 13d, a much quieter situation is observed under an upward refraction condition. In such a case, the quiet zone is located between wind turbine #2 and #3. It should be emphasized that uniform inflow boundary condition is applied in the flowfield simulations of Figure 13c,d. However, the un-uniform or down/up-ward atmospheric conditions, which means atmospheric altitude-dependent variation of the sound speed, are used in the ANPropagation computations.

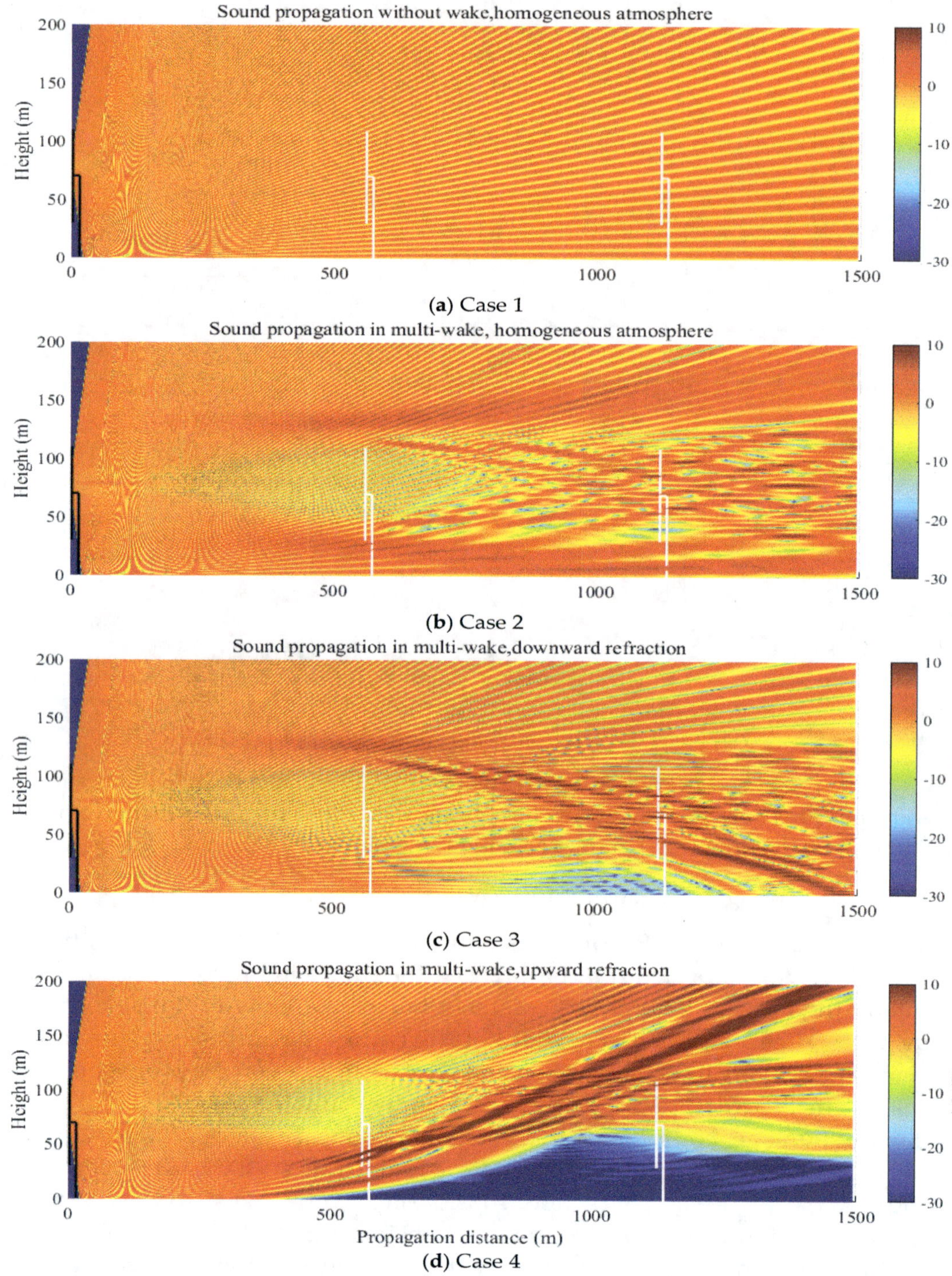

(**a**) Case 1
(**b**) Case 2
(**c**) Case 3
(**d**) Case 4

Figure 13. ANPropagation from the first ANSource: (**a**) Case 1, homogeneous atmospheric condition, no wake effect; (**b**) Case 2, homogeneous atmospheric condition with multi-wakes; (**c**) Case 3, downward refraction and multi-wakes; (**d**) Case 4, upward refraction and multi-wakes.

Manifesting the noise received by humans, the RSPL or ΔL at $h = 2$ m is extracted from Figure 13 and shown in Figure 14.The sound attenuation ΔL of Case 1 starts to vary linearly with the propagation distance at around $x = 400$ m, because the reflection effect becomes weaker after that. In contrast, the ΔL of Case 2, 3, and 4 vary obviously even after a long distance. In Cases 2 and 3, ΔL fluctuates over distance due to the wake effect and the strong downward refraction respectively. Whereas in Case 4, a more significant RSPL loss is shown which is caused by the upward refraction. It should be mentioned that the ΔL values might be quite different when the observer height changes. From this comparison, it can be concluded that ANPropagation can be greatly influenced by the environmental conditions. What is more, the change of SPL in the near downstream is much smaller than that at far downstream ($x > 400$ m), which shows the significance of long distance wind turbine noise predictions under complex atmospheric conditions. The near field ($x < 400$ m) simulations and measurements only belong to ANSource study which is not identical to the long-path ANPropagation.

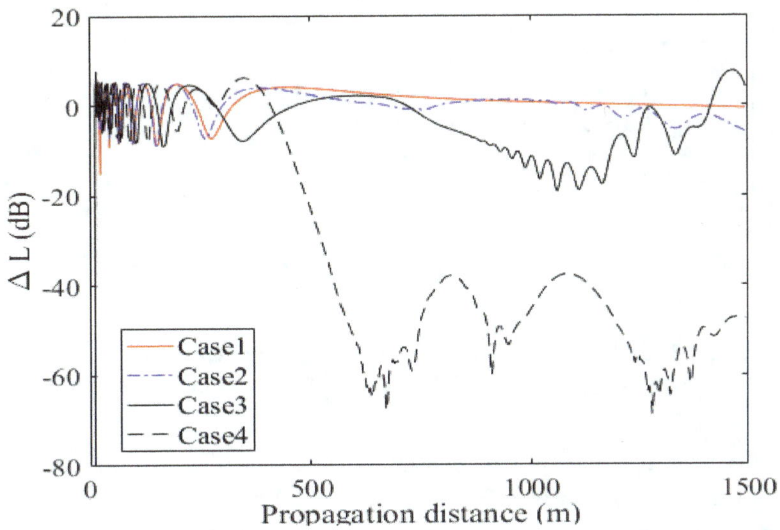

Figure 14. The RSPL at 2 m height extracted from four cases of Figure 13.

Result shown in Figure 14 represents the ΔL solely due to propagation which is computed at a fixed frequency of $f = 300$ Hz. Repeating the calculation for many frequencies and extracting the data at a receiver 1500 m away and 2 m height, a full spectrum of ΔL will be obtained. Using Equation (12) and considering all the effects (including ΔL), the SPL at receiver's location under Case 1 and Case 3 is gathered and shown in Figure 15. Besides, the spectrum of ANSource and at receiver's location (without ΔL) are also depicted in Figure 15. In this work, the ANSource is considered in an 1/3-octave band from 20 Hz to 5000 Hz. Since a minimal eight grid points are needed to resolve a sound wave, the grid number for high-frequency simulations increases sharply if the same domain size is adopted. For instance, there is a 100-fold increase in the grid number for frequency of 3000 Hz compared with that of 300 Hz. To reduce the computational effort, frequencies above 5000 Hz are not considered here. Another reason is that the noise reduction due to air absorption, $-\alpha D$ in Equation (12), is much larger than the other factors at large frequency ($f > 5000$ Hz). The SPL at the source location is shown with black line with asterisk and labeled as 'Source: Lp' in Figure 15. The black lines with stars represent the SPL at receiver (2 m height at $x = 1500$ m) without considering the propagation loss ΔL in Equation (12). The blue line (Case 1) and red line (Case 3) consider all the factors, including geometric attenuation, air absorption and ΔL, in the long-path ANPropagation simulations. The big deviation between the two black lines demonstrates that the geometrical attenuation ($10 \log_{10} 4\pi D^2$) contributes most to the ANPropagation loss. The even larger reduction of noise, when $f > 1000$ Hz, means that air absorption has a larger effect at high frequencies. The differences between blue/red lines and the black line with stars implies the value of ΔL. The value of blue and red lines are larger than that of black line with stars, which means positive ΔL near low frequency (f = 30 Hz). However, Case 1 and Case 3 show opposite ΔL around f = 200 Hz which implies the influences of wake and downward refraction.

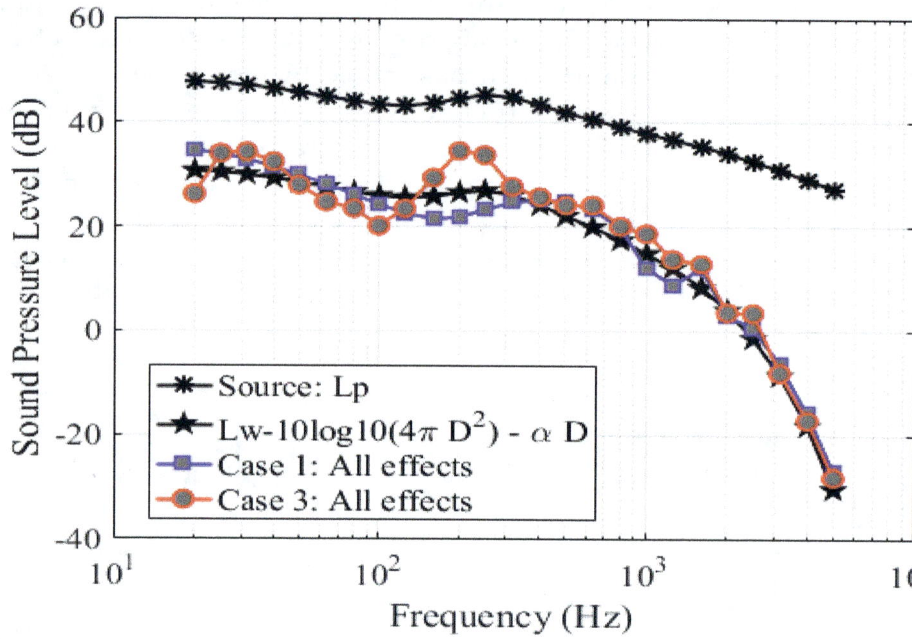

Figure 15. SPL of wind turbine #1 (Source: Lp) received at 1500 m distance (2 m height): with geometric attenuation and air absorption, Case 1 and Case 3 with all propagation effects.

Similar as the previous simulations, the following study focuses on the noise propagation area between 1120 m and 2500 m, and a higher frequency f = 1000 Hz is selected. In this region, the three individual wind turbine noise sources propagate and superimpose on each other. In Figure 16a, the results cover the entire domain behind the last wind turbine. The solution differs from Figure 13a in the sense that it is the RSPL from all the three wind turbines. In Figure 16b, the uniform sound profile at the inlet boundary is affected by the multi-wake (see Figure 12a), such that the resulted RSPL is enhanced around wake center. With the prescribed sound profile of Figure 12b, the RSPL is refracted downwards, as shown in Figure 16c. Similarly, using the sound profile given in Figure 12c, the resultant sound propagation is refracted upwards, see Figure 16d.

Figure 17 better illustrates the superposition of the sound pressure propagates from each wind turbine, which is a decomposition of Figure 16d. In the first figure, noise is generated from wind turbine #1 and propagates through the wake flow. The other two wind turbines are depicted with white lines, which imply that they are currently excluded for noise propagation. In the next simulation, shown in Figure 17b, ANPropagation from wind turbine #2 is individually considered, which excluded noise propagation from wind turbine #1 and #3. Similarly, wind turbine #3 is then considered for ANPropagation, which does not take account of the two upstream wind turbine noise propagations. The total RSPL is the summation of sound pressure following the logarithmic sumation rule, which is shown in Figure 17d.

Figure 18 shows the RSPL or ΔL extracted from a line at 2 m high. The overall sound pressure variation for different cases is smaller as compared to the previous results shown in Figure 14. This is due to the mixing of three individually simulated propagation field. Among all the test cases, Case 3 can be regarded as the worst-case scenario. For example, around the distance of x = 1800 m, there is an obvious increase of SPL due the strong downward refraction at this region, which is also observed in Figure 16c at x = 1800 m.

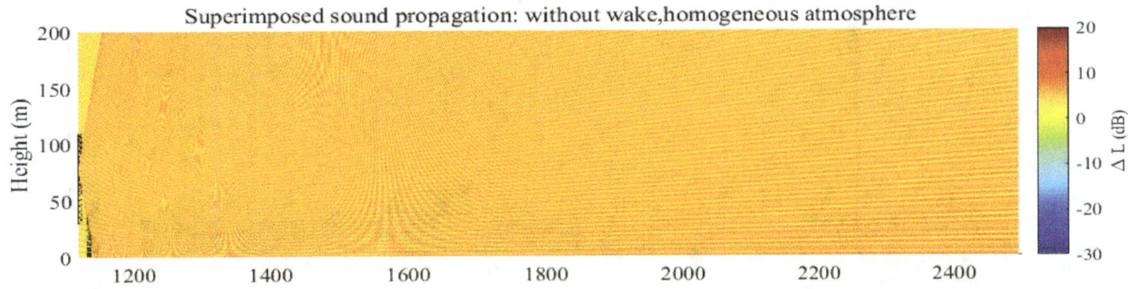

(a) Case 1

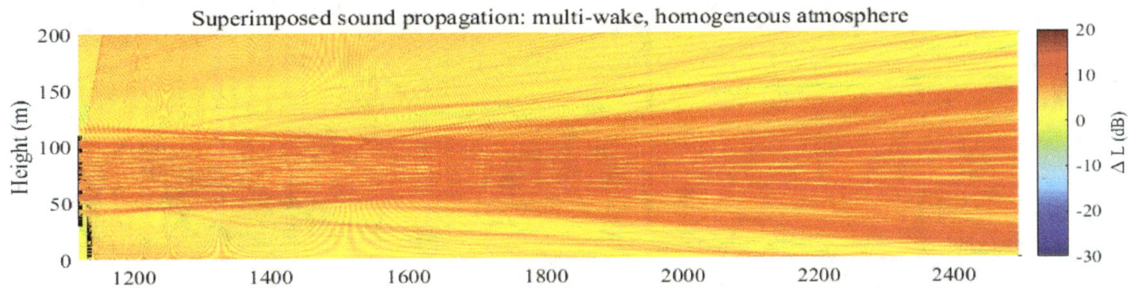

(b) Case 2

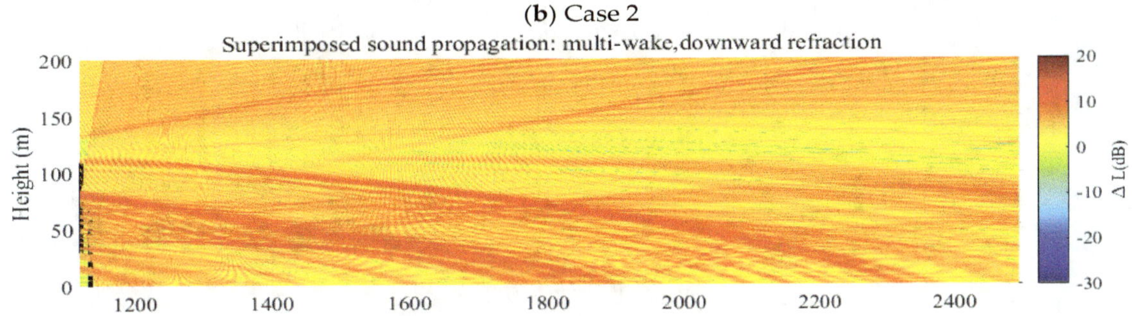

(c) Case 3

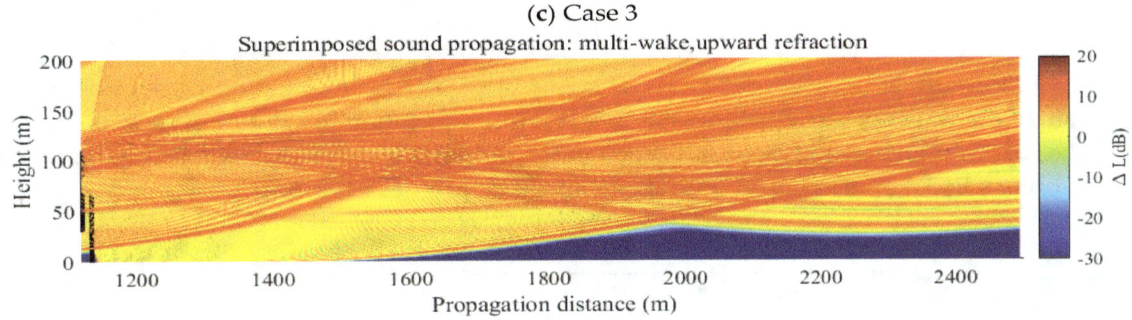

(d) Case 4

Figure 16. ANPropagation from all ANSources: (**a**) Case 1, homogeneous atmospheric condition, no wake effect; (**b**) Case 2, homogeneous atmospheric condition with multi-wakes; (**c**) Case 3, downward refraction and multi-wakes; (**d**) Case 4, upward refraction and multi-wakes.

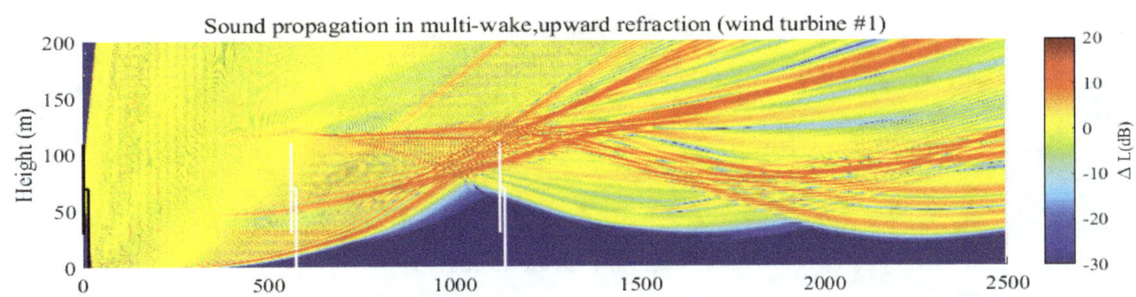

(**a**)

Figure 17. *Cont.*

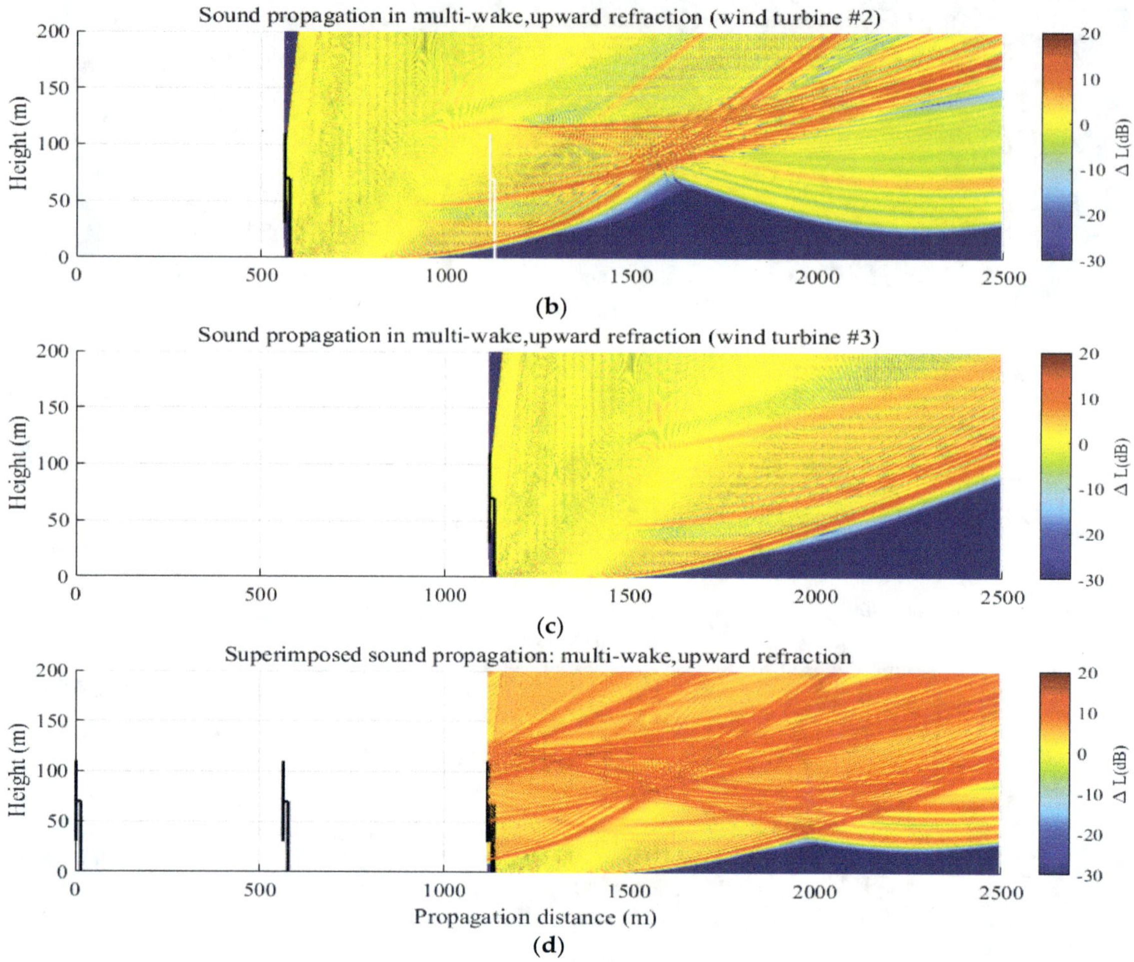

Figure 17. RSPL simulated at f = 1000 Hz and $b = -3$: (**a**) ANPropagation by wind turbine #1 downstream of the wakes of wind turbines # 1, 2, 3; (**b**) ANPropagation by wind turbine #2 downstream of the wakes of wind turbines #2, 3; (**c**) ANPropagation by wind turbine #3 downstream of the wakes of wind turbine #3; (**d**) superimposed RSPL from all wind turbines #1, 2, 3.

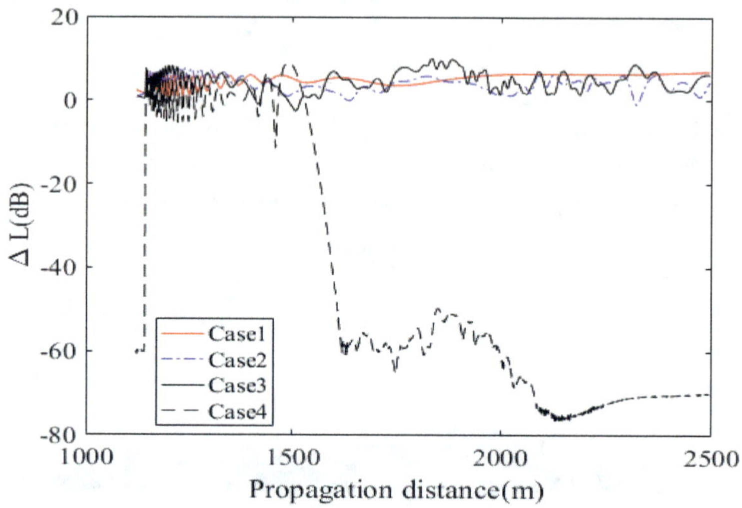

Figure 18. The RSPL at 2 m height extracted from four cases of Figure 16.

Similar to the procedures previously mentioned for Figure 15, Figure 19a–c show the noise spectra from wind turbines #1, 2, and 3, respectively. The difference is that the three wind turbines are individually considered for its noise generation and propagation at the given frequencies (20 Hz–5000 Hz). Due to the presence of velocity deficit in the wake, the inflow conditions of the two downstream wind turbines are slightly different, and thus the ANSources of the three wind turbines are slightly different from each wind turbine. In each figure, the original ANSource is shown on top of the other noise spectra. Similar to the situations in Figure 15, geometrical attenuation plays an primary role in the large noise reduction for ANPropagation at $f < 1000$ Hz, and the noise absorption by air has a strong effect at $f > 1000$ Hz. As expected, Case 3 shows more distinct characteristics for the frequencies. In Figure 19, the receiver is fixed at $x = 2500$ m, $z = 2$ m, therefore the sound pressure is largely attenuated for wind turbine #1 due to the longest propagation distance. A similar behavior is seen for wind turbines #2 and #3, where the noise level is the highest for wind turbine #3 due to the shortest relative distance from source to receiver. Figure 19a–c also show that at different relative propagation distances 2500 m (#1), 1940 m (#2) and 1380 m (#3), the distribution of SPL maintains some similarity. The differences of these sound spectra come from the combined effect of wake flow and the propagation distance. Finally, the total noise received from the three wind turbines can be got through logarithmically adding the sound spectra from individual wind turbines. In Figure 19d, the sources Lp1, Lp2, and Lp3 represent the ANSource level of the three wind turbines. The major difference is seen at the middle frequency range, which is due to the different input velocity data obtained from flow simulation. Case 1 and Case 3 all show that the most significant influence comes from wind turbine #3, which is not a surprise. To highlight more detailed difference in the noise spectra, a few typical values are compared in Table 1. At $f = 20$ Hz, the summated SPL from three turbines (Case 3) is 37.92 dB which is 3 dB higher than that from wind turbine #3 (Case 3). This difference becomes smaller when frequency increases. This implies that the low frequency noise generated from the far upstream wind turbines still plays an important role. On the other hand, if a simple sound propagation algorithm is applied, such as shown in the last column of Table 1, a large error is observed in a frequency range below 1000 Hz. It indicates that the high frequency noise will be heavily attenuated over long propagation distance except in special cases such as sound propagation over water (hard surface), stronger downward refraction under special climate conditions.

Table 1. Comparisons of SPL at some selected frequencies

	Case 3 (Figure 19c)	**Case 3 (Figure 19d)**	$L_w - 10\log_{10}4\pi D^2 - \alpha D$ **(Figure 19d)**
20 (Hz)	34.64 dB	37.92 dB	33.89 dB
100 (Hz)	12.61 dB	15.05 dB	28.53 dB
250 (Hz)	26.25 dB	28.33 dB	27.57 dB
1000 (Hz)	17.10 dB	17.99 dB	18.00 dB

Finally, it is interesting to predict the noise spectra at different receivers. The receiver height is again fixed at $z = 2$ m, and the horizontal distances measured from wind turbine #1 are: $x = [1620, 1820, 2020, 2220, 2420]$ m. The noise spectra are shown in Figure 20 and marked with different colors. The general trend is that the noise level decreases as observer distance increases. However, it does not always hold true at some distances and frequencies. For example, at $x = 1820$ m (22.75D), except at low frequencies, increase of noise level is observed as compared to the $x = 1620$ m case. The simulations show that the low frequency noise might be the major wind turbine noise source in the farfield. Since the spectra are obtained in farfield, the overall SPLs are quite low. According to the wind turbine noise regulations in many countries (such as 45 dB), the calculated noise evidently fulfill the limit. From the results, it is seen that SPLs over 1000 Hz is below 20 dB. For future studies, it is indicated that more focus should be paid on long-path ANPropagation at low frequencies.

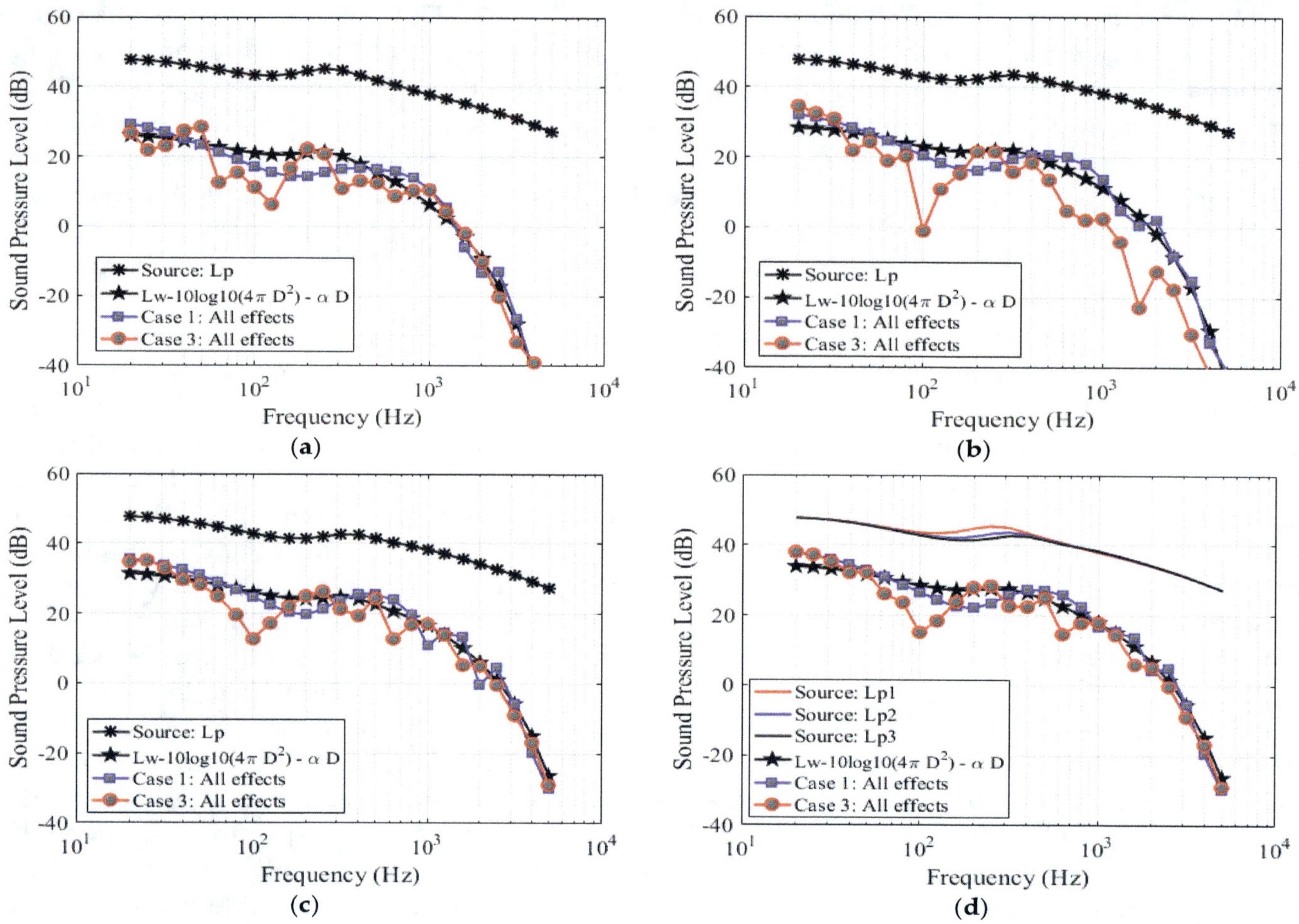

Figure 19. Receiver is fixed at $x = 2500$ m away from wind turbine #1, and the receiver height is 2 m. (**a**) SPL from wind turbine #1 (Source: Lp) to the receiver; (**b**) SPL from wind turbine #2 (Source: Lp) to the receiver; (**c**) SPL from wind turbine #3 (Source: Lp) to the receiver; (**d**) Total SPL from wind turbines #1, 2, 3 (Source: Lp1, Lp2, Lp3).

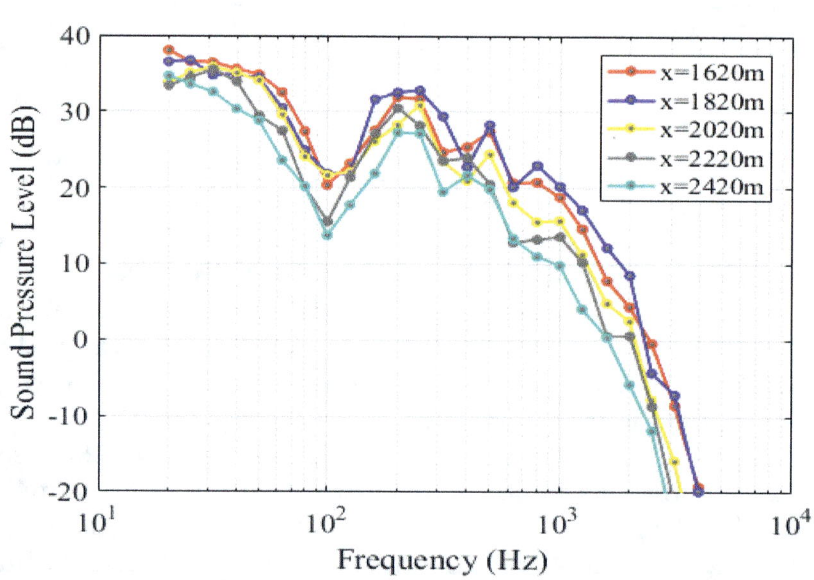

Figure 20. Total SPL from wind turbines #1, 2, 3 received at various distance.

4. Conclusions

Combing the flowfield aerodynamic simulation and the aeroacoustics computation of multi-wind turbines is the focus of this paper. To achieve this objective, ANSource modeling, flow simulation and ANPropagation are individually described and then integrated. The numerical computations are carried out using the in-house developed tool EllipSys. The ANSource modeling and ANPropagation modeling are implemented in the platform of EllipSys. The steady CFD simulations are performed to obtain the flowfield with a relatively efficient 2D-AD solver. The ANSource of wind turbine is calculated using the flowfield information from CFD computations. Then the ANPropagation is simulated which needs the data from both CFD simulations and ANSource computations. Results from different case studies show that a long-path ANPropagation are greatly influenced by different flow conditions, atmosphere conditions and wake condition. An overall trend is that the ANSource from wind turbines is broadband, but the received noise level at long travelling distances from multi-wind turbines is complicated. The noise spectra show that the low frequency noise from a wind turbine cluster might be an important issue to be further investigated. It has been shown in the results that at a distance over 1 km (about 12 rotor diameter), the high frequency components are largely attenuated, say at $f > 1000$ Hz. In a large wind farm with superimposition of noise sources from multi-wind turbines, the computational load will be heavy if high frequency noise is simulated. The ray theory, as compared in Sections 3.1–3.2, is a high-frequency approach therefore the low frequency solutions are not very accurate if strong refraction and diffraction are represented. The present numerical approach has same accuracy over all frequency range. However, to solve high frequency ANPropagation, higher grid density is required, thus increasing computational time. Fortunately, results presented in Figures 18 and 19 indicate that it is the low frequency noise components that contribute to the long range wind turbine ANPropagation. The receiver distances shown in Figure 19 ranges from 20 to 30 rotor diameters. It is clear that at such a large distance, attenuation of low frequency noise is still not significant.

Author Contributions: Conceptualization, W.J.Z. and Z.S.; Software, W.J.Z. and E.B.; Validation, Z.S. and W.J.Z.; Formal analysis, Z.S. and W.J.Z.; Resources, E.B., W.Z.S. and J.N.S.; Data curation, Z.S., W.J.Z., and J.C.; Writing—original draft preparation, Z.S. and W.J.Z.; Writing—Z.S., W.J.Z., W.Z.S., J.N.S., J.C., and H.Y.; Visualization, Z.S. and W.J.Z.; Supervision, W.J.Z.; Project administration, W.J.Z.; Funding acquisition, W.J.Z.

Acknowledgments: The authors wish to express acknowledgement to the National Nature Science Foundation of China under grant number 11672261.

References

1. Glauert, H. *Airplane Propellers. Aerodynamic Theory*; Springer: Berlin/Heidelberg, Germany, 1935; pp. 169–360.
2. Sun, Z.Y.; Chen, J.; Shen, W.Z.; Zhu, W.J. Improved blade element momentum theory for wind turbine aerodynamic computations. *Renew. Energy* **2016**, *96*, 824–831. [CrossRef]
3. Sørensen, J.N.; Shen, W.Z. Numerical Modeling of Wind Turbine Wakes. *J. Fluids Eng.* **2002**, *124*, 393. [CrossRef]
4. Sørensen, J.N.; Shen, W.Z.; Munduate, X. Analysis of wake states by a full-field actuator disc model. *Wind Energy* **1998**, *1*, 73–88. [CrossRef]
5. Lighthill, M.J. On Sound Generated Aerodynamically. I. General Theory. *Proc. R. Soc. Lond.* **1952**, *211*, 564–587. [CrossRef]
6. Curle, N. The Influence of Solid Boundaries upon Aerodynamic Sound. *Proc. R. Soc. Lond.* **1955**, *231*, 505–514. [CrossRef]
7. Williams, J.E.; Hall, L.H. Aerodynamic sound generation by turbulent flow in the vicinity of a scattering half plane. *J. Fluid Mech.* **2006**, *40*, 657–670. [CrossRef]
8. Hardin, J.C.; Pope, D.S. An acoustic/viscous splitting technique for computational aeroacoustics. *Theor. Comput. Fluid Dyn.* **1994**, *6*, 323–340. [CrossRef]

9. Zhu, W.J.; Shen, W.Z.; Sørensen, J.N. High-order numerical simulations of flow-induced noise. *Int. J. Numer. Methods Fluids* **2011**, *66*, 17–37. [CrossRef]
10. Zhu, W.J.; Shen, W.Z.; Barlas, E.; Bertagnolio, F.; Sørensen, J.N. Wind turbine noise generation and propagation modeling at DTU Wind Energy: A review. *Renew. Sustain. Energy Rev.* **2018**, *88*, 133–150. [CrossRef]
11. Salomons, E.M. *Computational Atmospheric Acoustics*; Springer Science & Business Media: Dordrecht, The Netherlands, 2001; ISBN 978-1-4020-0390-5.
12. Ostashev, V.E.; Wilson, D.K. Acoustics in Moving Inhomogeneous Media. *J. Acoust. Soc. Am.* **1999**, *105*, 2067. [CrossRef]
13. Blancbenon, P.; Dallois, L.; Juvé, D. Long Range Sound Propagation in a Turbulent Atmosphere Within the Parabolic Approximation. *Acta Acust. United Acust.* **2001**, *87*, 659–669. [CrossRef]
14. Brooks, T.F.; Pope, D.S.; Marcolini, M.A. *Airfoil Self-Noise and Prediction*; National Aeronautics and Space Administration: Washington, DC, USA, 1989.
15. Zhu, W.J.; Heilskov, N.; Shen, W.Z.; Sørensen, J.N. Modeling of Aerodynamically Generated Noise From Wind Turbines. *J. Sol. Energy Eng.* **2005**, *127*, 517–528. [CrossRef]
16. Lau, A.S.H.; Kim, J.W.; Hurault, J.; Vronsky, T. A study on the prediction of aerofoil trailing-edge noise for wind-turbine applications. *Wind Energy* **2017**, *20*, 233–252. [CrossRef]
17. Lee, S.; Lee, D.; Honhoff, S. Prediction of far-field wind turbine noise propagation with parabolic equation. *J. Acoust. Soc. Am.* **2016**, *140*, 767–778. [CrossRef] [PubMed]
18. Heimann, D.; Käsler, Y.; Gross, G. The wake of a wind turbine and its influence on sound propagation. *Meteorol. Z.* **2011**, *20*, 449–460. [CrossRef]
19. Barlas, E.; Zhu, W.J.; Shen, W.Z.; Kelly, M.; Andersen, S.J. Effects of wind turbine wake on atmospheric sound propagation. *Appl. Acoust.* **2017**, *122*, 51–61. [CrossRef]
20. Barlas, E.; Zhu, W.J.; Shen, W.Z.; Dag, K.O.; Moriarty, P. Consistent modelling of wind turbine noise propagation from source to receiver. *J. Acoust. Soc. Am.* **2017**, *142*, 3297. [CrossRef]
21. Delany, M.E.; Bazley, E.N. Acoustical properties of fibrous absorbent materials. *J. Acoust. Soc. Am.* **2005**, *48*, 105–116. [CrossRef]
22. Jónsson, G.B.; Jacobsen, F. A Comparison of Two Engineering Models for Outdoor Sound Propagation: Harmonoise and Nord2000. *Acta Acust. United Acust.* **2008**, *94*, 282–289. [CrossRef]
23. Plovsing, B.; Kragh, J. *Nord2000-Comprehensive Outdoor Sound Propagation Model. Part 2: Propagation in an Atmosphere with Refraction*; Delta Acoustics & Vibration Report; DELTA: Copenhagen, Denmark, 2006.
24. Mikkelsen, R. Actuator Disc Methods Applied to Wind Turbines. Ph.D. Thesis, Technical University of Denmark, Copenhagen, Denmark, 2003.
25. Michelsen, J.A. *Basis3D-A Platform for Development of Multiblock PDE Solvers*; Technical Report; University of Denmark: Copenhagen, Denmark, 1992.
26. Sørensen, N. *General Purpose Flow Solver Applied over Hills*; Technical Report; Risø National Laboratory: Roskilde, Denmark, 1995.
27. Tian, L.L.; Zhu, W.J.; Shen, W.Z.; Zhao, N.; Shen, Z.W. Development and validation of a new two-dimensional wake model for wind turbine wakes. *J. Wind Eng. Ind. Aerodyn.* **2015**, *137*, 90–99. [CrossRef]
28. Wu, Y.T.; Porté-Agel, F. Modeling turbine wakes and power losses within a wind farm using LES: An application to the Horns Rev offshore wind farm. *Renew. Energy* **2015**, *75*, 945–955. [CrossRef]
29. Sarmast, S.; Segalini, A.; Mikkelsen, R.F.; Ivanell, S. Comparison of the near-wake between actuator-line simulations and a simplified vortex model of a horizontal-axis wind turbine. *Wind Energy* **2016**, *19*, 471–481. [CrossRef]
30. Troldborg, N.; Sørensen, J.N.; Mikkelsen, R. Numerical simulations of wake characteristics of a wind turbine in uniform inflow. *Wind Energy* **2010**, *13*, 86–99. [CrossRef]
31. Troldborg, N.; Zahle, F.; Réthoré, P.E.; Sørensen, N.N. Comparison of wind turbine wake properties in non-sheared inflow predicted by different computational fluid dynamics rotor models. *Wind Energy* **2015**, *18*, 1239–1250. [CrossRef]
32. Tian, L.L.; Zhu, W.J.; Shen, W.Z.; Sørensen, J.N.; Zhao, N. Investigation of modified AD/RANS models for wind turbine wake predictions in large wind farm. *J. Phys.* **2014**. [CrossRef]
33. Tian, L.L.; Zhu, W.J.; Shen, W.Z.; Song, Y.L.; Zhao, N. Prediction of multi-wake problems using an improved Jensen wake model. *Renew. Energy* **2017**, *102*, 457–469. [CrossRef]

34. Zhu, W.J.; Shen, W.Z.; Sørensen, J.N.; Yang, H. Verification of a novel innovative blade root design for wind turbines using a hybrid numerical method. *Energy* **2017**, *141*, 1661–1670. [CrossRef]
35. Wilson, D.K.; Pettit, C.L.; Ostashev, V.E. Sound propagation in the atmospheric boundary layer. *Acoust. Today* **2015**, *11*, 44–53. [CrossRef]

4

Aerodynamic Force and Comprehensive Mechanical Performance of a Large Wind Turbine during a Typhoon based on WRF/CFD Nesting

Shitang Ke [1,2,*], Wenlin Yu [1], Jiufa Cao [2] and Tongguang Wang [2]

1 Department of Civil Engineering, Nanjing University of Aeronautics and Astronautics, 29 Yudao Road, Nanjing 210016, China; nuaayuwenlin@163.com

2 Jiangsu Key Laboratory of Hi-Tech Research for Wind Turbine Design, Nanjing University of Aeronautics and Astronautics, Nanjing 210016, China; caojiufa98@163.com (J.C.); tgwang@nuaa.edu.cn (T.W.)

* Correspondence: keshitang@163.com

Featured Application: The main research conclusions of this paper could provide references for load prediction of similar wind turbine systems during the occurrence of typhoon, and deepen the understanding on meso/microscale wind field nesting mechanism.

Abstract: Compared with normal wind, typhoons may change the flow field surrounding wind turbines, thus influencing their wind-induced responses and stability. The existing typhoon theoretical model in the civil engineering field is too simplified. To address this problem, the WRF (Weather Research Forecasting) model was introduced for high-resolution simulation of the Typhoon "Nuri" firstly. Secondly, the typhoon field was analyzed, and the wind speed profile of the boundary layer was fitted. Meanwhile, the normal wind speed profile with the same wind speed of the typhoon speed profile at the gradient height of class B landform in the code was set. These two wind speed profiles were integrated into the UDF (User Defined Function). On this basis, a five-MW wind turbine in Shenzhen was chosen as the research object. The action mechanism of speed was streamlined and turbulence energy surrounding the wind turbine was disclosed by microscale CFD (Computational Fluid Dynamics) simulation. The influencing laws of a typhoon and normal wind on wind pressure distribution were compared. Finally, key attention was paid to analyzing the structural response, buckling stability, and ultimate bearing capacity of the wind turbine system. The research results demonstrated that typhoons increased the aerodynamic force and structural responses, and decreased the overall buckling stability and ultimate bearing capacity of the wind turbine.

Keywords: typhoon; wind turbine; meso/microscale; aerodynamic force; mechanical performance

---

1. Introduction

A large wind turbine structure is more flexible [1] and has a more prominent wind-induced dynamic effect [2]. It is a typical wind-sensitive structure. Wind load is the control load of its structural design [3,4]. In particular, the wind-induced failure of large wind turbine structures occur frequently during strong typhoons [5,6]. For example, the Typhoon "Cuckoo" in 2003 affected the normal operation of wind power plants in south China, which halted 13 wind turbines. The Typhoon "Saomai" in 2006 landed on Zhejiang, and caused different degrees of failure of 20 units in the Cangnan Wind Power Plant. In 2014, the Typhoon "Rammasun" landed at Hainan Province, which caused the collapse of—and serious damage to—the 1.5-MW wind turbine with a 77-m blade diameter in the wind power plant. Compared with normal wind, typhoons form a more complicated near-ground wind field. The wind field characteristics of high turbulence, changeable direction, and great changes of shearing wind speed will intensify the fluctuation of the wind turbine structure. Under this circumstance, airflow movement on the near-wall surface is disorganized, and the surface pressure is changed significantly, which further caused significant changes in the wind-induced responses and the stability of the wind turbine system [7,8]. For these reasons, a contrastive analysis of aerodynamic performances and comprehensive performances of the large wind turbine system during typhoons and normal wind was of important theoretical significance and engineering values.

For studies concerning wind turbine performance during typhoon, Takeshi et al. [9] studied influences of blade, wind direction, and the relative position of a doorway on the collapse of wind turbines according to wind turbine collapse data and measured typhoon speed in Japan. Tang et al. [10] and Chou and Tu [11] analyzed stresses on the wind turbine structure under typhoon load by the simplified static analysis method. Li et al. [12] analyzed typhoon characteristics and the damages of different parts of wind turbines. They pointed out that the design of different parts of a wind turbine has to strengthen the structural strength and pay attention to the yaw of the wind turbine under typhoon load. Wang et al. [6] and Luo et al. [13] carried out a systematic analysis on wind-induced static/dynamic responses of the tower in the wind turbine system during a typhoon by combining a theoretical deduction and fluctuating wind spectrum. Some suggestions regarding the design of the typhoon resistance of the tower body in the wind turbine system were given. Lian et al. [14] and Zhou et al. [15] carried out a systematic analysis on the aerodynamic load on blades of the wind turbine system under typhoon loads by combining a theoretical analysis and measured data. They established a typhoon-resistance aerodynamic blade design optimization model for wind turbines. Cao [16] and An [17] analyzed the flow field and pressure field and structural stresses of the wind turbine under typhoon loads by theoretical studies and CFD numerical simulation. A new typhoon-resistance wind turbine design philosophy and structural design was proposed. Utsunomiya et al. [18] and Ma et al. [19] carried out a statistical analysis on the dynamic response and mooring linear tension response of a submarine platform of an offshore wind turbine under typhoon loads. They attached high attentions to the aerodynamic shape design and typhoon resistance characteristics of blades.

Most of the above studies concerning the typhoon resistance of a wind turbine are based on theoretical analysis and measured data. The theoretical systems in these studies were too simplified to reflect the specific wind profile and landing attenuation effect of a mesoscale typhoon field. Nowadays, mesoscale models have been well applied in case analyses of typhoons, early warnings, and disaster forecasts. Common models include Weather Research Forecasting (WRF), MM5 (Mesoscale Model Version 5), and ETA. Among these models, MM5 enjoyed widespread applications in earlier studies, because it was studied and used earlier. However, it has been never adopted by the NCEP (National Centers for Environmental Prediction), on the grounds that its dynamical framework is obsolete, and its procedures are not fairly standard. Utilized by the NCEP for operational forecast, ETA can hardly promptly assimilate the outstanding research findings of scientific research institutes and universities, although its simulation results are reliable. As a result, its promotion is restricted. Comparatively, the WRF model based on fluid dynamics and thermodynamics can not only effectively simulate wind speed, pressure, and temperature in a typhoon field, it can also comprehensively consider the evolution process, strong specificity, and attenuation characteristics of typhoons. In spite of its parametric schemes and the more complicated setting of some of its parameters, it takes more complete and accurate environmental parameters into account. Therefore, WRF is used most widely among all of the existing models. However, due to the large sphere of scope (hundreds of kilometers) of typhoons, grid resolution of the typhoon field is generally at the magnitude of kilometers. However, the overall dimension of a large wind turbine system is only at the magnitude of hectometers. To predict the aerodynamic load at blade edges accurately, it has to deepen into the near-ground boundary layer. The minimum grid size of the near-wall surface is generally lower than 10^{-2} m. The mesoscale model in WRF fails completely at this magnitude. Hence, it will be helpful for observing the airflow motion in a typhoon field by a more refined simulation of microscale typhoon fields based on the mesoscale WRF. WRF and Computational Fluid Dynamics (CFD), which are new-generation mesoscale and microscale forecast models respectively, are employed most widely as technologies for high-resolution and refined wind field forecasts. In the literature [20,21], refined numerical simulations were conducted on the bridge sites of mountainous areas by multi-scale coupling. The results suggested that the results of WRF could be more adaptable to inlet boundaries during numerical simulations of these wind fields by CFD. Thus, the existing problem about the mean wind speed at the inlet during the numerical simulations was successfully solved. In the literature [22–24], the complicated wind fields of lakes on the underlying surface were simulated meticulously by coupling WRF and CFD. According to the simulation results, WRF–CFD coupling better eliminates the mean deviation and the root mean squared error than WRF. Besides, it reflects the atmospheric characteristics of actual microscale topographical changes more accurately. Therefore, the meso/microscale nesting WRF/CFD has to be used for the high-precision simulation of a multi-scale flow field of large wind turbines under typhoon loads in order to solve various problems, such as the high-precision transition of parameters and flowing structure, multilayer and multi-scale grid nesting, multi-time scale control, and multi-scale mutation [25].

To solve these problems, the mesoscale WRF model was introduced for high temporal spatial resolution simulation of Typhoon "Nuri". After the typhoon field information was analyzed, the wind speed profile in the simulation region was acquired based on the nonlinear least square fitting. On this basis, the normal wind speed profile with the same wind speed of the typhoon speed profile at the gradient height of the class B landform was set. Later, the aerodynamic distribution on the surface of a five-MW horizontal axis wind turbine in a wind power plant in Shenzhen (China) under typhoon loads and normal wind was studied by microscale CFD numerical simulation. Finally, the mechanical performance and stability performance of a large wind turbine system under typhoon loads and normal wind were compared by combining them with the finite element method.

2. Mesoscale Typhoon Field Simulation and Analysis

2.1. WRF Model

WRF is the mesoscale weather research and forecasting system that was developed by the America Environmental Forecasting Center, National Center of Atmospheric Research of the United States and Rainstorm Analysis and Forecasting Center of University of Central Oklahoma [25]. The Euler equation model based on a non-static force equilibrium was used as the WRF–ARW (Advanced Research WRF) model of the dynamic framework. The horizontal computational domain adopted the Arakawa-C meshing scheme [26] (Figure 1a). The Arakawa-C grid can express scalar and vector parameters simultaneously, which is conducive to increase the accuracy of a high-resolution simulation. Besides, the Arakawa-C grid has good dispersion properties and conservativeness. The vertical computational domain applied the Euler mass coordinate η along the terrain (Figure 1b):

$$\eta = (p_h - p_{ht})/\mu, \quad \mu = p_{hs} - p_{ht} \tag{1}$$

where p_h is the air pressure at the target point p_{ht} is the air pressure of the top layer, which is a constant, and p_{hs} is the near-ground air pressure.

The WRF model is transferable when it is run on the Linux system. With considerations to the influences of physical processes such as water vapor, long-wave radiations, short-wave radiations, cumulus cloud, and the underlying surface, the WRF model can simulate airflow, air pressure, and wind field characteristics in a large region. The results can be used as the input boundary conditions for CFD numerical simulation.

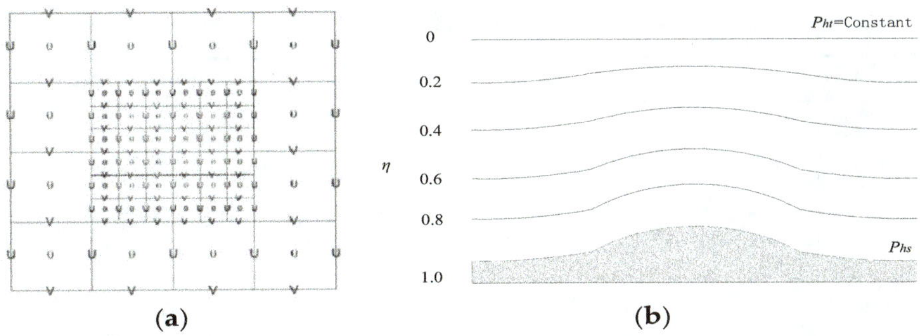

Figure 1. Meshing of the Weather Research Forecasting (WRF) model. (**a**) Horizontal grids (space between coarse and thin grids is 3:1); (**b**) Vertical grids.

2.2. Selection of Physical Scheme and Parameter Settings

The #12 Typhoon "Nuri" in 2008 was chosen as the simulation object. Based on the information provided by the Typhoon Website of China Meteorological Administration, it landed on coastal regions of Saigon, Hong Kong on 16:55 on 22 August and the maximum wind near the center was 12-level (33 m/s). On the same day, it landed again at Nanlang Town, Zhongshan City, Guangdong Province at 22:10, and the maximum wind near the center was eight-level (25 m/s). For effective numerical simulation of typhoon path as well the air pressure field and wind speed field, the simulation covers the whole process from the first landing at Saigon, Hong Kong to the second landing at Zhongshan City, Guangdong Province. The initial atmospheric conditions and time-dependent boundary conditions, which were applied in the WRF model, were all based on the FNL (Final Global Data Assimilation System) [27–30]. It covered 27 atmospheric layers. The sampling interval was set at 6 h, and the spatial resolution was 1° × 1°. The WRF simulation region had 37 vertical layers, and the air pressure at the top layer was set 5000 Pa.

The simulation center chose a district in north Shenzhen. Typhoon "Nuri" ran through this region at 20:00 on 22 August. With comprehensive considerations to data demands and computational conditions, the three-dimensional (3D) nesting program was used to calculate typhoon field in the WRF model. The horizontal mesh resolution was 13.5 km, 4.5 km, and 1.5 km, respectively. The longitude and latitude of center of the simulation region was (114.1° E, 22.5° N). The coarse grid space D01 in the external surface was 13.5 km, and the number of grids was 211 × 211. The grid space D02 in the second layer was 4.5 km, and the number of grids was 217 × 217. The grid space D03 in the inner layer was 1.5 km, and the number of grids was 241 × 241. The map projection applied the Lambert program. A total of 37 vertical layers were set, forming upper-sparse and lower-dense stratification patterns. There were 19 layers below 1000 m. The simulation domain is shown in Figure 2. The altitudes of different vertical layers are shown in Table 1.

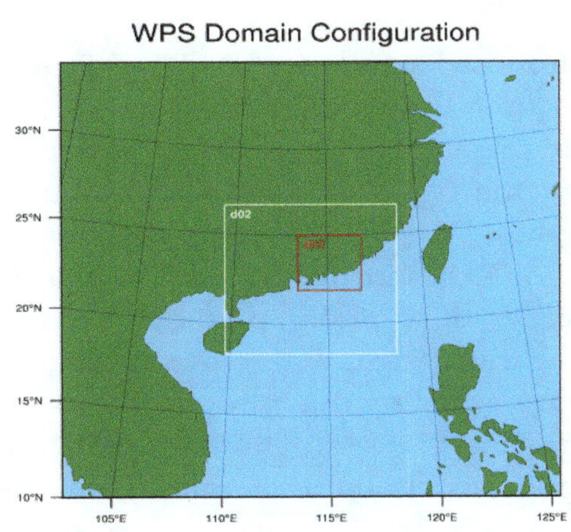

Figure 2. Simulation domain of the Typhoon "Nuri".

Table 1. Altitudes of different vertical layers.

Layers	Altitude/m	Layers	Altitude/m
0	0	19	997.85
1	5.41	20	1441.03
2	21.64	21	1864.44
3	43.28	22	2383.35
4	64.96	23	2929.90
5	86.67	24	3508.54
6	108.42	25	4124.07
7	130.20	26	4780.99
8	152.01	27	5484.65
9	179.34	28	6243.02
10	217.70	29	7067.05
11	272.71	30	8489.99
12	350.13	31	10,070.22
13	439.15	32	11,332.60
14	528.81	33	12,805.37
15	619.11	34	14,592.36
16	710.08	35	17,014.78
17	824.83	36	19,617.33
18	987.29	37	——

The hub height of the wind turbine is generally lower than 150 m, which is below the thickness of the boundary layer. The hub height is significantly influenced by airflow mass, humidity, and heat transmission in the boundary layer [31–35]. Therefore, the settings of the physical parameterization scheme can affect the simulation accuracy of boundary layer of the typhoon field directly. In this paper, the physical parameters of WRF model are listed in Table 2.

Table 2. Physical parameter settings of the WRF model.

WRF Parameters	Main Zone (D01)	Nesting Zone (D02)	Nesting Zone (D03)
Horizontal resolution	13.5 km	4.5 km	1.5 km
Integral time step		180 s	
Microphysical process scheme		Lin	
Long-wave radiation		RRTM	
Short-wave radiation		Dudhia	
Near-ground layer scheme		Monin-Obukhov	
Land surface process scheme		Noah	
Planet boundary layer scheme		MYJ	
Cumulus convective parameterization scheme		Kain-Fritsch	

2.3. Simulation Results and Discussions

To verify the validity of the proposed simulation model of Typhoon "Nuri", the typhoon path and minimum sea level pressure throughout the simulation are shown in Figure 3. The simulated path was drawn according to the minimum pressure in the cyclonic center of the WRF model. The measured path was the best path provided by the Typhoon Website of the China Meteorological Administration, and it could also refer to some literature cites [36,37]. The simulated path and the measured path were stable, and both paths moved toward the northwest. In early simulations, the simulated path was close to the measured path. However, the typhoon path deflected gradually as time went on, which might be caused by the intensifying initial error with the increase of simulation time. The evolution of typhoon intensity was mainly represented by the minimum sea level pressure. The simulated value and measured value of minimum sea level pressure in the simulation period differed slightly, but the distribution trend was basically consistent. Both the simulated value and the measured value increased slowly with time, and increased sharply in the late simulation period. At this moment, the typhoon became weak and disappeared gradually.

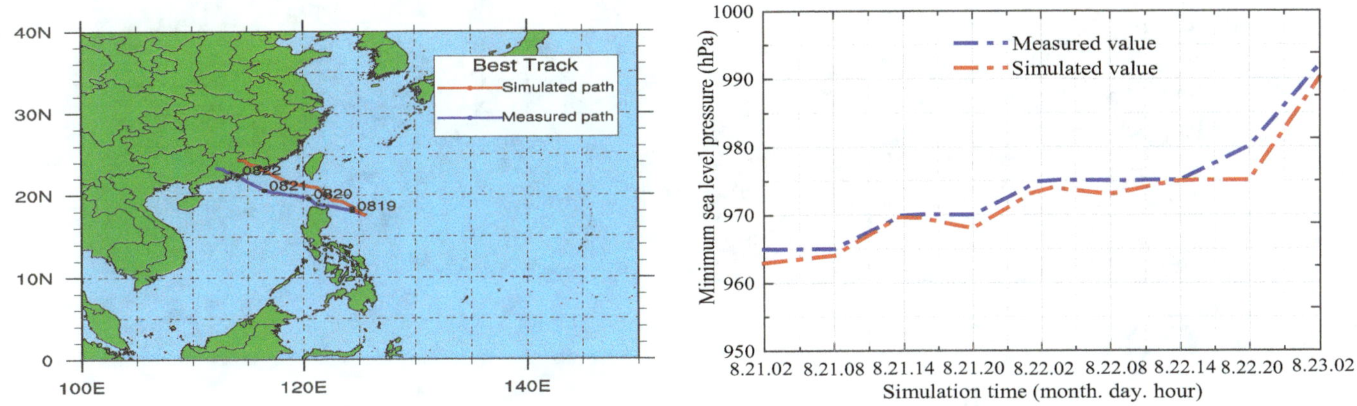

(a) (b)

Figure 3. Typhoon path and minimum sea level pressure throughout the simulation. (**a**) Typhoon path; (**b**) Minimum sea level pressure.

Simulation results and measured results in the simulation regions at typical height for the landing of the typhoon at 20:00 on 22 August are shown in Figures 4 and 5. These nephograms are conducive to analyzing the wind field information when the Typhoon "Nuri" ran through the center of the simulation region. It can be concluded from the analysis that:

(1) The typhoon had a low-pressure center. The land underlying the surface was rougher than the ocean after landing, thus cutting the heat source. Hence, the upflow was weakened, while the surrounding airflows continued radiation toward the typhoon center, which gradually further increased the air pressure. After comparison, it was found that the simulated pressure distribution agreed well with the measured results.

(2) Wind speed increased due to the invasion of the peripheral nephsystem of the typhoon. The wind speed increased continuously as the typhoon approached. The wind speed closer to the center was the higher.

(3) Influenced by the downward transmission of atmospheric heats, the MYJ boundary layer scheme contributed a high computational accuracy of turbulent fluctuation fluxes, such as heats and momentum in the boundary layer of the region. This implied that the low atmosphere heats were mainly carried into the boundary layer by the upper atmosphere through high temperature, which caused the hybrid heating of the low atmosphere.

(4) Cloud precipitation close to the typhoon belongs to the long-term convective precipitation. The large rainfall volume in this region is related to the typhoon strength and the central distance. The Noah pavement scheme can simulate the surface flux and airflow convergence field well.

(5) There is a minor difference between the simulation results and the measured results of the typhoon, but the distribution trend is basically consistent. In the central area of the typhoon simulation, the ranges of wind pressure, wind speed, temperature, and rainfall were 860–920 Pa, 22–26 m/s, 294–297 K, and 40–48 mm/h, respectively. These data are close to the measured data on the website of the National Meteorological Center. The maximum error is lower than 15%.

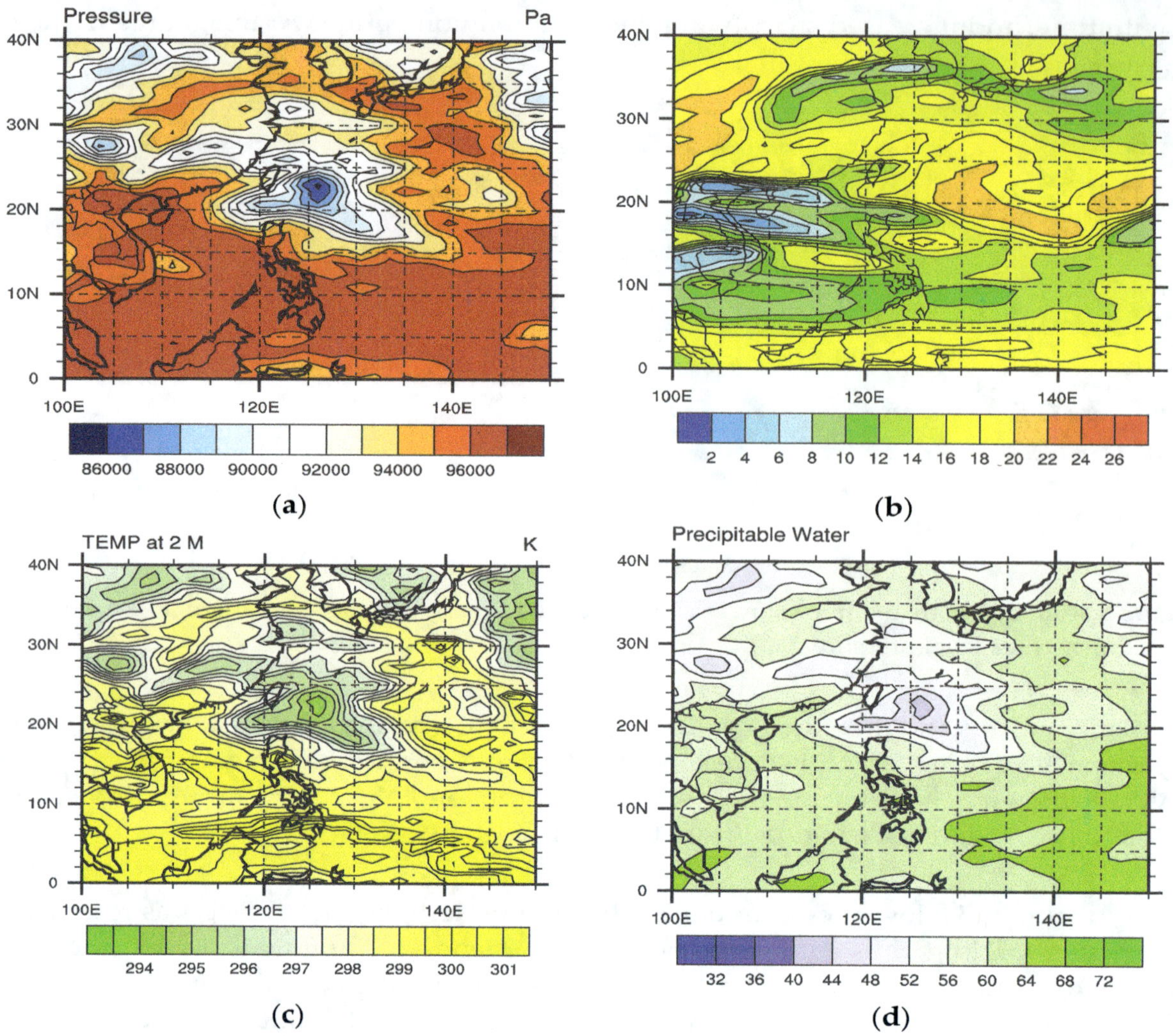

Figure 4. Simulation results for landing of typhoon. (**a**) Air pressure nephogram; (**b**) Wind speed nephogram; (**c**) Temperature nephogram; (**d**) Rainfall nephogram.

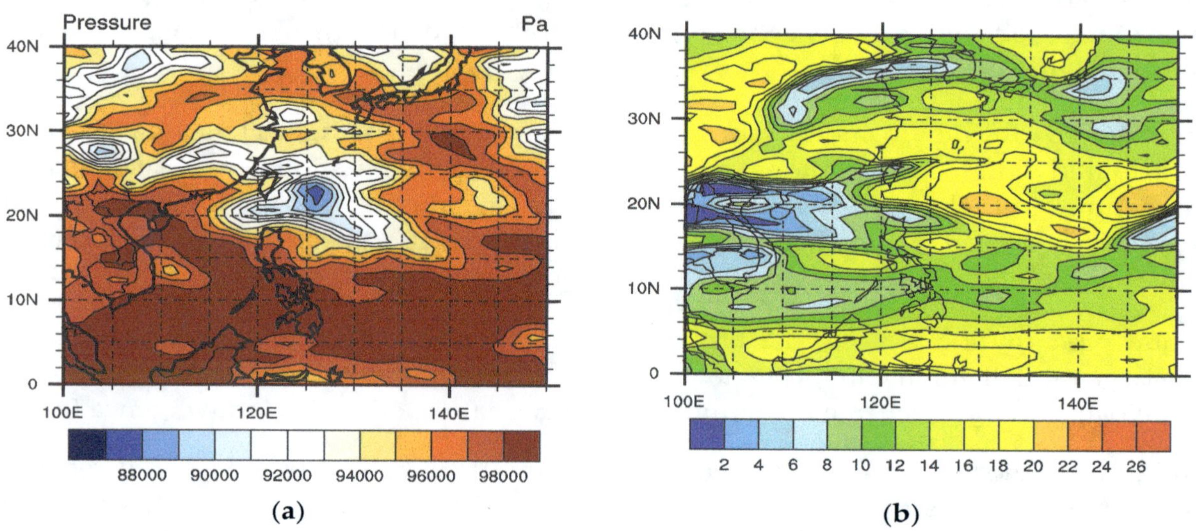

Figure 5. *Cont.*

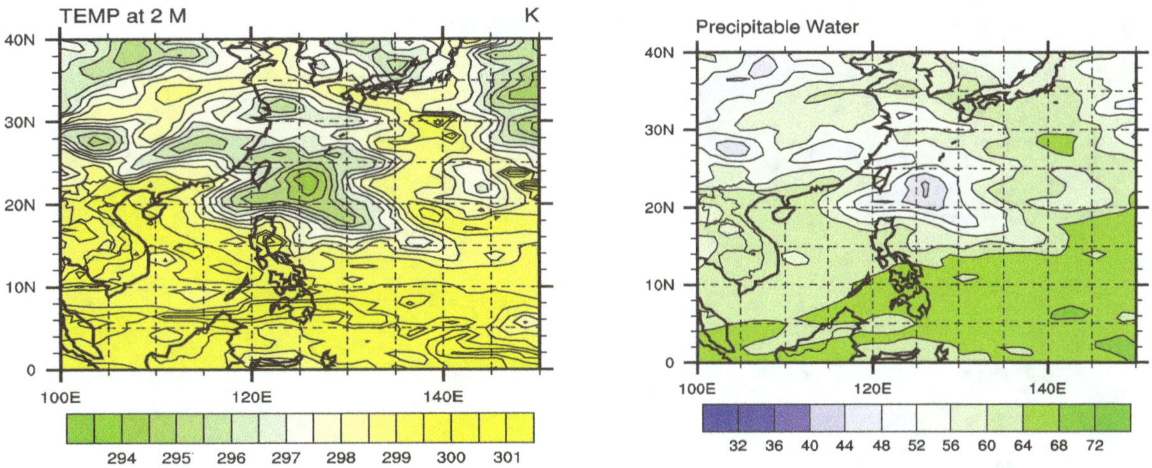

(c) (d)

Figure 5. Measured results for landing of typhoon. (**a**) Air pressure nephogram; (**b**) Wind speed nephogram; (**c**) Temperature nephogram; (**d**) Rainfall nephogram.

The wind speed streamlines before, at, and after the landing of the typhoon are shown in Figure 6. No strong convection was observed before the landing of the typhoon. However, it turned to a southeaster gradually at the landing of the typhoon. After the landing, surrounding airflows continued to radiate toward the typhoon center. Wind speed streamlines are disorganized, without a regular development trend.

The wind speed profiles close to the typhoon center at different moments are shown in Figure 7a. According to analysis, wind speed distributes uniformly in the near-ground region at different moments, but wind speed distributes disorderly at high altitude. The typhoon speed at 20:00 on 22 August was significantly lower than those at other moments. This might be because the typhoon had landed at Shenzhen at this moment, but it was on the sea at other moments. The near-ground wind speed in the center of the simulation region and the corresponding nonlinear least square fitting curves are shown in Figure 7b. The fitting value of the wind speed profile index was 0.118, which was smaller than the surface roughness index (0.15) of class B landform under normal wind, indicating the good fitting effect in the near-ground typhoon field (simulation accuracy reached 95.97%). The typhoon speed at 10 m of height was high, and increased slowly with the increase of height. Moreover, it defined that the typhoon field at 350 m altitude of class B landform in the code [38] was equal to the wind speed in the normal wind field for the purpose of qualitative and quantitative analyses of the difference between the normal wind field and the typhoon field. The near-ground typhoon profile and the normal wind profile were integrated into the UDF and used as the initial boundary conditions in the follow-up microscale CFD numerical simulation.

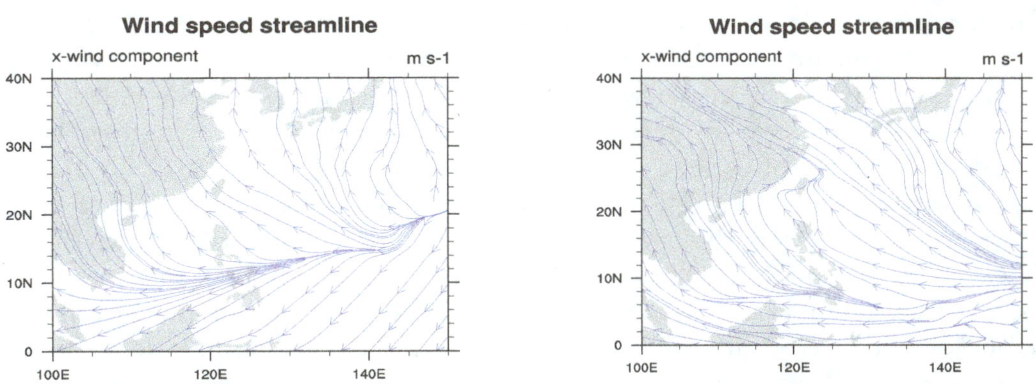

(a) (b)

Figure 6. *Cont.*

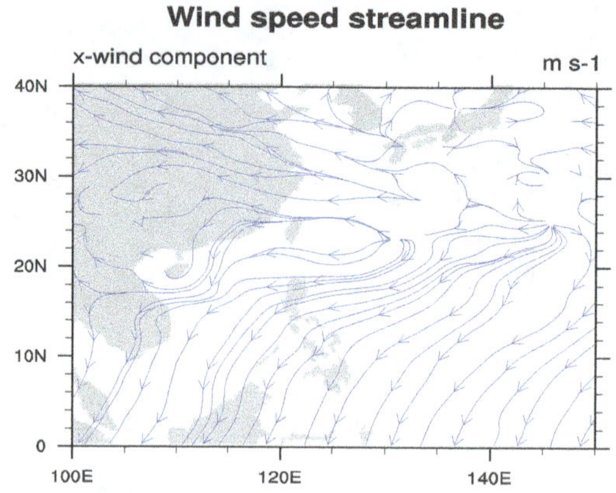

(c)

Figure 6. Wind speed streamlines before, at, and after landing of the typhoon. (**a**) Before landing (14:00 on 22 August); (**b**) At landing (20:00 on 22 August); (**c**) After landing (02:00 on 23 August).

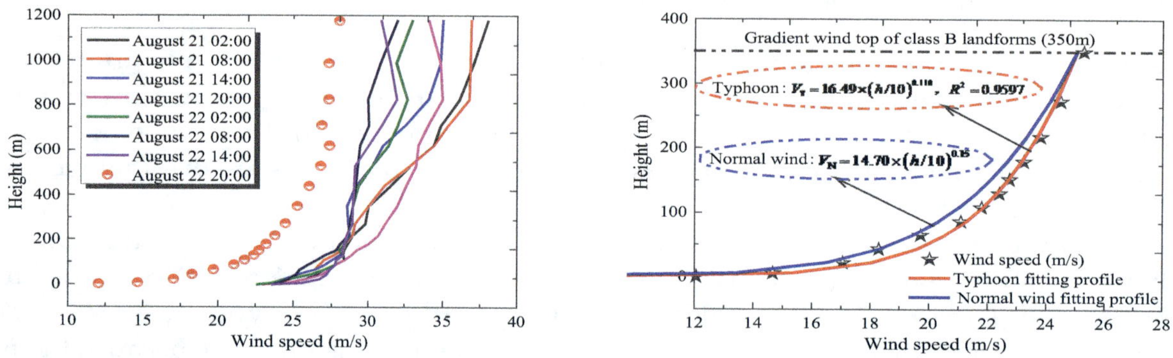

(a) (b)

Figure 7. Wind speed profile close to the typhoon center and near-ground wind speed and fitting curves in the center of the simulation region at different moments. (**a**) Wind speed profile close to the typhoon center at different moments; (**b**) Near-ground wind speed and fitting curves in the center of the simulation region.

3. Microscale CFD Numerical Simulation

3.1. Brief Introduction to the Project

The main structural design parameters and model of a five-MW windward horizontal axis three-blade wind turbine in a wind power plant in Shenzhen, China are shown in Table 3. This five-MW wind turbine was developed in the Jiangsu Key Laboratory of Hi-Tech Research for Wind Turbine Design of Nanjing University of Aeronautics and Astronautics [39]. The tower body was a long thickness-variable structure. The upper wall thickness was 40 mm, and the bottom wall thickness was 90 mm. The nacelle size was 18 m (Length) × 6 m (Width) × 6 m (Height). The dip angle of the blade was 5°, and the cut-out speed was 25 m/s. The blades were distributed uniformly along the circumferential direction at an angle of 120°, and the blade length was 60 m. The parameters of the different blade sections along the span are listed in Table 4. The airfoil types of the blade were NH02_40, NH02_35, NH02_30, NH02_25, NH02_21, NH02_18, and NH02_15. Given the coordinates, chord lengths, and installation angles of the discrete points on the different blade profile sections in the wind turbine blades, the three-dimensional (3D) coordinates of different blade profile sections were gained through coordinate transformation. Finally, the 3D model of the blade was constructed.

Specific steps of coordinate transformation included [40]:

(1) The coordinates of discrete points on different blade profile sections were calculated from relevant software. The normalized coordinates (x_0, y_0) of different blade elements were transformed to the coordinates (x_1, y_1) of discrete point under the coordinate system 1:

$$(x_1, y_1) = c \times (x_0, y_0) \tag{2}$$

(2) The coordinates were further transformed to the coordinates $(x_2,\ y_2)$ under the coordinate system 2:

$$(x_2,\ y_2) = (x_1,\ y_1) - (X,\ Y) \tag{3}$$

where (X, Y) are the coordinates of the pressure center.

(3) The coordinates were further rotated to the actual spatial coordinates (x, y, z):

$$x = \sqrt{x_2^2 + y_2^2} \times \cos\left(\arctan\frac{y_2}{x_2} + \theta\right) \tag{4}$$

$$y = \sqrt{x_2^2 + y_2^2} \times \sin\left(\arctan\frac{y_2}{x_2} + \theta\right) \tag{5}$$

$$z = r \tag{6}$$

where θ is the installation angle, and r is the turning radius of the blade elements.

The inflow angle is the included angle between the plane of rotation and inflow speed in the traditional theory of blade momentum. However, the calculation formula of the inflow angle during the blade design stage of the wind turbine according to the Baez limit is:

$$\cot\varphi = \frac{3}{2}\lambda \tag{7}$$

where λ is the tip velocity ratio.

According to the relationship between the tip velocity ratio and the power coefficient, the optimal tip velocity ratio can be determined, and the inflow angle of each blade element is:

$$\cot\varphi = \frac{3}{2}\lambda_r \tag{8}$$

where λ_r is the optimal tip velocity ratio.

The tower body, blade, and nacelle models were constructed successively. Finally, the 3D physical model of a large wind turbine was formed by Boolean operation.

Table 3. Main structural design parameters and model of the five-MW wind turbine.

Parameters	Numerical Value	3D Blade Model	3D Wind Turbine Model
Tower height	124 m		
Radius at tower top	3.0 m		
Radius at tower bottom	3.5 m		
Thickness at tower top	0.04 m		
Thickness at tower bottom	0.09 m		
Blade length	60 m		
Nacelle size	18 m × 6 m × 6 m		
Cut-out speed	25 m/s		
Yaw rotation speed	0.5°/s		

Table 4. Blade parameters of wind turbine.

Position/%	Blade Span/m	Chord Length/m	Installation Angle/°	Blade Pitch Angle/°	Position/%	Blade Span/m	Chord Length/m	Installation Angle/°	Blade Pitch Angle/°
5	3	2.9	0.823	37.14	55	33	1.95	0.169	−0.293
10	6	3.66	0.64	26.672	60	36	1.75	0.156	−1.072
15	9	4.41	0.507	19.069	65	39	1.58	0.144	−1.736
20	12	4.56	0.414	13.692	70	42	1.42	0.134	−2.31
25	15	4.25	0.346	9.83	75	45	1.27	0.125	−2.81
30	18	3.91	0.296	6.976	80	48	1.12	0.118	−3.25
35	21	3.59	0.258	4.802	85	51	0.98	0.111	−3.64
40	24	3.05	0.229	3.103	90	54	0.83	0.105	−3.987
45	27	2.63	0.205	1.742	95	57	0.69	0.099	−4.299
50	30	2.29	0.186	0.63	100	60	0.54	0.095	−4.58

3.2. Computational Domain and Meshing

To assure the full development of the wake flow of the wind turbine [40,41], the computational domain was determined as 12D × 5D × 5D (flow direction X × extension direction Y × vertical direction Z, and D is the diameter of the wind wheel). The wind turbine was placed 3D to the entrance of the computational domain. To achieve satisfying computational efficiency and accuracy [42,43], due to the complicated blade appearance, the hybrid grid discrete form was applied to divide the whole computational domain into the internal and external parts. The local encrypted region covered the wind turbine model, which used the non-structured meshing. Non-structured meshing produces a tetra/mixed tetrahedral mesh in ICEM CFD automatically by Robust (Octree). The peripheral region had a regular shape that used the high-quality structured meshing. The grid number, grid quality, and windward pressure coefficient of the tower body under different meshing schemes are shown in Table 5. Grid quality includes the minimum orthogonal quality of the grids and grid skewness. The minimum orthogonal quality of the grids shall be larger than 0.1, and it is best to be higher than 0.2. No negative volume is allowed. The grid skewness shall be lower than 0.95, and it is suggested to be lower than 0.9 [44]. It can be seen from Table 5 that with the increase of the total number of grids, the grid quality and windward pressure coefficient converged gradually. No significant differences were observed in grid quality and computational results under a 9.3 million meshing program and a 28.4 million meshing program. Considering computational accuracy and efficiency, this paper chose the 9.3 million meshing program. When processing flow in the near-wall region, the standard wall function was applied to connect the physical variables on the wall and the unknown variables in the turbulence core region directly. In this paper, the cooperation of the standard wall function and standard k-ε model (which will be introduced in Section 3.3) can increase the simulation accuracy more significantly. Essentially, the flowing conditions in the turbulence core region were solved by the k-ε model. The flow in the wall surface was not solved. Instead, the physical parameters (e.g., speed) of this region were solved by a semi-empirical formula directly. Since the near-wall flow was processed by the standard wall function in this paper, no encryption of the wall surface was needed during meshing, and it only has to place the first internal node in the logarithmic law region. The calculated results demonstrated that the $y+$ of the model wall could assure bottom grids in the logarithmic law, which met common engineering requirements. The meshing schemes of the overall computational domain and model are shown in Figure 8. Both the number of grids and grid quality met the requirements of calculation.

Table 5. Grid number, grid quality, and windward pressure coefficient on the tower body under different meshing schemes.

Meshing Schemes	1	2	3	4	5
Total number of grids	1.1 million	4.5 million	7.4 million	9.3 million	28.4 million
Minimum orthogonal quality of grids	0.13	0.36	0.53	0.60	0.64
Grid skewness	0.95	0.87	0.82	0.74	0.71
Windward pressure coefficient	0.92	0.88	0.85	0.80	0.79

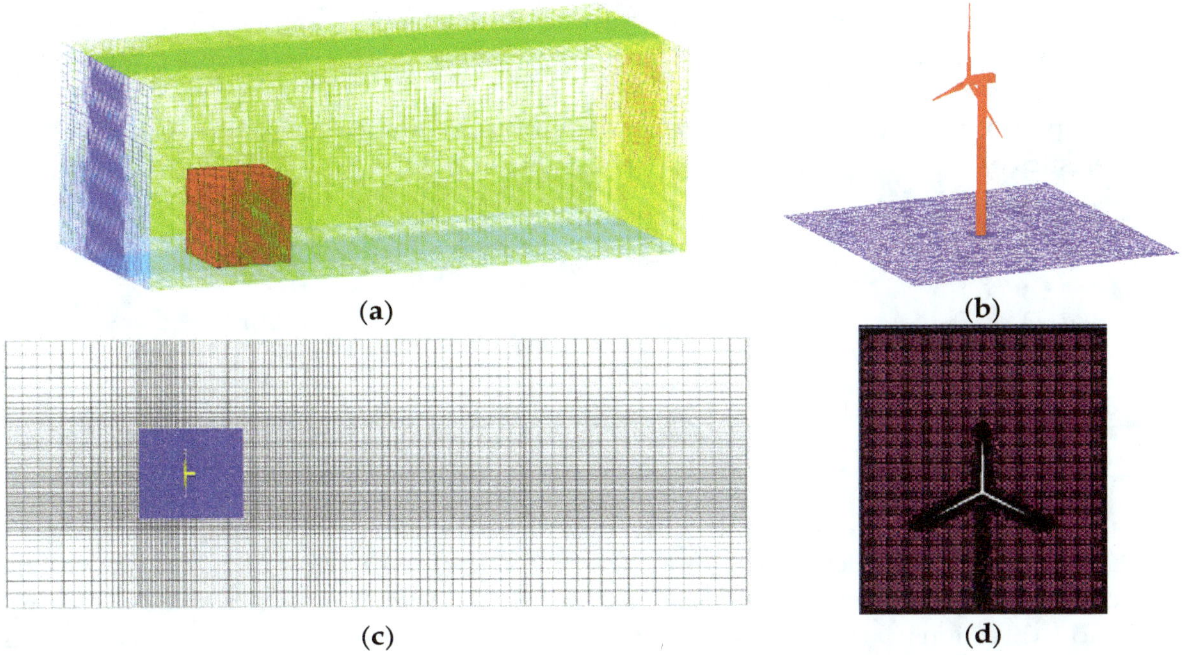

Figure 8. Computational domain and meshing schemes. (**a**) Meshing of the overall computational domain; (**b**) Meshing of local computational domain; (**c**) Meshing of the x-y plane; (**d**) Meshing of the encrypted y-z plane.

3.3. Selection of Turbulence Model

The 3D steady implicit algorithm was used in this paper. The air wind field chose the incompressible flow field and the turbulence model chose the standard k-ε model [45]. The speed vector and pressure coupling problem in the momentum equation was solved by the SIMPLEC algorithm. The turbulent kinetic energy, turbulent dissipation term, and momentum equation in the calculation all adopted the second-order upwind discretion, and the calculation residual error of the governing equation (Navier-Stokes(N-S) equation) was set 1 × 10−6. During the simulation, the enhanced wall surface function model was started to assure that the logarithmic distribution of the bottom grids was true.

The continuity equation and momentum equation of the viscous incompressible N-S equation based on Reynolds average are described by Equations (9)–(13).

$$\frac{\partial \mu_i}{\partial x_i} = 0 \tag{9}$$

$$\frac{\partial(\rho\overline{\mu_i})}{\partial t} + \frac{\partial(\rho\overline{\mu_i}\overline{\mu_j})}{\partial x_j} = -\frac{\partial(\overline{p})}{\partial x_i} + \frac{\partial}{\partial x_j}(\mu\frac{\partial\overline{\mu_i}}{\partial x_j} - \rho\overline{\mu_i'\mu_j'}) \tag{10}$$

where $i, j = 1, 2$; ρ is the air density, which is generally 1.225 kg/m^3L and μ is the dynamic coefficient of viscosity, which is generally 1.7894×10^{-5} kg/(m·s)

$$\frac{\partial(\rho_m k)}{\partial t} + \frac{\partial(\rho_m k\mu_j)}{\partial x_j} = P - \rho_m\varepsilon + \frac{\partial}{\partial x_j}[(\mu + \frac{\mu_t}{\sigma_k})\frac{\partial k}{\partial x_j}] \tag{11}$$

$$\frac{\partial(\rho_m\varepsilon)}{\partial t} + \frac{\partial(\rho_m\xi\mu_j)}{\partial x_j} = C\xi_1\frac{\xi}{k}P_t - C\xi_2\rho_m\frac{\xi^2}{k} + \frac{\partial}{\partial x_j}[(\mu + \frac{\mu_t}{\sigma_k})\frac{\partial\xi}{\partial x_j}] \tag{12}$$

$$\mu_t = \frac{C_\mu\rho_m k^2}{\xi} \tag{13}$$

where k and ε are the turbulence energy and the turbulence dissipation rate, P_t is the generating term of turbulence energy, and μ_t is the viscosity coefficient of turbulence. The model constants are $C_{\zeta 1} = 1.44$, $C_{\zeta 2} = 1.92$, $\sigma_\zeta = 1.3$, $\sigma_k = 1.0$, and $C_\mu = 0.09$.

Under the premise of meeting the above mass and momentum conservation equations, the basic flow characteristics of turbulence are described by Equations (14)–(16).

$$\frac{\partial H}{\partial t} + \frac{\partial(\mu H)}{\partial x} + \frac{\partial(vH)}{\partial y} + \frac{\partial(wH)}{\partial z} = 0 \tag{14}$$

$$\frac{\partial(\mu H)}{\partial t} + \frac{\partial}{\partial x}\left|u^2 H + \frac{1}{2}gH^2\right| + \frac{\partial(\mu v H)}{\partial y} = Hfv - gH\frac{\partial Z_b}{\partial x} - \frac{H}{\rho}\frac{\partial p}{\partial x} + \frac{\tau_{ax} - \tau_{bx}}{\rho} \tag{15}$$

$$\frac{\partial(vH)}{\partial t} + \frac{\partial(\mu v H)}{\partial x} + \frac{\partial}{\partial y}\left|v^2 H + \frac{1}{2}gH^2\right| = -Hfu - gH\frac{\partial Z_b}{\partial y} - \frac{H}{\rho}\frac{\partial p}{\partial y} + \frac{\tau_{ay} - \tau_{by}}{\rho} \tag{16}$$

where u, v, and w are components of the speed vector $u(z)$ on the x, y, and z directions; H is the reference height; g is the gravitational acceleration; w is the angular velocity of ground rotation; φ is the geological latitude of the calculation point; Z_b is the elevation of ground; τ_{ax} and τ_{ay} are the wind stress components along the x and y directions; τ_{bx} and τ_{by} are the ground frictional force components along the x and y directions; p is the pressure difference; and ρ is the air density. f can be expressed as:

$$f = 2w \sin \varphi \tag{17}$$

The k-ε model is a turbulence model that is applicable to the calculation of high Reynolds numbers. It can predict the fully developed turbulence flow well and it is suitable to the simulation of the surrounding flow fields of similar megawatt wind turbines [46].

3.4. Settings of Boundary Conditions

The inlet of the computational domain was used as the speed inlet boundary, and the outlet was the pressure outlet boundary. Two side walls and the top surface used the symmetric boundaries. The wind turbine and ground set the wall boundary. The overlapping surface of the local and peripheral computational domain was set as the interface. The wind field computational domain and its boundary conditions are shown in Figure 9. Wind velocity profiles were set as the indexes of the atmospheric boundary layer in the computational domain, which are based on the profiles of typhoon and normal wind that are shown in Figure 7b of the revised manuscript, respectively. The velocity profiles of the typhoon and normal wind were determined by formulas (18) and (19) as follows:

$$V_{\mathrm{T}} = 16.49 \times (h/10)^{0.118} \tag{18}$$

$$V_{\mathrm{N}} = 14.70 \times (h/10)^{0.15} \tag{19}$$

where, V_{T} and V_{N} were velocity of typhoon and normal wind at the height of h. They equaled to 16.49 and 14.70 respectively at the height of 10 m. 0.118 was the index on the velocity profiles of typhoon fit by nonlinear least squares, while 0.15 was the roughness index of surface on Type B landforms as specified by Chinese Code [38].

In this study, turbulence of wind fields near the ground was determined by Formulas (20) and (21) under typhoon and normal wind respectively.

$$I_{\mathrm{T}} = 0.20 \times (h/10)^{-0.118} \tag{20}$$

$$I_{\mathrm{N}} = 0.14 \times (h/10)^{-0.15} \tag{21}$$

where 0.20 was the nominal turbulence of the typhoon "Nuri" at the height of 10 m, which was determined based on the synchronous monitoring results of the typhoon [37], and 0.14 was the nominal turbulence on Type B landforms at the height of 10 m, as specified by Chinese code [38].

In typhoon field simulation, fluid parameters (e.g., mean wind profile, scale of turbulence, turbulence energy, turbulence integral scale, and specific dissipation rate) in the fluctuating wind field were integrated into Fluent by UDF based on the above k-ε model and fitting typhoon profile (Figure 10). This was to accomplish the simulation of the typhoon field in a meso/microscale nesting large wind turbine.

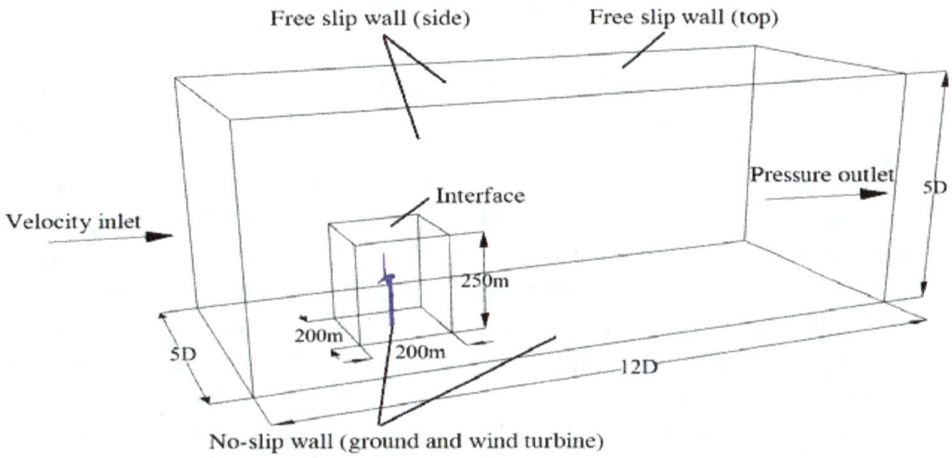

Figure 9. Boundary conditions of the computational domain.

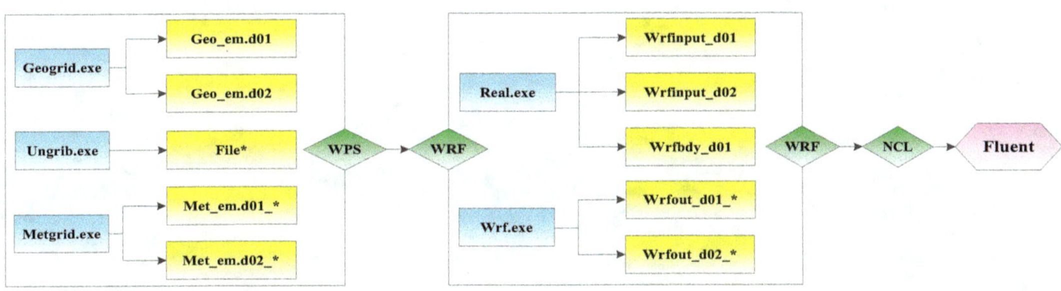

Figure 10. Procedures of WRF-CFD (Computational Fluid Dynamics) computation and nesting.

3.5. Analysis of Aerodynamic Results

The surface pressure coefficient on the wind turbine tower is a main index for evaluating the aerodynamic performances of a wind turbine system. Figure 11 shows the typical cross-section circumferential curve of the pressure coefficient for a wind turbine tower under normal wind conditions, and compares the coefficient with the code value [38]. The analysis suggested that the numerical simulation results of the wind turbine tower under normal wind conditions were generally in accord with the circumferential distribution laws and numerical values specified by Chinese codes. They were only lower than the value specified by the codes out of wind, but the maximum error was below 10%. This demonstrated that the numerical simulation results of CFD were effective.

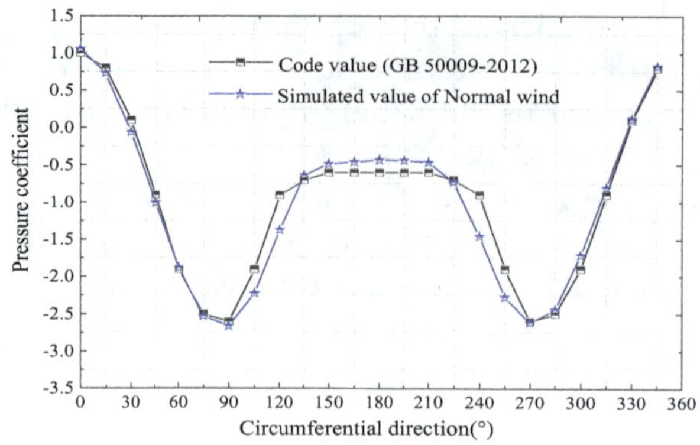

Figure 11. Numerical simulation results under normal wind conditions versus value specified by Chinese code.

The wind speed streamlines and turbulence energy distribution on typical sections of the tower body under normal wind and typhoon loads are shown in Figures 12 and 13. H was the height of the tower body. The wind speed streamlines presented relatively consistent distribution under normal wind and typhoon loads. The wind speed surrounding the tower body and blade under the typhoon loads was higher than that under normal wind. The inflow split at 0° of the windward side of the tower body, thus generates reflux and different sizes of eddies on the leeside of the tower body. Additionally, the inflow moved along the blade surface and ran through the front and rear edges of the blades, which formed small-sized eddies on the blade leeside. At the same time, the turbulence energy surrounding the tower body and blades under typhoon loads was significantly stronger than that under normal wind. Wakes on the blade and tower body developed the large-scaled eddy increment region. Besides, the interaction between the blade and the tower body was significant on the middle and upper sections of the tower body.

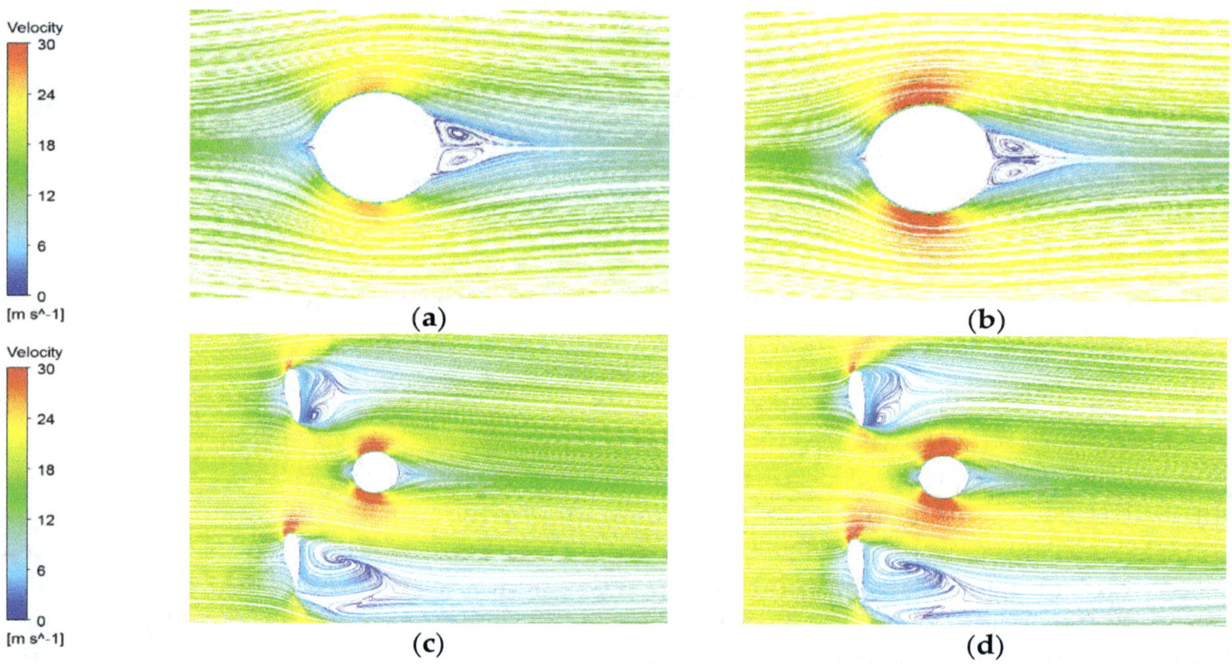

Figure 12. Wind speed streamlines at a typical section of the tower body under normal wind and typhoon loads. (**a**) Normal wind (0.3H); (**b**) Typhoon (0.3H); (**c**) Normal wind (0.8H); (**d**) Typhoon (0.8H).

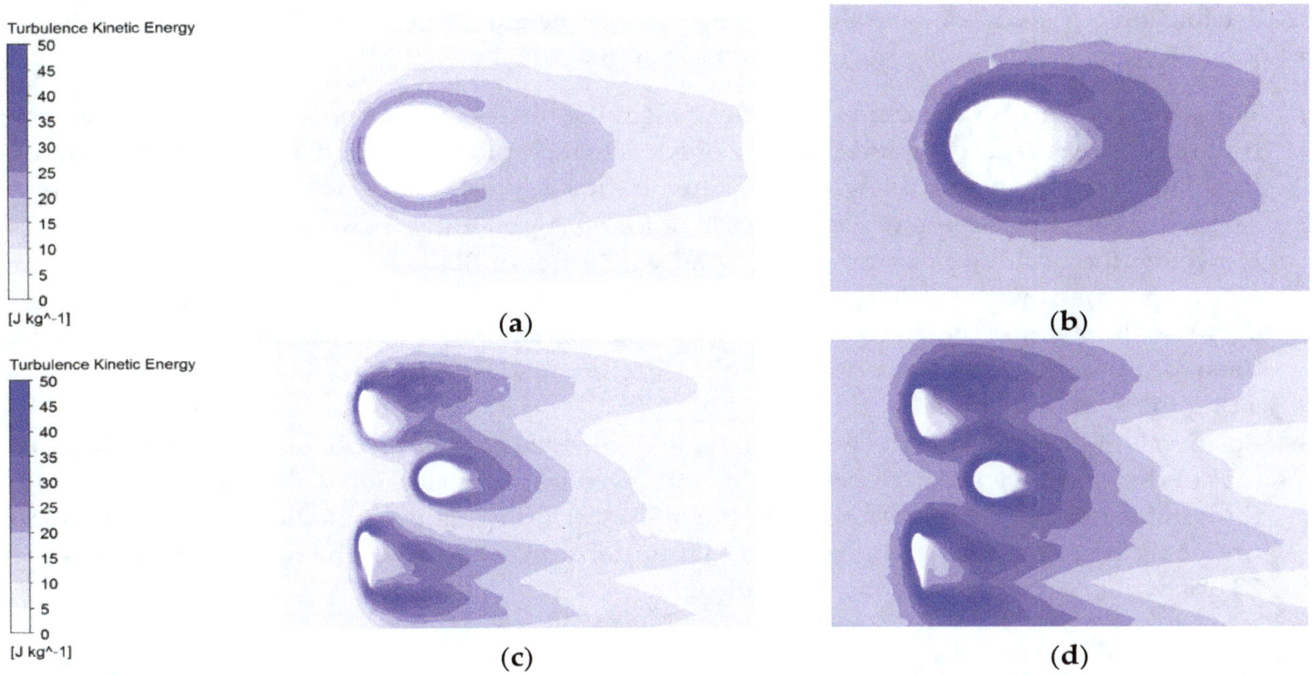

Figure 13. Turbulence energy distribution at a typical section of the tower body under normal wind and typhoon loads. (**a**) Normal wind (0.3H); (**b**) Typhoon (0.3H); (**c**) Normal wind (0.8H); (**d**) Typhoon (0.8H).

The distribution curves of the wind pressure coefficient at the typical section of the tower body under normal wind and typhoon loads are shown in Figure 14. Under normal wind and typhoon loads, the circumferential wind pressure coefficient on the tower body was symmetric along the windward side. With the increase of circumferential angle, it decreased firstly and then increased until it stabilized close to the leeside. However, extremum positive pressure on the windward side and extremum negative pressure on the lateral surface under the typhoon were significantly higher compared with those under normal wind. The maximum increase of the wind pressure coefficient on the windward side was 61.1%, and the maximum increase of the extreme negative pressure on the lateral surface was 44.9%. Both were achieved at bottom of the tower body.

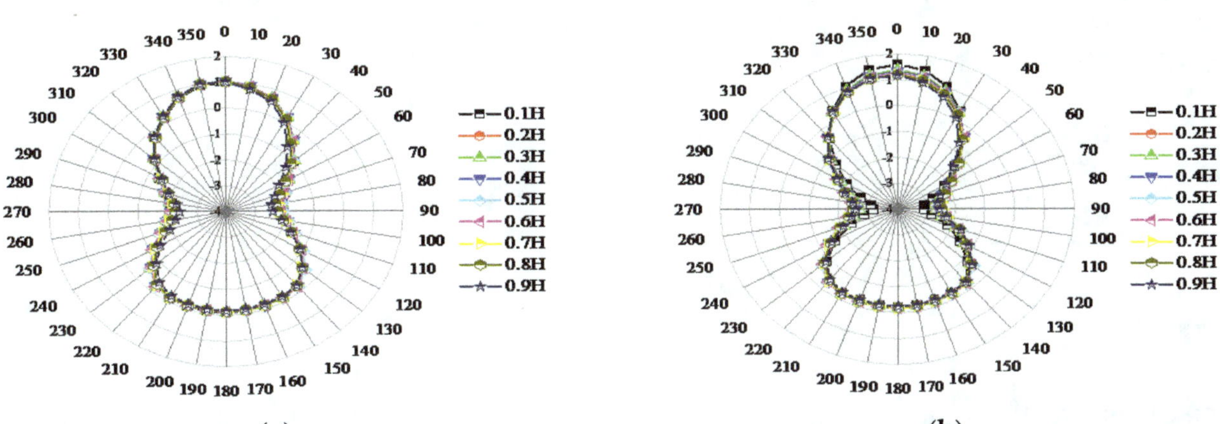

Figure 14. Distribution curves of wind pressure coefficient on tower body under normal wind and typhoon loads. (**a**) Normal wind; (**b**) Typhoon.

The straight-up blade was defined Blade A. The remaining two clockwise rotating blades were Blade B and Blade C. Wind pressure coefficient nephograms on the windward side and leeside of blade as well as mean along the blade span under normal wind and typhoon loads are shown in Figure 15.

The distribution curves of the overall wind pressure coefficient of the blades under normal wind and typhoon loads are shown in Figure 16. It can be seen that:

(1) The wind pressure on the windward side of all of the blades was positive under normal wind and typhoon loads. Negative wind pressure only occurred at the front blade edges and rear edges of the blade root. However, the wind pressure on the leeside of blades was basically negative.

(2) The wind pressure on the windward side of Blade A generally increased along the blade span. However, the positive pressure on the windward sides of Blade B and Blade C changed slightly along the span, without irregular changes. Negative pressure on the leeside of the blades decreased firstly, and then increased along the span direction. If these two effects were superposed, the overall wind pressure coefficient of the different blades decreased firstly, and then increased along the span direction.

(3) Positive wind pressure on the windward side of the blades under typhoon loads was higher than that under normal winds. The negative wind pressure on the leeside was slightly different from that under normal winds. As a result, the overall wind pressure coefficient on different blades under typhoon loads was higher than that under normal winds. The maximum increase was 23.6%, which occurred at the tip of Blade C.

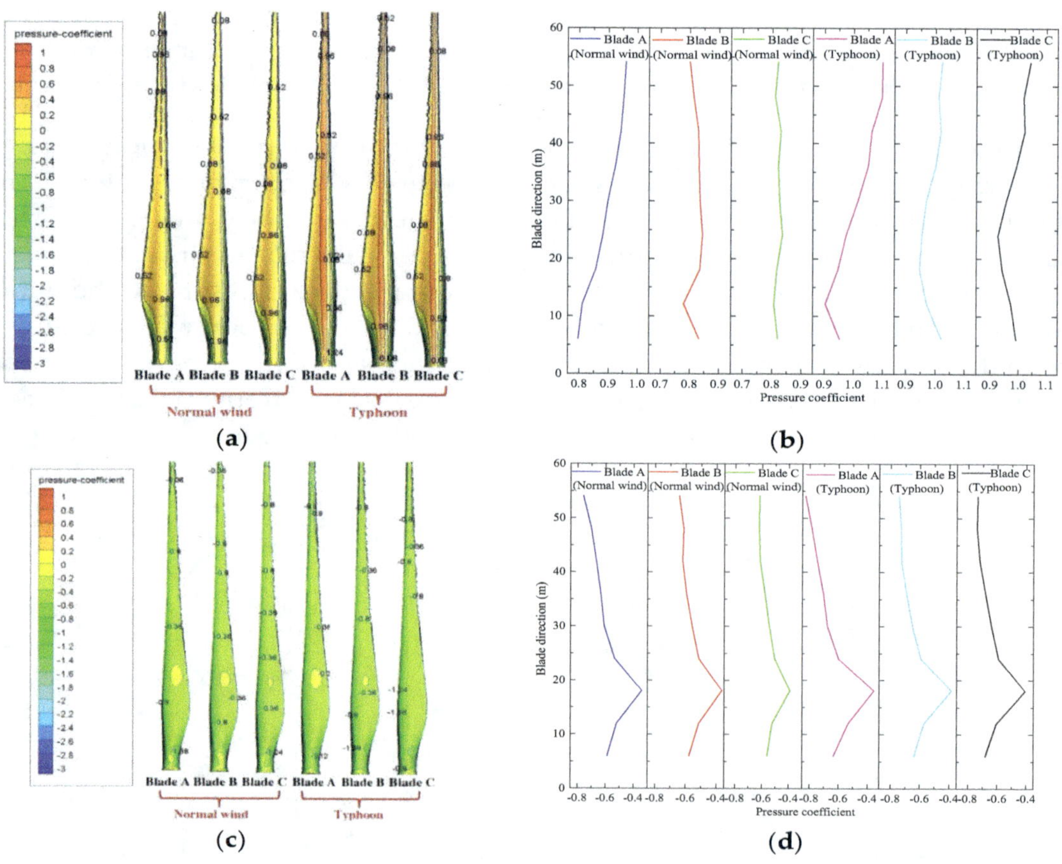

Figure 15. Wind pressure coefficient nephograms on the windward and leeside of blades as well as the mean along the blade under normal wind and typhoon loads. (**a**) Wind pressure coefficient nephogram on the windward side; (**b**) Wind pressure coefficient curves on the windward side; (**c**) Wind pressure coefficient nephogram on the leeside; (**d**) Wind pressure coefficient curves on the leeside.

The distribution curves of the lift coefficient and drag coefficient on the different height sections of the tower body under normal wind and typhoon loads are shown in Figure 17. The lift–drag coefficient ratio is shown in Table 6. The calculation formulas of the across-wind lift coefficient (C_L) and downwind drag coefficient (C_D) are shown as follows [4]:

$$C_L = \frac{\sum_{i=1}^{n} C_{Pi} A_i \sin \theta_i}{A_T} \quad (22)$$

$$C_D = \frac{\sum_{i=1}^{n} C_{Pi} A_i \cos \theta_i}{A_T} \quad (23)$$

where Cpi is the mean wind pressure coefficient at the measuring point i on the tower body, A_i is the pressure coverage area at the measuring point i, θ_i is the included angle between the pressure at measuring point i and the wind axis, and A_T is the projection area of the overall structure along the wind axis.

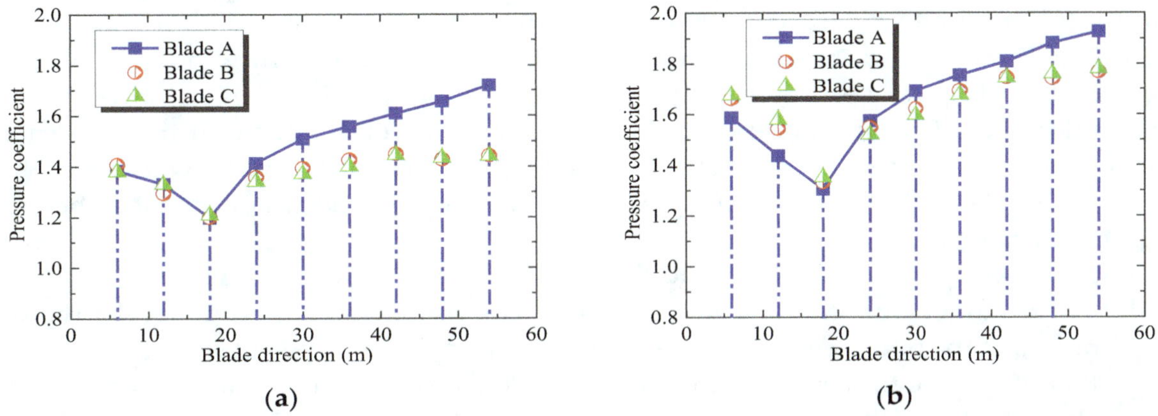

Figure 16. Distribution curves of overall wind pressure coefficient on blades under normal wind and typhoon loads. (**a**) Normal wind; (**b**) Typhoon.

The following can be seen from Figure 17 and Table 6. (1) The lift coefficient was negative under normal wind, but it was alternatively positive and negative under typhoon loads. However, the distribution law of the lift coefficient was consistent under normal wind and typhoon loads, and the numerical value was small; (2) The drag coefficient was mainly in the range of 0.35~0.4 under normal winds, and it was in the range of 0.45~0.65 under typhoon loads. The drag coefficient at different heights under the typhoon loads increased dramatically than that under the normal wind. The maximum was about 0.18, which was at the tower top. This was mainly caused by the large disturbance of the tower top to blades.

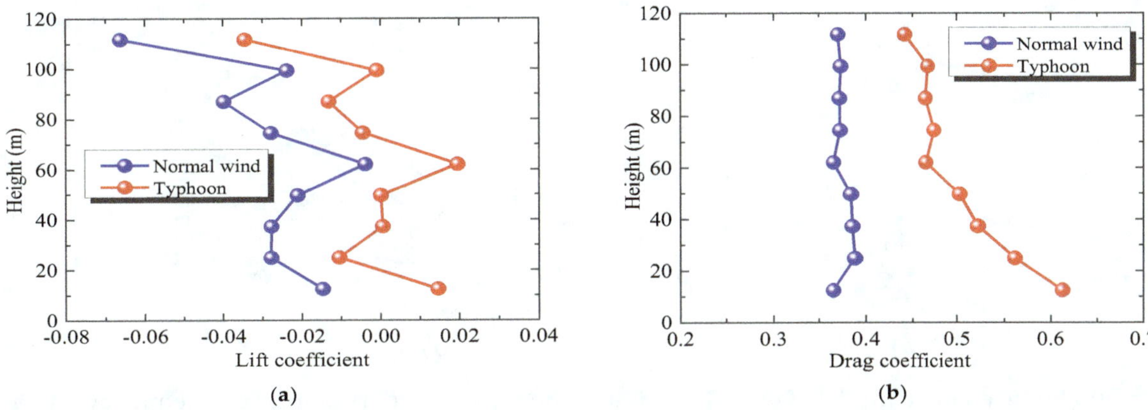

Figure 17. Distribution curves of lift coefficient and drag coefficient of the tower body under normal wind and typhoon loads. (**a**) Lift coefficient. (**b**) Drag coefficient.

Table 6. Lift–drag coefficient ratio of the tower body under normal wind and typhoon loads.

C_L/C_D	Height of the Tower Body								
	12.4 m	**24.8 m**	**37.2 m**	**49.6 m**	**62 m**	**74.4 m**	**86.8 m**	**99.2 m**	**111.6 m**
Normal wind	−0.040	−0.071	−0.072	−0.055	−0.011	−0.074	−0.107	−0.064	−0.179
Typhoon loads	0.024	−0.019	0.001	0.001	0.042	−0.010	−0.028	−0.002	−0.078

4. Response Analysis of the Wind Turbine Coupling Structure

4.1. Finite Element Modeling and Dynamic Characteristics

The tower–blade integrated finite element coupling model was constructed by the large universal finite element analysis software ANSYS. The tower body and blade were simulated by Shell63 unit, whereas the nacelle and internal structure were simulated as a whole by using Beam 189 units. The round raft basis unit was Solid65. The interaction between the foundation and the basis was simulated by a Combin14 unit. After finite element models of different wind turbine parts were constructed, a coupling connection may cause the relative slippage of models under loads, which was attributed to the different unit types of different parts. Therefore, the multipoint restriction unit coupling command was used to connect different parts and construct the tower–blade integrated simulation model. According to the principle of efficiency and the accuracy balance, the model was divided into 4122 units.

The natural frequency of vibration and the mode of vibration of the wind turbine were solved by a block Lanczos algorithm after modeling [47,48]. The finite element model of the wind turbine, the first 100 orders of the frequency distribution, and the mode of vibration at typical orders are shown in Figures 18 and 19. The tower–blade integrated model had a low natural frequency of vibration. The fundamental frequency was 0.137 Hz, and the first 10 orders of frequency were lower than 1 Hz. At the same time, the natural frequency of the vibration of the structure was low and distributed densely. The structural frequency achieved approximately linear growth with the number of vibration orders, but the growth rate decreased after 90 orders. In addition, three blades in the tower–blade integrated model made front-to-back movement at the first, fifth, and 10th order. The blades made complicated front-to-back and left-to-right swings at the 30th order and 50th order, which were accompanied by the bending deformation of the tower body. According to the multi-order mode of vibration, the low-order mode of vibration in the tower–blade integrated model was mainly manifested by the front-to-back and left-to-right swing of the blades. Under the high order of vibration, the tower body and blades developed deformations and buckling failure.

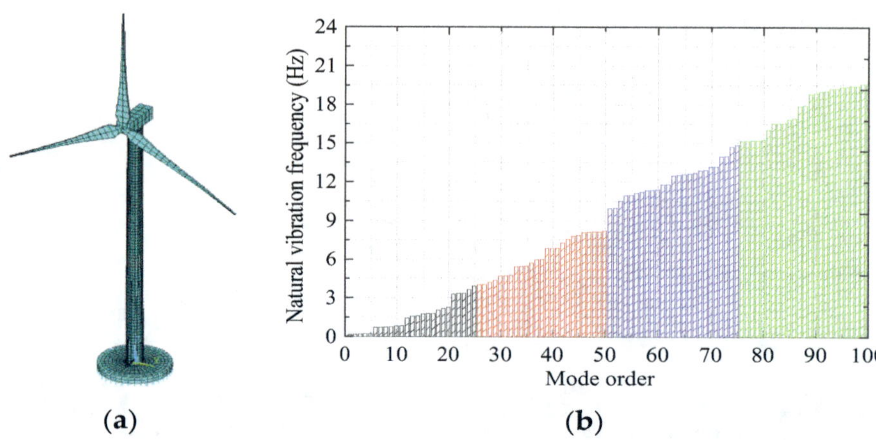

Figure 18. Finite element model of wind turbine and the first 100 orders of inherent frequency. (**a**) Finite element model; (**b**) First 100 orders of inherent frequency.

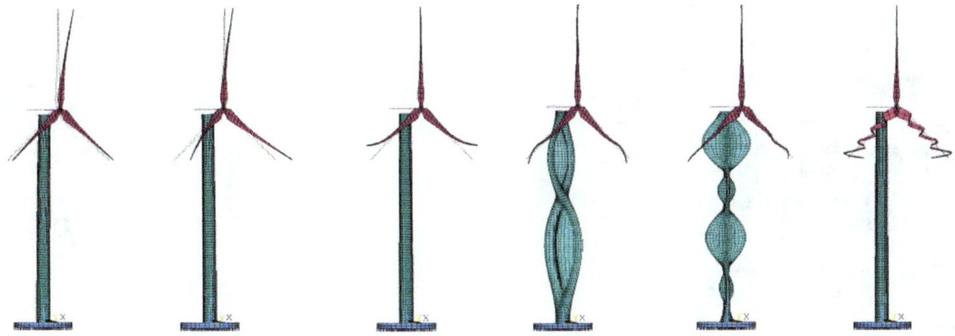

(a) (b) (c) (d) (e) (f)

Figure 19. Mode of vibration of the tower–blade coupling model at different orders. (**a**) First order; (**b**) Fifth order; (**c**) 10th order; (**d**) 30th order; (**e**) 50th order; (**f**) 100th order.

4.2. Stress Analysis on the Tower Body

The 3D distributions of the radial displacement of the tower body and distributions of internal stress responses at the tower bottom under normal wind and typhoon loads are shown in Figures 20 and 21. It can be seen that: (1) the radial distribution of the tower body presented relatively similar distribution patterns under normal wind and typhoon loads. The radial displacement increased gradually going up of the tower body, forming two extreme regions on the windward side and leeside. The maximum positive and negative displacements were at 180° and 0°; (2) Different wind field mainly affected the radial displacement at the middle and upper parts of the tower body. Under the typhoon loads, the maximum radial displacement of the lower tower body was 0.117 m, which was increased by 28.6% compared with that under normal wind; (3) Internal stresses at the tower bottom were symmetric along the windward side under normal wind and typhoon loads. Despite the slight difference in the meridian axial force at the tower bottom, other internal stresses presented basically consistent circumferential distribution laws under normal wind and typhoon loads. All of the internal stresses achieved the maximum value close to the circumferential 330° of the tower body; (4) Extreme internal stresses at the tower bottom under typhoon loads were stronger than those under normal wind loads. The maximum meridian axial force, maximum shearing force, maximum circumferential bending moment, and maximum meridian bending moment were increased by 36.1%, 34.8%, 34.2%, and 46.5%, respectively. All occurred close to the circumferential 330°.

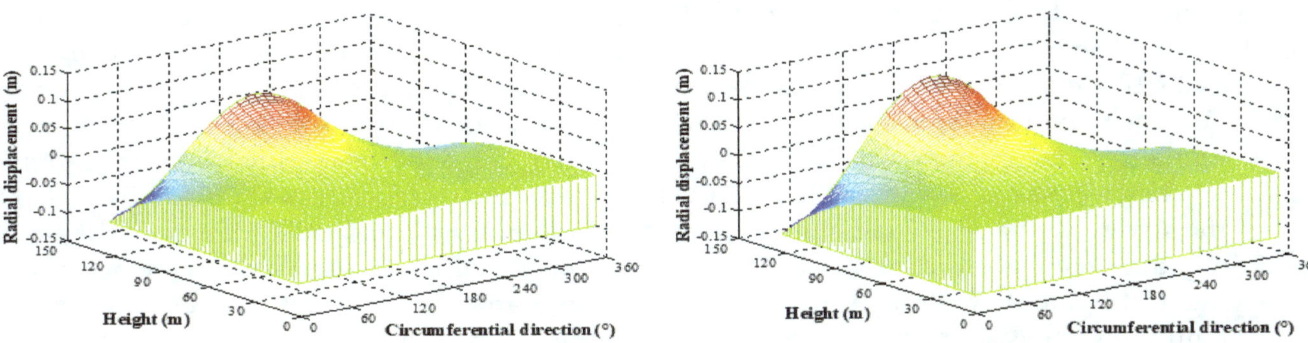

Figure 20. 3D radial displacement of tower body under normal wind and typhoon loads. (**a**) Normal wind; (**b**) Typhoon.

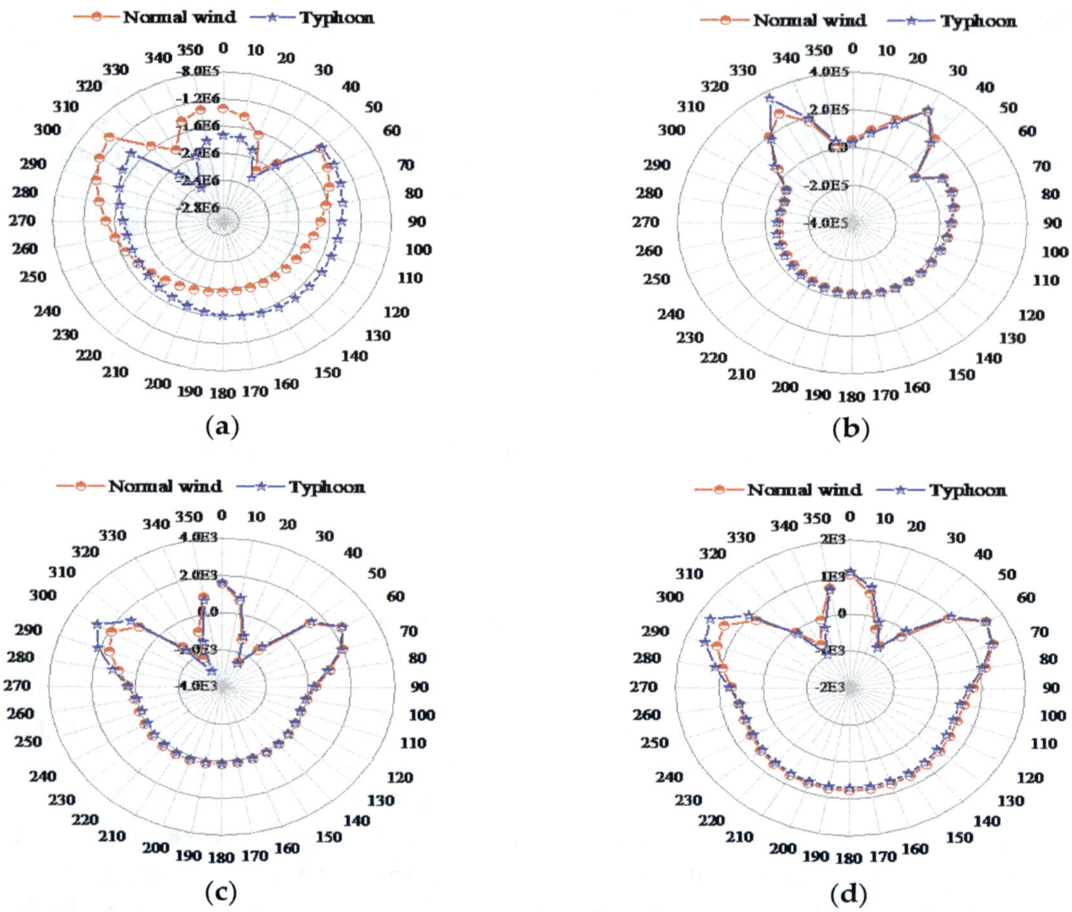

Figure 21. Internal stress responses at the tower bottom of the wind turbine system under normal wind and typhoon loads. (**a**) Meridian axial force Ty/(N); (**b**) Shearing force Txy/(N); (**c**) Circumferential bending moment Mx/(N·m); (**d**) Meridian bending moment My/(N·m).

4.3. Stress Analysis of Blades

The internal stress responses at blade roots and downwind displacement distribution along the blade span under normal wind and typhoon loads are shown in Figures 22 and 23. Downwind displacements at the blade tips are shown in Table 7. It was found that: (1) under normal wind and typhoon loads, the internal stresses at the root of Blade A were significantly higher than those of Blade B and Blade C. The numerical values of Blade B and Blade C were similar. Moreover, the internal stresses at the blade roots under typhoon loads were significantly higher than those under normal wind loads. The maximum increase of the shearing force was 41.3%. The maximum increase of the circumferential bending moment was 39.6%. The maximum increase of the meridian bending moment was 39.5%. They were all achieved by Blade C; (2) The downwind displacements of Blade B and Blade C along the span were basically same under normal wind and typhoon loads. However, they were significantly different from those of Blade A. With the extension along the blade span, the downwind displacement of Blade A was higher than those of Blade B and Blade C, and then became smaller. Next, the numerical values that were close to the middle of the blades were the same. However, the numerical values that were close to the middle of Blade A exceeded those of Blade B and Blade C. The numerical values of the three blades were equal at about 35 m of the blade span under normal wind, and at about 30 m of the blade span under typhoon loads; (3) The downwind displacements of the blades were larger under typhoon loads than those under normal wind loads. The maximum increase of downwind displacement was 49.3%, which was achieved at the tip of Blade A.

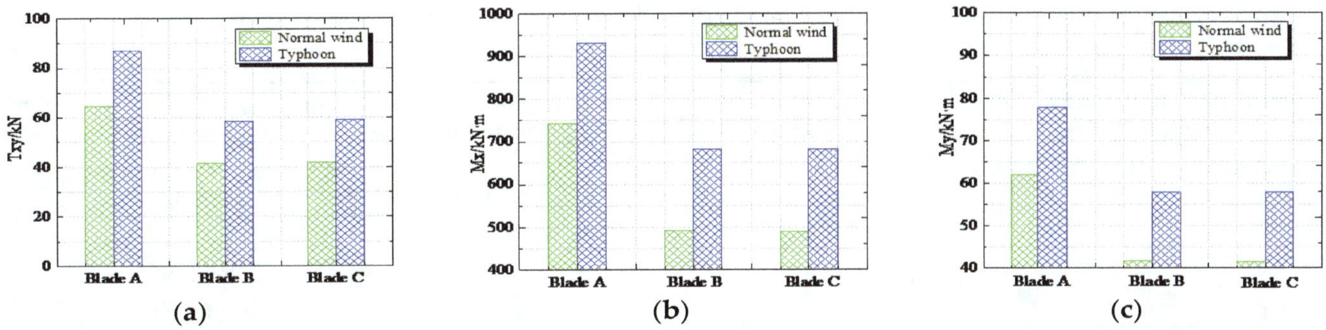

Figure 22. Internal stress responses at the blade roots of the wind turbine under normal wind and typhoon loads. (**a**) Shearing force (Txy); (**b**) Circumferential bending moment (Mx); (**c**) Meridian bending moment (My).

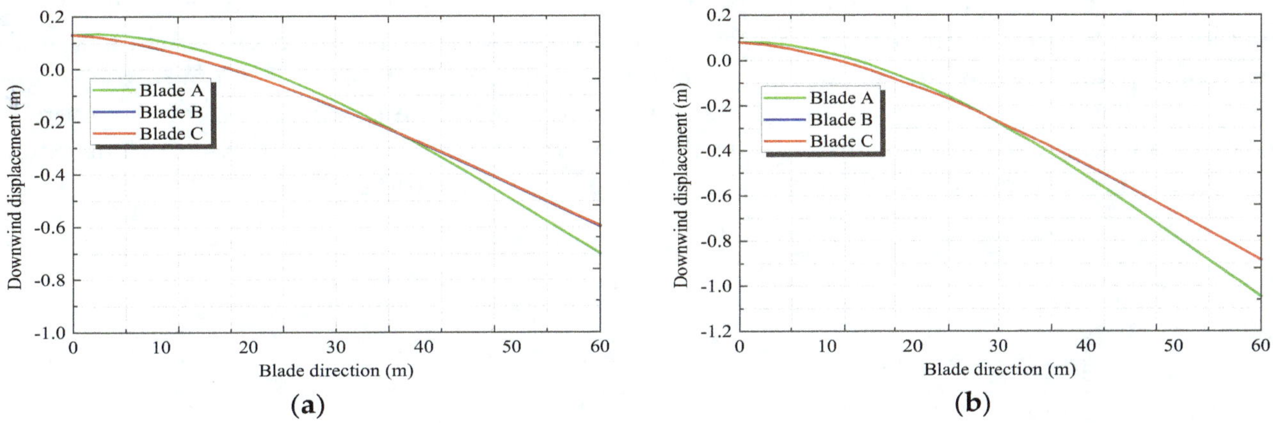

Figure 23. Distribution curves of downwind displacements of three blades under normal wind and typhoon loads. (**a**) Normal wind; (**b**) Typhoon.

Table 7. Downwind displacements at blade tips under normal wind and typhoon loads.

Blade A		Blade B		Blade C	
Normal Wind	**Typhoon**	**Normal Wind**	**Typhoon**	**Normal Wind**	**Typhoon**
−0.701 m	−1.047 m	−0.599 m	−0.885 m	−0.596 m	−0.884 m

5. Wind-Induced Stability Analysis of Wind Turbine System

5.1. Buckling Stability Analysis

The buckling mode and eigenvalues of the tower-blade integrated model under normal wind and typhoon loads are shown in Table 8. Based on comparison, the critical wind speed was higher than the designed wind speed under normal wind and typhoon loads, meeting the design requirements. The buckling position of the wind turbine was at the tower top. The buckling displacement and critical wind speed of the large wind turbine system were significantly influenced by different wind fields. The buckling coefficient, critical wind speed, and maximum displacement under typhoon loads were increased by 1.3%, 30.9%, and 28% compared with those under normal wind loads.

Table 8. Comparison of buckling mode and eigenvalues of wind turbine system under normal wind and typhoon loads.

Buckling Eigenvalue under Normal Wind		Buckling Mode under Normal Wind	Buckling Mode under Typhoon Loads	Buckling Eigenvalue under Typhoon Loads	
Buckling coefficient	2.955			Buckling coefficient	2.992
Critical wind speed/(m/s)	24.12			Critical wind speed/(m/s)	31.57
Maximum displacement/m	0.025			Maximum displacement/m	0.032

5.2. *Ultimate Bearing Capacity*

Changes in the buckling displacement of the wind turbine system with wind speed under normal wind and typhoon loads are shown in Figure 24. It was found that with the leveling loading of the wind speed, the buckling displacement of the wind turbine system under normal wind and typhoon loads increased firstly, and then decreased, and finally increased slightly. It reached the peak at about 20 m/s of the reference wind speed, and reached the minimum at about 40 m/s of the reference wind speed. The above trend was due to the "reverse effect" [46,49]. Moreover, the typhoon effect would significantly decrease the ultimate bearing capacity of the wind turbine system. The buckling displacement was increased by 38.5% at 40 m/s of the reference wind speed.

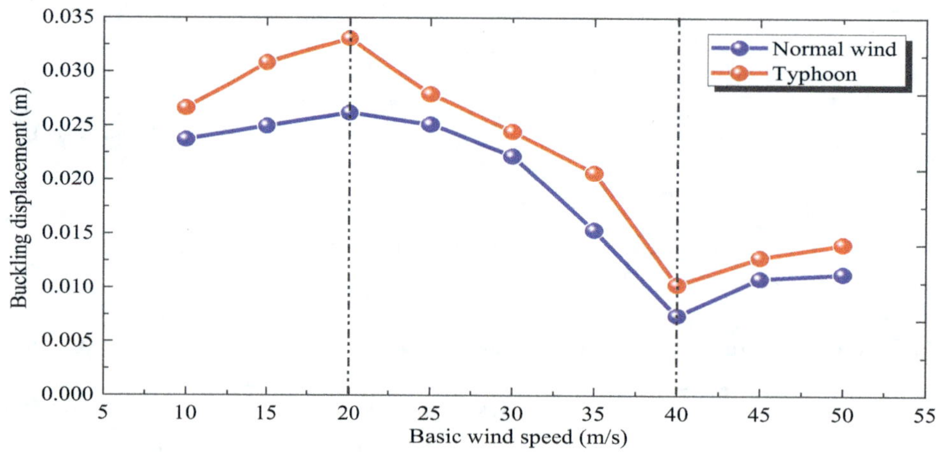

Figure 24. Changes of the buckling displacement of the wind turbine with wind speed under normal wind and typhoon loads.

6. Summary and Conclusions

High-resolution simulation of the Typhoon "Nuri" was accomplished based on the mesoscale WRF model. Later, differences regarding the aerodynamic force, wind-induced responses, buckling stability, and ultimate bearing capacity of the wind turbine system under normal wind and typhoon loads were analyzed by microscale CFD numerical simulation and the finite element method. Some major conclusions could be drawn:

- A high-resolution simulation of Typhoon "Nuri" was carried out through the mesoscale WRF model. The typhoon profile index that was based on the least square fitting was 0.118, which was about 20% smaller than the wind profile index under normal wind loads. In this paper, the meso/microscaled nested downscaling method can effectively simulate a three-dimensional typhoon field of similar structures and provide effective input loads for follow-up comprehensive stress analysis.

- The flow field distribution surrounding the wind turbine is basically similar under normal wind and typhoon effects. However, the airflow under typhoon conditions showed higher wind speed and turbulence compared with those under normal wind. Positive pressure on windward surface of the tower body, negative pressure on the crosswind surface of the tower body, the overall pressure coefficient of the blade, the resistance coefficient of the tower body, and the coupling structural response all increased significantly. The buckling stability and ultimate bearing capacity of the coupling system decreased significantly.
- The pressure coefficient on the windward surface of the tower body under typhoon conditions increased by 61.1% to the maximum extent. The maximum increment of the negative pressure extremum on the crosswind surface was 44.9%. Both were observed at the bottom of the tower body. The typhoon influenced the pressure coefficient on the windward surface of the blades significantly, but it only affected the leeside surface slightly. The maximum increment of the overall pressure coefficient of the blade was 23.6%, which occurred at the tip of Blade C. Under typhoon conditions, the resistance coefficient of the tower body increased significantly, and the maximum increment, 67.5%, occurred at the bottom of the tower body.
- Under typhoon conditions, the radial displacement at the tower top of the wind turbine increased by 28.6%. The increments of the maximum meridian axial force, maximum shearing force, maximum circumferential bending moment, and maximum meridian bending moment were 36.1%, 34.8%, 34.2%, and 46.5%, respectively. All were observed close to the circumferential 330°. The maximum increments of the shearing force, circumferential bending moment, and meridian bending moment at the blade root under typhoon conditions were 41.3%, 39.6%, and 39.5%, which were observed on Blade C. The maximum increment of downwind displacement at the blade tip occurred on Blade A, which reached 49.3%.
- Under typhoon conditions, the buckling coefficient, critical wind speed, and maximum buckling displacement of the wind turbine system increased by 1.3%, 30.9%, and 28%, respectively. The wind turbine system achieved the maximum buckling displacement when the basic wind speed was about 20 m/s, and the minimum buckling displacement at about 40 m/s of the basic wind speed. When the basic wind speed was about 40 m/s, the buckling displacement under typhoon conditions increased by 38.5%.

Author Contributions: Conceptualization, S.K.; Data curation, S.K.; Formal analysis, W.Y.; Funding acquisition, S.K.; Investigation, J.C.; Methodology, T.W.; Software, W.Y.; Writing—original draft, S.K.; Writing—Review & Editing, S.K.

Acknowledgments: The authors would like to thank Yaojun Ge and Yukio Tamura for valuable discussions on this article. The authors are sincerely grateful to the reviewers for their valuable review comments, which substantially improved the paper.

References

1. Murtagh, P.J.; Basu, B.; Broderick, B.M. Along-wind response of a wind turbine tower with blade coupling subjected to rotationally sampled wind loading. *Eng. Struct.* **2005**, *27*, 1209–1219. [CrossRef]
2. Ke, S.T.; Ge, Y.J.; Wang, T.G.; Cao, J.F.; Tamura, Y. Wind field simulation and wind-induced responses of large wind turbine tower-blade coupled structure. *Struct. Des. Tall Spec. Build.* **2015**, *24*, 571–590. [CrossRef]
3. Hansen, M.O.L.; Sørensen, J.N.; Voutsinas, S.; Sørensen, N.; Madsen, H.A. State of the art in wind turbine aerodynamics and aeroelasticity. *Prog. Aerosp. Sci.* **2006**, *42*, 285–330. [CrossRef]
4. Ke, S.T.; Yu, W.; Wang, T.G.; Zhao, L.; Ge, Y.J. Wind loads and load-effects of large scale wind turbine tower with different halt positions of blade. *Wind Struct. Int. J.* **2016**, *23*, 559–575. [CrossRef]
5. Utsunomiya, T.; Sato, I.; Yoshida, S.; Ookubo, H.; Ishida, S. Dynamic response analysis of a floating offshore wind turbine during severe typhoon event. In Proceedings of the ASME 2013 32nd International Conference on Ocean, Offshore and Arctic Engineering, Nantes, France, 9–14 June 2013; p. V008T09A032.

6. Wang, Z.Y.; Zhang, B.; Zhao, Y.; Liu, G. H.; Jiang, J.Q. Study on the vibration response of wind turbine tower under typhoons. *Acta Energ. Sol. Sin.* **2013**, *34*, 1434–1442. (In Chinese)
7. Chen, X.; Xu, J.Z. Structural failure analysis of wind turbines impacted by super typhoon Usagi. *Eng. Fail. Anal.* **2016**, *60*, 391–404. [CrossRef]
8. Ishihara, T.; Yamaguchi, A.; Takahara, K.; Mekaru, T.; Matsuura, S. An analysis of damaged wind turbines by typhoon Maemi in 2003. In Proceedings of the 6th Asia-Pacific Conference on Wind Engineering, Seoul, Korea, 12–14 September 2005; pp. 1413–1428.
9. Takeshi, I.; Van, P.P.; Fujino, Y.; Takahara, K.; Mekaru, T. A field test and full dynamic simulation on a stall regulated wind turbine. In Proceedings of the Sixth Asia-Pacific Conference on Wind Engineering (APCWE-VI), Seoul, Korea, 12–14 September 2005; Volume 48, pp. 599–612.
10. Tang, W.L.; Yuan, Q.; Han, Z.H. Withstanding typhoon design of wind turbine tower. *Acta Energ. Sol. Sin.* **2008**, *29*, 422–427. (In Chinese)
11. Chou, J.S.; Tu, W.T. Failure analysis and risk management of a collapsed large wind turbine tower. *Eng. Fail. Anal.* **2011**, *18*, 295–313. [CrossRef]
12. Li, Z.Q.; Chen, S.J.; Ma, H.; Feng, T. Design defect of wind turbine operating in typhoon activity zone. *Eng. Fail. Anal.* **2013**, *27*, 165–172. [CrossRef]
13. Luo, Y.; Chen, Q.; Ren, D.; Li, X.; Liao, Y. The light-weight design of tower for under typhoon wind turbine based on the typhoon control strategy. In Proceedings of the International Conference on Sensors, Mechatronics and Automation, Zhuhai, China, 12–13 November 2016.
14. Lian, J.; Jia, Y.; Wang, H.; Liu, F. Numerical study of the aerodynamic loads on offshore wind turbines under typhoon with full wind direction. *Energies* **2016**, *9*, 613. [CrossRef]
15. Zhou, P.Z.; Long, S.H.; Chen, R. Optimal design of aerodynamic shape of wind turbine blade against typhoon based on genetic algorithm. *J. Changsha Univ. Technol. (Nat. Sci.)* **2015**, *12*, 89–94. (In Chinese)
16. Cao, Z. Design of a vertical axis wind turbine to resist typhoon. *Appl. Energy Technol.* **2015**, *35*, 50–52. (In Chinese)
17. An, Y. *Study on the Performance of Large Double Runner Wind Turbine against Typhoon*; South China University of Technology: Guangzhou, Guangdong, 2011. (In Chinese)
18. Utsunomiya, T.; Sato, I.; Yoshida, S.; Ookubo, H.; Ishida, S. At-sea experiment of a spar-type floating wind turbine (behavior during typhoon attack). *Wind Energy* **2015**, *39*, 86–97.
19. Ma, Z.; Li, W.; Ren, N.X.; Ou, J. The typhoon effect on the aerodynamic performance of a floating offshore wind turbine. *J. Ocean Eng. Sci.* **2017**, *35*, 87–99. [CrossRef]
20. Lundquist, J.K.; Mirocha, J.D.; Chow, F.K.; Kosovic, B.; Lundquist, K.A. Simulating atmosphere flow for wind energy applications with WRF-LES. *Univ. North Texas* **2008**, *7*, 281–295.
21. Shen, L.; Han, Y.; Dong, G.C.; Cai, C.S.; Zheng, J.R. Numerical simulation of wind field on bridge site located in mountainous area and gorge based on WRF. *China J. Highw. Transp.* **2017**, *30*, 104–113. (In Chinese)
22. Papanastasiou, D.K.; Melas, D.; Lissaridis, I. Study of wind field under sea breeze conditions; an application of WRF model. *Atmos. Res.* **2010**, *98*, 102–117. [CrossRef]
23. Dimitrov, N.; Natarajan, A.; Mann, J. Effects of normal and extreme turbulence spectral parameters on wind turbine loads. *Renew. Energy* **2017**, *10*, 1180–1193. [CrossRef]
24. Li, J.; Song, X.P.; Cheng, X.L.; Hu, F.; Zhu, R. Dynamical simulation of wind flow from synoptic scale to turbine scale. *Acta Energ. Sol. Sin.* **2015**, *36*, 806–811. (In Chinese)
25. Maruyama, T.; Tomokiyo, E.; Maeda, J. Simulation of strong wind field by non-hydrostatic mesoscale model and its applicability for wind hazard assessment of buildings and houses. *Hydrol. Res. Lett.* **2010**, *16*, 40–44. [CrossRef]
26. Skamarock, W.C. A description of the advanced research WRF version 3. *NCAR Tech.* **2005**, *113*, 7–25.
27. Cha, D.H.; Wang, Y.A. Dynamical Initialization Scheme for Real-time Forecasts of Tropical Cyclones Using the WRF Model. *Mon. Weather Rev.* **2013**, *141*, 964–986. [CrossRef]
28. Gan, C.M.; Hogrefe, C.; Mathur, R.; Pleim, J.; Xing, J.; Wong, D.; Gilliam, R.; Pouliot, G.; Wei, C. Assessment of the effects of horizontal grid resolution on long-term air quality trends using coupled WRF-CMAQ simulations. *Atmos. Environ.* **2016**, *132*, 207–216. [CrossRef]
29. Zou, Z.C.; Deng, Y.C.; Deng, Z.P. Analysis on the effect of different initial fields on wind field simulation of WRF model. *Renew. Energy* **2015**, *33*, 1048–1053. (In Chinese)
30. Pan, X.D.; Li, X. Research on the effect of horizontal resolution on WRF model—Taking Heihe river basin WRF simulation as an example. *Sci. Res. Inf. Technol. Appl.* **2011**, *2*, 126–137. (In Chinese)

31. Yang, J. *Numerical Simulation Study of Local Typhoon Wind Fields Nested by WRF and CFD*; Harbin Institute of Technology: Harbin, Heilongjiang, 2015. (In Chinese)
32. Miao, Y.; Liu, S.; Chen, B.; Zhang, B.; Wang, S.; Li, S. Simulating urban flow and dispersion in Beijing by coupling a CFD model with the WRF model. *Adv. Atmos. Sci.* **2013**, *30*, 1663–1678. [CrossRef]
33. Liu, Y.S.; Miao, S.G.; Zhang, C.L.; Cui, G.X.; Zheng, Z.S. Study on micro-atmospheric environment by coupling large eddy simulation with mesoscale model. *J. Wind Eng. Ind. Aerodyn.* **2012**, *s107–s108*, 106–117. [CrossRef]
34. Davis, C.; Wang, W.; Chen, S.S.; Chen, Y.; Corbosiero, K.; DeMaria, M.; Dudhia, J.; Holland, G.; Klemp, J.; Michalakes, J.; et al. Prediction of Landfalling Hurricanes with the Advanced Hurricane WRF Model. *Month. Weather Rev.* **2008**, *136*, 1990–2005. [CrossRef]
35. Lou, X.F.; Hu, Z.J.; Wang, P.Y.; Zhou, X.J. Introduction of mesoscale model cloud precipitation physics scheme. *J. Appl. Meteorol.* **2003**, *14*, 49–59. (In Chinese)
36. Yi, J.K. Diagnosis and analysis of the movement, landing site and intensity change of typhoon No. 0812 'Nuri'. In Proceedings of the 26th Annual Meeting of the Chinese Meteorological Society, Hangzhou, Zhejiang, 14 October 2009. (In Chinese)
37. Song, L.L.; Pang, J.B.; Jiang, C.L.; Huang, H.H.; Qin, P. Measurement and analysis of turbulence characteristics of 'Nuri' typhoon in Macao friendship bridge. *Chin. Sci. Technol. Sci.* **2010**, *40*, 1409–1419. (In Chinese)
38. *Load Code for the Design of Building Structures*; GB 50009-2012; The Ministry of Structure of the People's Republic of China: Beijing, China, 2012. (In Chinese)
39. Wang, X.H.; Ke, S.T. Study on aerodynamic performance of large wind turbine tower considering blade yawing and interference effect. *Chin. J. Electr. Eng.* **2018**, *38*, 4546–4554, 4655. (In Chinese)
40. Zhang, L.D.; Ren, L.C.; Lian, L.J.; Chen, R.S. Study on shape design and three-dimensional solid modeling of wind turbine blades. *Acta Energ. Sol. Sin.* **2008**, *29*, 1177–1180.
41. Luo, K.; Zhang, S.; Gao, Z.; Wang, J.; Zhang, L.; Yuan, R.; Fan, J.; Cen, K. Large-eddy simulation and wind-tunnel measurement of aerodynamics and aeroacoustics of a horizontal-axis wind turbine. *Renew. Energy* **2015**, *77*, 351–362. [CrossRef]
42. Kuo, J.Y.J.; Romero, D.A.; Beck, J.C.; Amon, C.H. Wind farm layout optimization on complex terrains—Integrating a CFD wake model with mixed-integer programming. *Appl. Energy* **2016**, *178*, 404–414. [CrossRef]
43. Chattot, J.J. Effects of blade tip modifications on wind turbine performance using vortex model. *Comput. Fluids* **2009**, *38*, 1405–1410. [CrossRef]
44. Jiang, F. *Advanced Applications and Case Study of Fluent*; Tsinghua University Press: Beijing, China, 2008. (In Chinese)
45. Genç, M.S. Numerical simulation of flow over a thin aerofoil at a high reynolds number using a transition model. *J. Mech. Eng. Sci.* **2010**, *1*, 1–10. [CrossRef]
46. Ke, S.T.; Xu, L.; Ge, Y.J. The aerostatic response and stability performance of a wind turbine tower-blade coupled system considering blade shutdown position. *Wind Struct. Int. J.* **2017**, *25*, 507–535.
47. Ke, S.T.; Wang, T.G.; Ge, Y.J.; Tamura, Y. Wind-induced responses and equivalent static wind loads of tower-blade coupled large wind turbine system. *Struct. Eng. Mech. Int. J.* **2014**, *52*, 485–505. [CrossRef]
48. Cárdenas, D.; Escárpita, A.A.; Elizalde, H.; Aguirre, J.J.; Ahuett, H.; Marzocca, P.; Probst, O. Numerical validation of a finite element thin-walled beam model of a composite wind turbine blade. *Wind Energy* **2012**, *15*, 203–223. [CrossRef]
49. Ke, S.T.; Wang, X.H. Wind vibration response and stability analysis of large-scale wind turbine systems considering the yaw and interference effects of blades. *J. Hunan Univ. (Nat. Sci. Ed.)* **2018**, *45*, 61–70. (In Chinese)

5

Experimental Study on Aerodynamic Characteristics of a Gurney Flap on a Wind Turbine Airfoil under High Turbulent Flow Condition

Junwei Yang [1,2,3], Hua Yang [1,2,*], Weijun Zhu [1,2], Nailu Li [1,2] and Yiping Yuan [4]

1 College of Electrical, Energy and Power Engineering, Yangzhou University, Yangzhou 225127, China; yangjunwei@yzu.edu.cn (J.Y.); wjzhu@yzu.edu.cn (W.Z.); nlli@yzu.edu.cn (N.L.)
2 New Energy Research Center, Yangzhou University, Yangzhou 225009, China
3 College of Hydraulic Science and Engineering, Yangzhou University, Yangzhou 225009, China
4 Jiangsu Key Laboratory of Hi-Tech Research for Wind Turbine Design, Nanjing University of Aeronautics and Astronautics, Nanjing 210016, China; yuanyiping@nuaa.edu.cn
* Correspondence: yanghua@yzu.edu.cn

Abstract: The objective of the current work is to experimentally investigate the effect of turbulent flow on an airfoil with a Gurney flap. The wind tunnel experiments were performed for the DTU-LN221 airfoil under different turbulence level (T.I. of 0.2%, 10.5% and 19.0%) and various flap configurations. The height of the Gurney flaps varies from 1% to 2% of the chord length; the thickness of the Gurney flaps varies from 0.25% to 0.75% of the chord length. The Gurney flap was vertical fixed on the pressure side of the airfoil at nearly 100% measured from the leading edge. By replacing the turbulence grille in the wind tunnel, measured data indicated a stall delay phenomenon while increasing the inflow turbulence level. By further changing the height and the thickness of the Gurney flap, it was found that the height of the Gurney flap is a very important parameter whereas the thickness parameter has little influence. Besides, velocity in the near wake zone was measured by hot-wire anemometry, showing the mechanisms of lift enhancement. The results demonstrate that under low turbulent inflow condition, the maximum lift coefficient of the airfoil with flaps increased by 8.47% to 13.50% (i.e., thickness of 0.75%), and the Gurney flap became less effective after stall angle. The Gurney flap with different heights increased the lift-to-drag ratio from 2.74% to 14.35% under 10.5% of turbulence intensity (i.e., thickness of 0.75%). However, under much a larger turbulence environment (19.0%), the benefit to the aerodynamic performance was negligible.

Keywords: wind tunnel experiment; wind turbine airfoil; turbulence; Gurney flap; aerodynamic characteristics

---

1. Introduction

Wind power generation technology has been maturely developed in the past decades. Airfoil is a basic element of a wind turbine blade, and its aerodynamic characteristics have a major influence on the wind energy conversion efficiency. Among the conventional rotor aerodynamic design strategy, the blade add-ons were of particular interest to further improve wind energy efficiency. Therefore, mounting flap to the airfoil trailing edge was one of the most feasible methods to improve the aerodynamic performance of wind turbines. In addition to power production, such a technique can also effectively reduce the aerodynamic loads both of wind turbine blades and tower. If a sophisticated controller was implemented, the flap can further reduce turbulence-induced fatigue loads, so that longer lifetime was guaranteed. After the pioneering work of Liebeck [1], a large number of studies have been conducted to explain the phenomena induced by the presence of this device. More recently, research objects were most focused on airfoil attached various shapes of flaps, such as Gurney flaps [2,3], triangular flaps [4], separate trailing edge flaps [5,6] and deformable trailing edge flaps [7,8]. For experiment tests, Zhang et al. [9] and Amini et al. [10] studied the aerodynamic effect of a Gurney flap based on the airfoil through wind tunnel experiments. T Lee et al. [11] carried out a wind tunnel test of the lift force and pitching moment coefficients of both trailing edge flaps and Gurney flaps with different shape parameters. Based on a 5 MW reference wind turbine, Chen et al. [12] designed and optimized a trailing edge flaps such that the blade mass can be further reduced but still maintain the desired power performance. Medina et al. [13] explored the flow mechanisms of a flap at a high deflection angle. When the flap works in a separated flow region, he provided some ideas for realizing instantaneous action or alleviating extra aerodynamic loads on wind turbines. Elsayed et al. [14] studied the flap tip vortexes and characterized the flow structures behind a flap in a low-speed wind tunnel by using particle image velocimetry. Little et al. [15] designed a trailing edge flap by using a single medium plasma driver which resulted in a higher lift force. Bergami et al. [16] designed an active controller of a trailing edge flap on a 5 MW reference wind turbine and proved that the flap could effectively control aerodynamic loads. Edward et al. [17] also conducted experiments on a flaps noise drop. According to the wind tunnel tests based on a full-size rotor, Straub et al. [18] found that flaps could also be used to control noise and vibration such that the noise generated by blade vortex interaction could be reduced by 6 dB. For numerical simulations and theoretical analyses, Traub et al. [19] fitted a semi-empirical equation by summarizing the performance of a large number of flaps. Lario et al. [20] numerically solved the unsteady flow field of Gurney flaps at high Reynolds numbers through the discontinuous Galerkin method. By analyzing dynamic characteristics of an airfoil with Gurney flaps through the numerical simulation, Li et al. [21] found that flaps were capable to reduce unsteady aerodynamic loads of wind turbines. Zhu et al. [22] numerically simulated the airfoil with trailing edge flaps using the immersed boundary method and found that flaps could be combined with the paddle movement to adjust the aerodynamic loads of a wind turbine airfoil. Ng et al. [23] investigated the trailing edge flap together with an aeroelastic analysis. It was noted that the trailing edge flap could be a smart device to control aeroelastic deformation.

All the above researches on the flaps were built on the uniform inflow of low turbulence intensity. In the past, there were few studies carried out by flow over flaps under high turbulence intensity [24,25], however, such a flow condition often occurs on wind turbines operating in a wind farm. Considering wind turbines operate in a turbulence environment, the conclusions obtained from the previous studies might not be accurate. Therefore, to simulate wind turbines under turbulence environment, active and passive wind tunnel turbulence generation methods can be used. The active technology includes a vibrating grille and multi-fan wind tunnel; the maximum turbulence intensity can reach more than 20% [26]. The passive control structure was relatively simple, which can be divided into grille, wedge, and rough square types among the passive turbulence generators, it was more convenient to construct grilles, which have gain very popular use. In this experiment, specific turbulence levels were passively controlled by a grille with proper grid size.

The investigations presented in this paper were focused on the coupled effects of Gurney flap and turbulence inflow. The Gurney flaps were experimentally investigated under various turbulence intensities and flap configurations. The desired turbulent flow passes the airfoil was achieved by changing the grille size as well as the distance between the grille and the airfoil model. The hot-wire anemometer was used to record the wind speed and turbulence intensity in a flow cross-section. The quantitative information obtained during the experiment includes: (1) turbulent field descriptions, (2) airfoil pressure coefficients, (3) lift-to-drag coefficients, (4) wake measurements. On that basis, the aerodynamic performance of Gurney flaps with different heights and thicknesses was researched. The values of lift and pitching-moment coefficients were obtained through the integration of surface pressures. The wake rake array was also used to determine the values of drag coefficient. Furthermore, the flow fields near the trailing edge of airfoil were tested which could further verify the reason for lift improvement. Finally, concluding remarks accompany the discussion of the experimental investigations.

2. Experimental Setup

The experiments were carried out in a wind tunnel located at Yangzhou University. It was a close-loop type wind tunnel which contains two experimental sections. The measurements were conducted in the smaller section with the cross-section parameters of 3 m × 1.5 m, and the length is 3 m. The operational wind speed range was 0~50 m/s and the calibrated maximum turbulence intensity in the free stream was 0.2%. The DTU-LN221 airfoil model [27], as shown in Figure 1, was adopted which has a chord length of 0.6 m and a span length of 1.5 m. The airfoil model was vertically placed in the test section. The bottom part of the airfoil section was connected by the rotating shaft, which was fixed on a rotational plate, as demonstrated in Figure 2. Therefore, the angle of attack can be remotely controlled via a shaft connection with a motor below the wind tunnel. The airfoil model was made of aluminum alloy where small taps were drilled on the surface of the middle section, with the location sketched in Figure 1. The arc length between the taps was approximately spaced every 2.5% of chord length starting from the leading edge to 90% of the trailing edge. The hollow plastic hoses were connected on the reverse side of the airfoil surface which has an outer diameter of 1.2 mm. Thus, a total of 77 pressure taps were arranged for the pressure measurement on the surface of the airfoil. The pressures on the airfoil surface were sensed by the electronic pressure acquisition system of PSI (Pressure Systems Inc., Hampton, VA, USA). The pressure system used in the experiment has a sampling frequency of 333.3 Hz, a measuring range of ±2.5 kPa and a measuring accuracy of ±0.05%. Figure 1 also shows the chord-wise position, so the lift and pitching-moment of the test airfoil were achieved by integrating the pressure and the position of the taps on the surface. The lift coefficient and pitching-moment coefficients were given by

$$\begin{aligned} C_l &= C_n \cos\alpha - C_t \sin\alpha \\ C_m &= \int_0^1 \left(C_{pl} - C_{pu}\right)(0.25 - \hat{x})\,d\hat{x} \end{aligned} \tag{1}$$

where

$$\begin{aligned} C_n &= \int_0^1 \left(C_{pl} - C_{pu}\right) d\hat{x} \\ C_t &= \int_{\hat{y}_{l\ max}}^{\hat{y}_{u\ max}} \left(C_{p,be} - C_{p,af}\right) d\hat{y} \end{aligned} \tag{2}$$

where C_l and C_m are the lift coefficient and pitching-moment coefficient, respectively; αrepresents the angle of attack; $\hat{x}$ is the relative chord length; $\hat{y}$ is the thickness value relative to the chord length; C_{pu} and C_{pl} are the pressure coefficient on the suction and pressure sides of the airfoil, respectively; C_n and C_t are the normal force coefficient and tangential force coefficient, respectively. $C_{p,be}$ and $C_{p,af}$ are the pressure coefficient before and after the maximum thickness of airfoil; $\hat{y}_{u\ max}$ and $\hat{y}_{l\ max}$ represent the maximum thickness values of the suction and pressure sides relative to chord length.

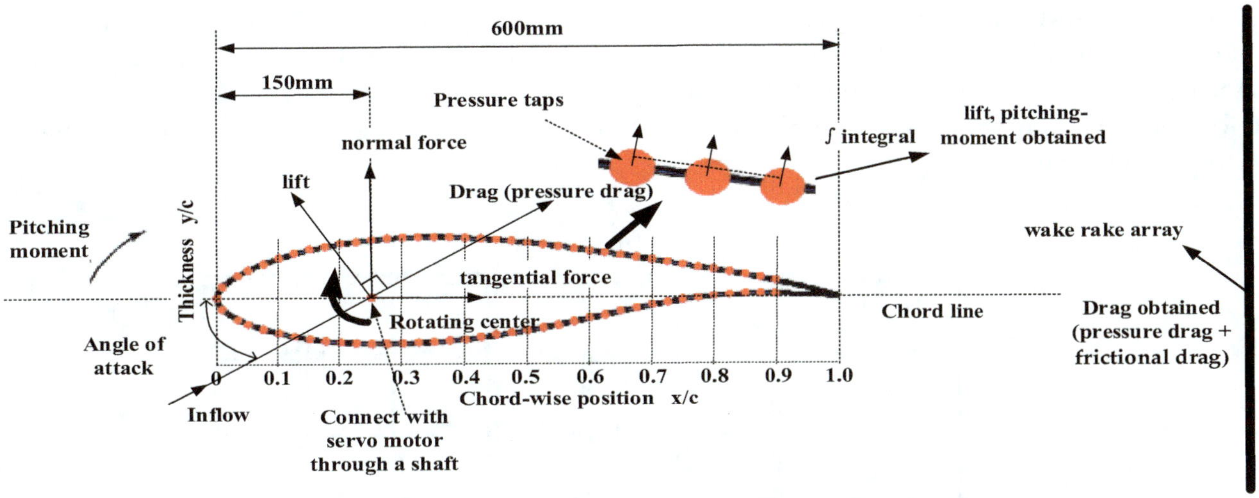

Figure 1. Schematic diagram of measuring position.

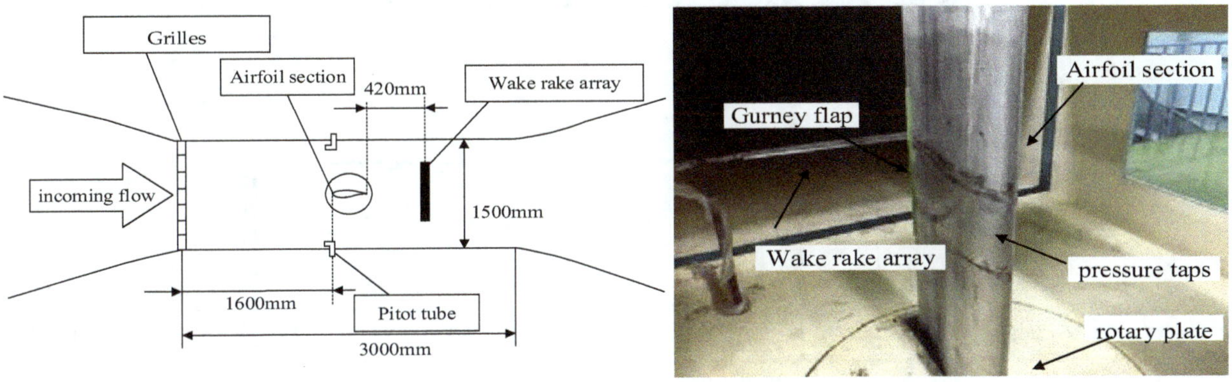

Figure 2. Experimental test device.

At a small angle of attack, the aerodynamic drag largely consists of frictional drag, while the data measured by surface pressure taps cannot accurately represent the drag force, so the drag can be measured by the momentum method more precisely. As shown in Figure 2, a wake rake array was placed at 0.7 chord length behind the trailing edge of the airfoil and at the same vertical level as the pressure taps. The measurement range of the wake probes was 80.8 cm, 102 total pressure pipes (with an outer diameter of 1.2 mm) and 4 static pressure pipes (with an outer diameter of 2 mm) were averagely arranged, with an 20 cm apart for the static pressure pipes. To prevent air leakage, all pressure tubes were connected by plastic hoses to the pressure measurement device. Besides, two Pitot tubes were installed at 1.6 m measured from the downstream of grilles where the free stream velocity was recorded. The drag coefficients were given as follows

$$C_d = \frac{2}{c}\int_w \sqrt{\frac{P_{01} - P}{P_0 - P_\infty}}\left[1 - \sqrt{\frac{P_{01} - P_\infty}{P_0 - P_\infty}}\right]ds \tag{3}$$

where C_d is the drag coefficient; c is the chord length, w is the range of integration; s is the coordinate along the thickness direction of the wake rake array; P_∞ and P_0 are the static pressure and total pressure measured by the Pitot tubes; P and P_{01} are the static pressure and total pressure measured by the wake rake array. It should be noted that there is a total pressure loss along the Pitot tubes to the wake rake array, therefore the total pressure loss should be added to each total pressure measuring point of the wake rake array.

Figure 3 presents the schematic diagram of the grille geometry. The grilles assembled with many squared alloys with geometry specified by four parameters a, b, c and d. These small squares were bolted together and on top of the grille the rubber pads were attached. During the measurements, two types of grilles were implemented. The dimensions of the two grilles were provided in Table 1. The width of the longitudinal grille and the transverse grille were denoted by a and c, respectively; the width and the height of the spacing were denoted by b and d, respectively, the thickness along the flow direction was e.

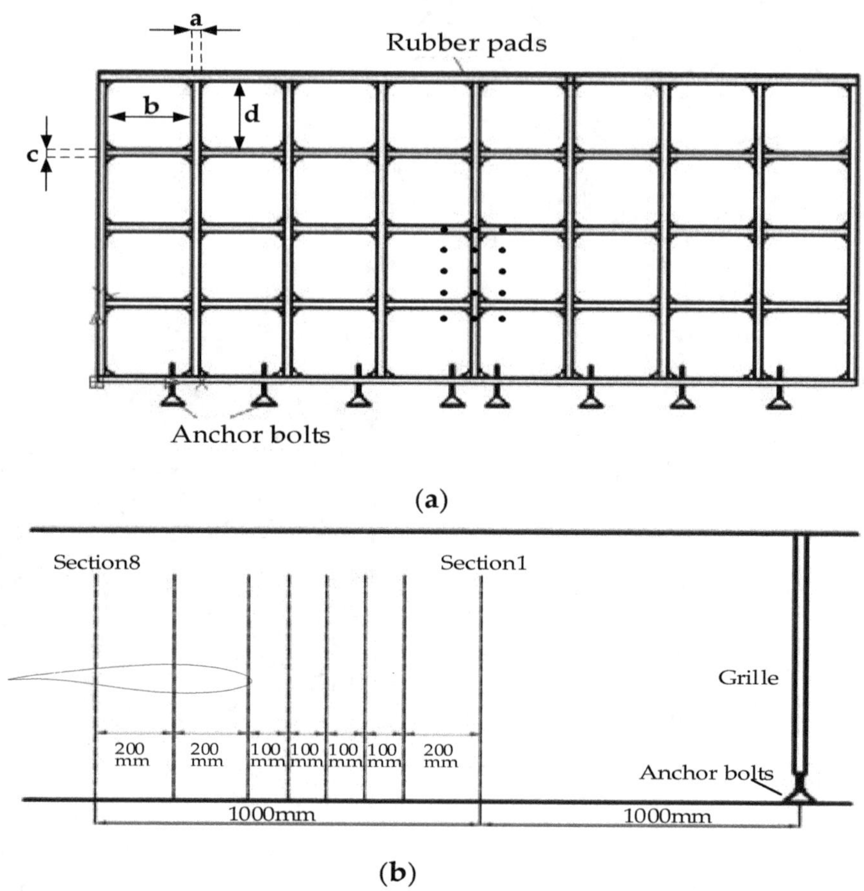

Figure 3. Schematic diagram of experimental grille. (**a**) Front view; (**b**) side view.

Table 1. Dimensions of the experiment grilles.

	a (cm)	b (cm)	c (cm)	d (cm)	e (cm)
Scheme 1	3	34	3	32	3
Scheme 2	6	31.4	6	29.8	3

Behind the grille, there were eight measurement sections designed for velocity recording. At each of the section, 15 measurement points were uniformly spaced. The center of the topmost measurement point corresponds to the geometrical center of the wind tunnel section. The horizontal and vertical spacing of measuring points were 10 cm, as displayed by the black spots in Figure 3a. In addition, we set eight measurement sections one to two meters behind the grille, as shown in Figure 3b. The wind speed was measured through a Dantec hot-wire probe (type 55P61). The sampling frequency was 5 kHz, and the test wind speeds were consistent with that in the airfoil experiment. Figure 4 displays the device of turbulence measurement. Besides, wake measurements in this paper were carried out on an electronically controlled traverser consisting of a two-dimensional probe, as displayed in Figure 5.

Figure 4. Grille turbulence generated device.

Figure 5. Wake measurements.

The influence of the flap height and thickness were two key parameters which influence the overall aerodynamic performance of the airfoil. For this reason, a total of nine types of flaps were designed in the experiments, namely, the flap with heights h of 1%, 1.5% and 2% chord length (6 mm, 9 mm and 12 mm, respectively) and with relative thicknesses d of 0.25%, 0.5% and 0.75% chord length (1.5 mm, 3 mm and 4.5 mm, respectively). The diagram of the flap used in the experiment was given in Figure 6. As shown, the flap has an L shape; the bottom part has a width of 7 mm which was attached to the pressure side of the airfoil trailing edge. During the experiments, the flap was vertically attached near the position of 100% of the chord length, see Figure 2 with the specific experimental setup. The coordinate system behind the trailing edge was defined for subsequent wake measurements, also the x and y axis were set in the vertical and the forward directions, respectively. The black spots in the Figure 6 represented the hot-wire measuring positions set at downstream of the trailing edge.

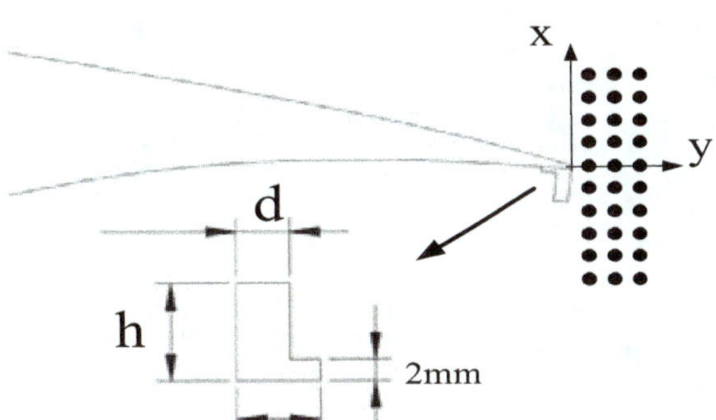

Figure 6. Flap geometry.

3. Experimental Results and Analysis

In the experiment, the free stream flow velocity was 20 m/s, and the corresponding Reynolds number based on the airfoil chord length was 0.8×10^6, although wind turbines often operate at wind speed below 20 m/s, but the magnitude of the Reynolds number was up to an order of 10^6. Therefore, the experimental value of the Reynolds number was chosen to approach an order of magnitude corresponding to those obtained from full-scale wind turbines. Two types of grilles generate different turbulence levels in the wind tunnel, such that the experiments were mainly separated into low and high turbulence cases. To be specific, a grille was placed at the upstream of the airfoil test section which results in different aerodynamic characteristics of flow over the airfoil with and without a Gurney flap.

3.1. Experimental Results of Grille Turbulence Generator

The measurements were first conducted for the two grilles. Figure 7 illustrates the time series of the instantaneous velocity at the central measuring point of the 6th section. As shown in Figure 7, the incoming wind became highly disturbed, for this reason, a statistical analysis was done for the flow field. The decay of turbulence intensity from the grilles towards the airfoil test section was reported in Figure 8a. Within the scope of the test, the averaged turbulence intensity was recorded along 8 downstream sections (each measured with 15 points). Besides, the margin of error in Figure 8a was expressed as the standard deviation of each section data. According to Figure 8a, the average turbulence intensity of downstream in scheme one dropped from 13.5% at the position of 1.0 m behind the grille to 9.4% at the position of 2.0 m; the average turbulence intensity of the downstream in scheme two was higher than scheme 1, which decreased from 27.7% at the position of 1.0 m behind the grille to 15.9% at the position of 2.0 m. The standard deviations of turbulence intensity of the two schemes drops quickly towards the test section. The change in turbulence level indicates that there were very high turbulence fluctuations just behind the grille which yielded very unstable flow field. The flow became gradually stabilized owing to the mutual dissipation over a propagation distance. To ensure a homogeneous and isotropic turbulent field, so the distance 1.6 m (from the leading edge of the airfoil to the grille) was selected for placing the leading edge of the airfoil. The turbulence intensity measured at the 6th section of the two schemes was 10.5% and 19.0%, respectively. Figure 8b illustrates the power spectral density (PSD) of the wind speed at the two grille schemes measured at the central measuring point of the 6th section (1.6 m measured from grille to the hot-wire probes). Obviously, the larger the turbulence intensity was generated, the larger the amplitude of the power spectral density was observed. Moreover, the power spectra amplitude in the downstream direction was consistent with the corresponding Karman spectra, and it dropped down rapidly after the frequency over 30 Hz, which suggests that the grille distance was more proportional to the turbulent kinetic energy.

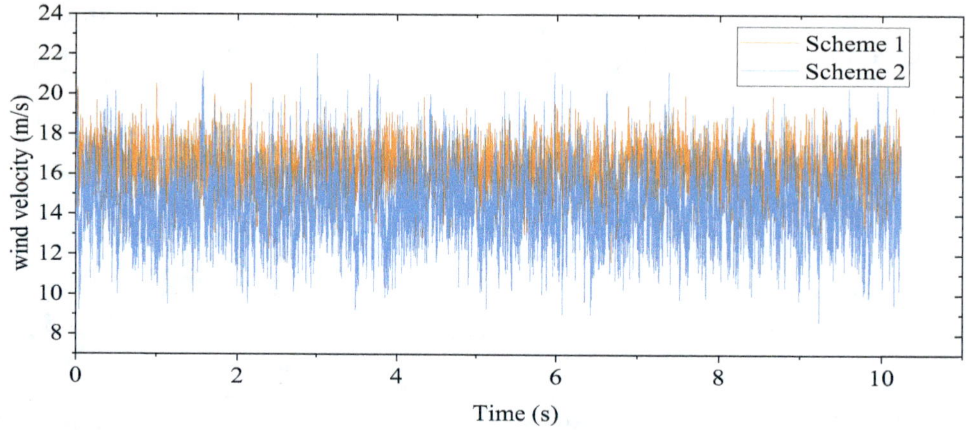

Figure 7. The time series of the instantaneous streamwise velocity (the 6th section, central measuring point).

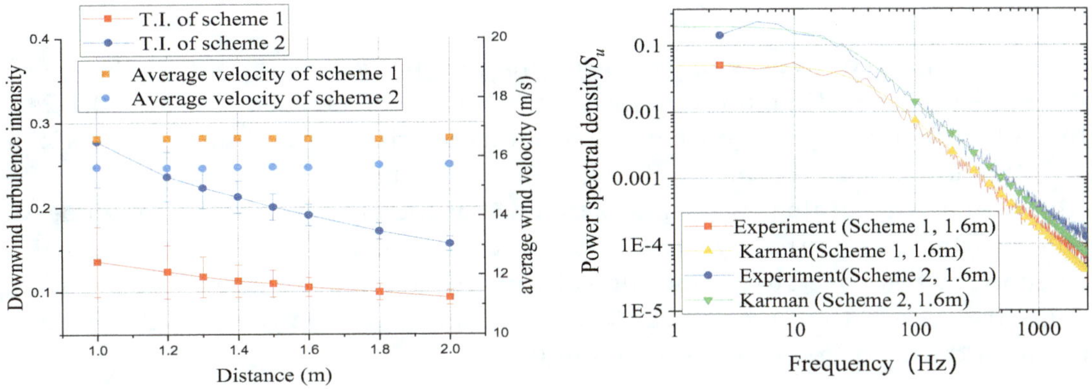

(a) (b)

Figure 8. Description of the flow field behind the grilles. (**a**) Turbulence intensity and average velocity (Re: 0.8×10^6); (**b**) Comparison of the PSD (1.6 m behind the grille).

3.2. Experimental Results of the Baseline Airfoil

To verify the accuracy of the experiment, the experimental data from the wind tunnel of LM Wind Power were selected as a reference [28]. The aerodynamic data of the DTU-LN221 airfoil under uniform inflow were measured in the LM wind tunnel experiment at the approximate Reynolds number of 1.5×10^6. The lift and drag data were selected as a cross-validation case. In consideration of the interference effect of wind tunnel wall, the maximum blockage ratio in this experiment was about 8.4%, so the experiment measured data of the airfoil were corrected by after the reference [29], and the experimental data analyzed below were all corrected. In order to perform a comparison, Figure 9 shows the comparison of experimental data of the DTU-LN221 airfoil under uniform inflow condition. To minimize the uncertainty, measurements were performed several times for comparison test at the approximate Reynolds number of 1.5×10^6, YZU1, YZU2, YZU3 represents the results of the repeated experiments. As shown in Figure 9, the lift and drag coefficients were very similar before the stall angle, while there were some discrepancies in the stall state. Consider the cause of flow separation, the deviation of results of drag is quite considerable, but the stall angles tested in the two wind tunnels were nearly the same. The reason for this may be that although the airfoils were of the same type, they were individually manufactured, for example the roughness of the airfoil surface may affect the experimental results. Besides, there were some differences in drag measurements, LM adopted the model surface pressure distribution in a certain range of the angle of attack, while YZU used the wake probes at all angles. Similar phenomena were seen in different wind tunnel laboratories [30–32]. In the following, for reason of comparisons, the measured data are averaged.

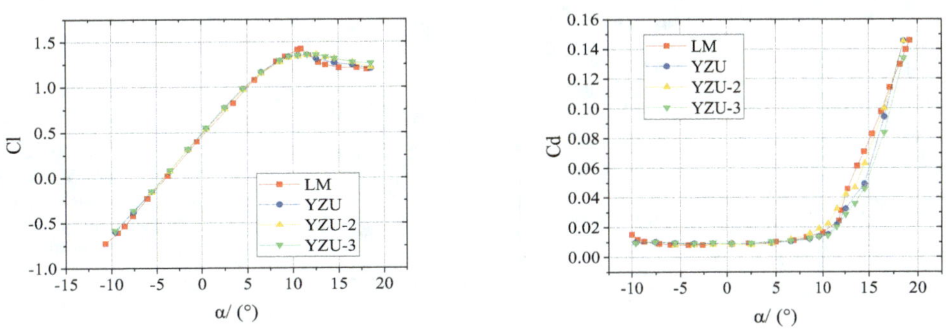

(a) (b)

Figure 9. Comparison of experimental data of the DTU-LN221 airfoil (uniform inflow, Re: 1.5×10^6). (**a**) Lift coefficient comparison; (**b**) drag coefficient comparison.

3.3. Experimental Results of the Gurney Flap under Uniform Inflow

This subsection starts to investigate the aerodynamic characteristics of the effect of Gurney flap in low turbulent inflow condition, the Reynolds number was decreased to 0.8×10^6. Seen from the Figure 10, three lift and drag curves were compared with the original at the angle of attack ranged from $-9.6°$ to $14.4°$. As shown, the lift coefficients increased with the increase of the height of the Gurney flap before the stall angle, while it was also realized that the effect of the Gurney flap was limited at a high angle of attack. The lift coefficient slopes and the stall angles of the flapped airfoil remain essentially unchanged compared to the baseline airfoil, but the flapped airfoil caused a leftward shift compared to the baseline airfoil. In order to show the changes of aerodynamic characteristics clearly, the following lift and drag differences were defined:

$$\Delta cl = \frac{\left|cl_{bs} - cl_{gf}\right|}{cl_{bs}} \times 100\% \tag{4}$$

$$\Delta cd = \frac{\left|cd_{bs} - cd_{gf}\right|}{cd_{bs}} \times 100\% \tag{5}$$

$$\Delta cld = \frac{\left|cld_{bs} - cld_{gf}\right|}{cld_{bs}} \times 100\% \tag{6}$$

where cl_{bs} and cd_{bs} are the lift and drag coefficients obtained from the baseline airfoil, respectively. cl_{gf} and cd_{gf} are the lift and drag coefficients obtained from the flapped airfoil. cld_{bs} and cld_{bs} are the lift-to-drag ratio obtained from the baseline airfoil and flapped airfoil.

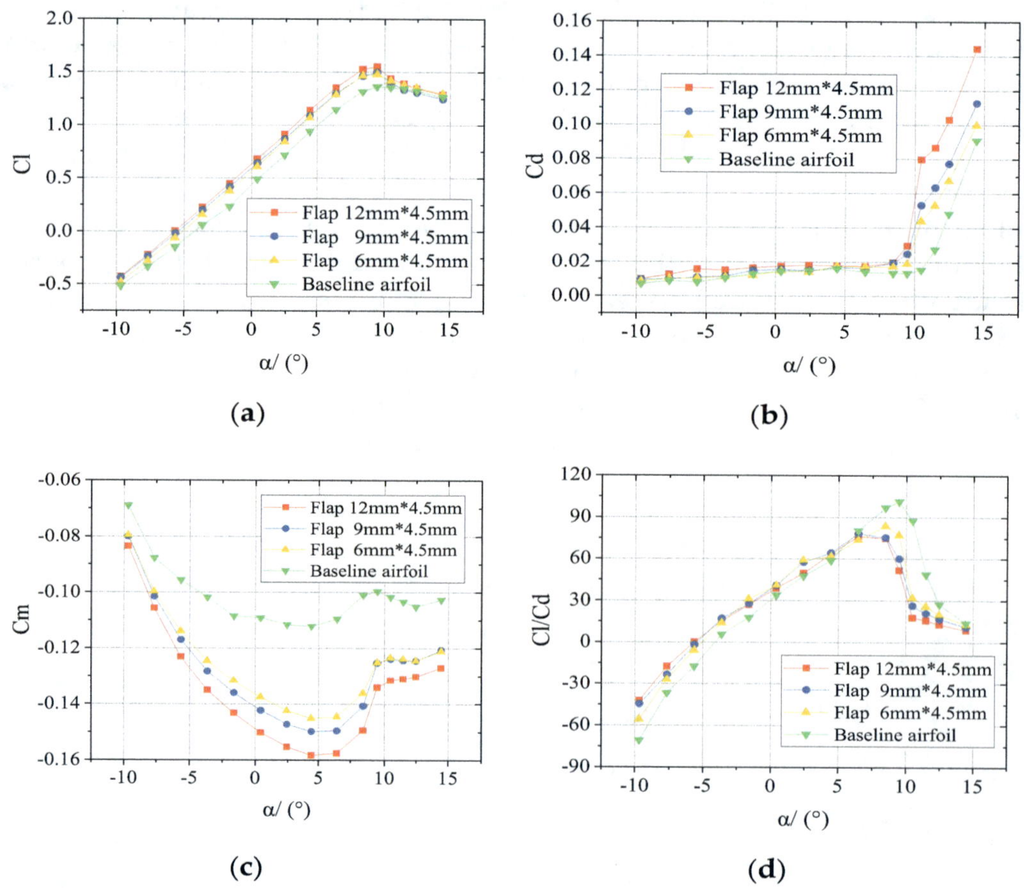

Figure 10. Variation of the aerodynamic characteristics with different heights of Gurney flaps (uniform flow, Re: 0.8×10^6). (**a**) Lift coefficient comparison; (**b**) drag coefficient comparison; (**c**) pitching-moment coefficient comparison; (**d**) lift-to-drag ratio comparison.

When the flap height changes to 6 mm, 9 mm and 12 mm, the maximum lift coefficients were increased by 8.47%, 9.56% and 13.50% at 9.4 degrees, respectively. Looking into the drag coefficients shown in Figure 10b, within the range of −9.6~10.4°, the drag coefficients show regular change where the frictional drag dominates as expected. The drag coefficients rise rapidly in the stall state, at this time the change in drag coefficient was dominated by pressure drag. For example, at an angle of attack of 12.4°, with the flap heights of 6 mm, 9 mm, 12 mm the drag coefficients were increased by 39.67%, 59.92% and 113.43%, respectively. According to Figure 10c, the Gurney flap generated a prominent increase in the pitching-moment coefficient compared to the baseline airfoil, the value increased with the heights of the Gurney flaps and reached negative peak values at approximately 4.4°. Based on Figure 10d, the lift-to-drag ratios were compared. When the angle of attack was less than 8.4°, the lift-to-drag ratios were all larger than the baseline airfoil. However, the maximum lift-to-drag ratios of the Gurney flap at three heights were smaller than the baseline airfoil. When the flap heights were 6 mm, 9 mm and 12 mm, the corresponding maximum lift-to-drag ratios were decreased by 17.16%, 22.79% and 24.47%, respectively. In the range after the stall angle, the presence of the Gurney flap reduces the lift-to-drag ratios and the higher the flap height, the more the lift efficiency decreases. When the angle of attack was 12.4°, for example, the lift-to-drag ratios of the flap with heights of 6 mm, 9 mm and 12 mm were decreased by 26.55%, 38.11% and 52.20%, respectively. Meanwhile, we observed the same trend from the other thicknesses of the Gurney flaps (i.e., of 1.5 mm), When the flap height changes to 6 mm, 9 mm and 12 mm, the maximum lift coefficients were increased by 9.27%, 9.78% and 14.08% at 9.4 degrees.

The following study in this section will be focused on the effect of the thickness of the Gurney flap. Figure 11 illustrates the changes in the aerodynamic characteristics of the baseline airfoil and airfoil with a height of 9 mm of Gurney flaps under uniform flow. As seen from the figure, by changing the thickness of the Gurney flap, the lift-to-drag curves of the flapped airfoils with different thicknesses almost overlaps, it seems that the flap thickness does not have obvious effect in the aerodynamic performance. The results imply that the small increase of the flap thickness does not change the degree of downward turning of the mean flow and recirculation flow around the trailing edge of the airfoil.

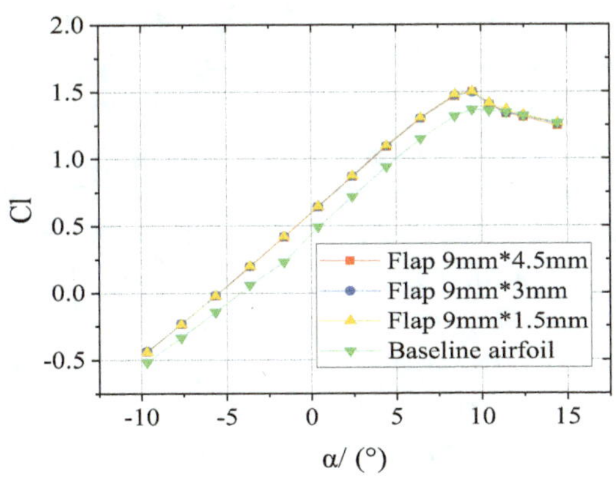

(a)

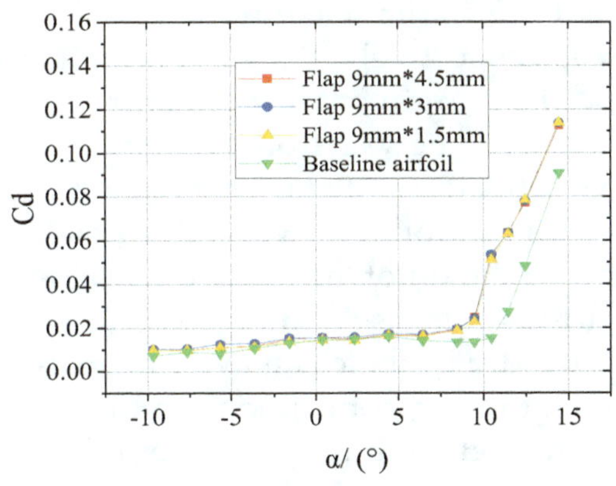

(b)

Figure 11. *Cont.*

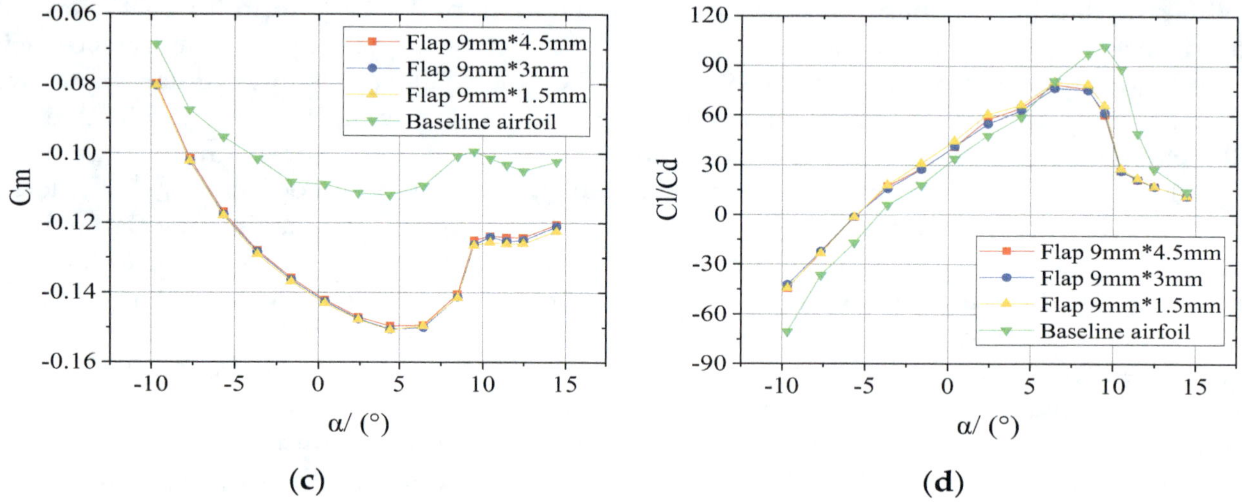

Figure 11. Variation of the aerodynamic characteristics with different thicknesses of Gurney flaps (uniform flow, Re: 0.8×10^6). (**a**) Lift coefficient comparison; (**b**) drag coefficient comparison; (**c**) pitching-moment coefficient comparison; (**d**) lift-to-drag ratio comparison.

3.4. Experimental Results of Gurney Flap under 10.5% Turbulence Intensity

This subsection starts to investigate the aerodynamic characteristics of the Gurney flap under a turbulent inflow with an intensity of 10.5%. Figure 12 shows the changes in the aerodynamic characteristics of the baseline airfoil and the Gurney flaps with a thickness of 4.5 mm when the turbulence intensity of the incoming flow was 10.5%. The measured angle of attack ranged from −9.6° to 20.4°. As shown, under the present turbulence intensity of 10.5%, all cases induced a stall delay. The stall angle was nearly delayed from the original 9.4° to about 16.4° under turbulent flow. In Figure 12a, with the increase of turbulence intensity, the maximum lift coefficient of the baseline airfoil reaches 1.611 and increases by 17.59% as compared with uniform inflow condition. An interesting observation was that in such a turbulence environment, the lift coefficients were not significantly different from each other under the current three flapped configurations. For the flap heights of 6 mm, 9 mm and 12 mm, the maximum lift coefficients were increased by 15.21%, 17.19% and 17.26% at 18.4 degrees, respectively. The increase rate was found larger than that under the uniform turbulence condition. The margin of error in Figure 12a was expressed as the variance of the time-averaged lift coefficient. As can be seen from the Figure 12a, the pressure uncertainty increased with the increase of the angle of attack, and the lift coefficient fluctuations caused by the pressure fluctuations were the most obvious pronounced when the attack angle reaches the stall angle. According to Figure 12b, the increase in the lift of the flapped airfoil was at a cost of increasing drag. The drag coefficient of the flapped airfoils was also non-linearly raised in the stall state. For example, when the angle of attack was 20.4°, the drag coefficients were increased by 12.50%, 37.50% and 44.53%, respectively. As can be seen from Figure 12c, compared with the baseline airfoil, the Gurney flap also generated a prominent increase in the pitching-moment coefficient, the negative peak values delay to about 6.4°compared with the uniform inflow condition. As demonstrated in Figure 12d, the peak value of the lift-to-drag ratio becomes less as compared with the uniform flow condition in all cases. Unlike the low turbulence condition, the maximum lift-to-drag of Gurney flap with different heights slightly increased relative to the baseline airfoil. The maximum lift-to-drag raised 14.35%, 9.15% and 2.74% respectively when the flap heights were 6 mm, 9 mm, and 12 mm, respectively, with the maximum lift-to-drag ratio occurred at the smallest flap height. Besides, in contrast to the situations of 1.5 mm thickness of the Gurney flaps, the Gurney flaps also achieved a lift increase under 10.5% turbulence intensity. The maximum lift coefficient was increased by 14.77%, 14.83%, and 16.44% for height equals to 6 mm, 9 mm and 12 mm, respectively. The two thickness configurations both had maximum lift-to-drag efficiency at 6 mm height, for the three heights, respectively.

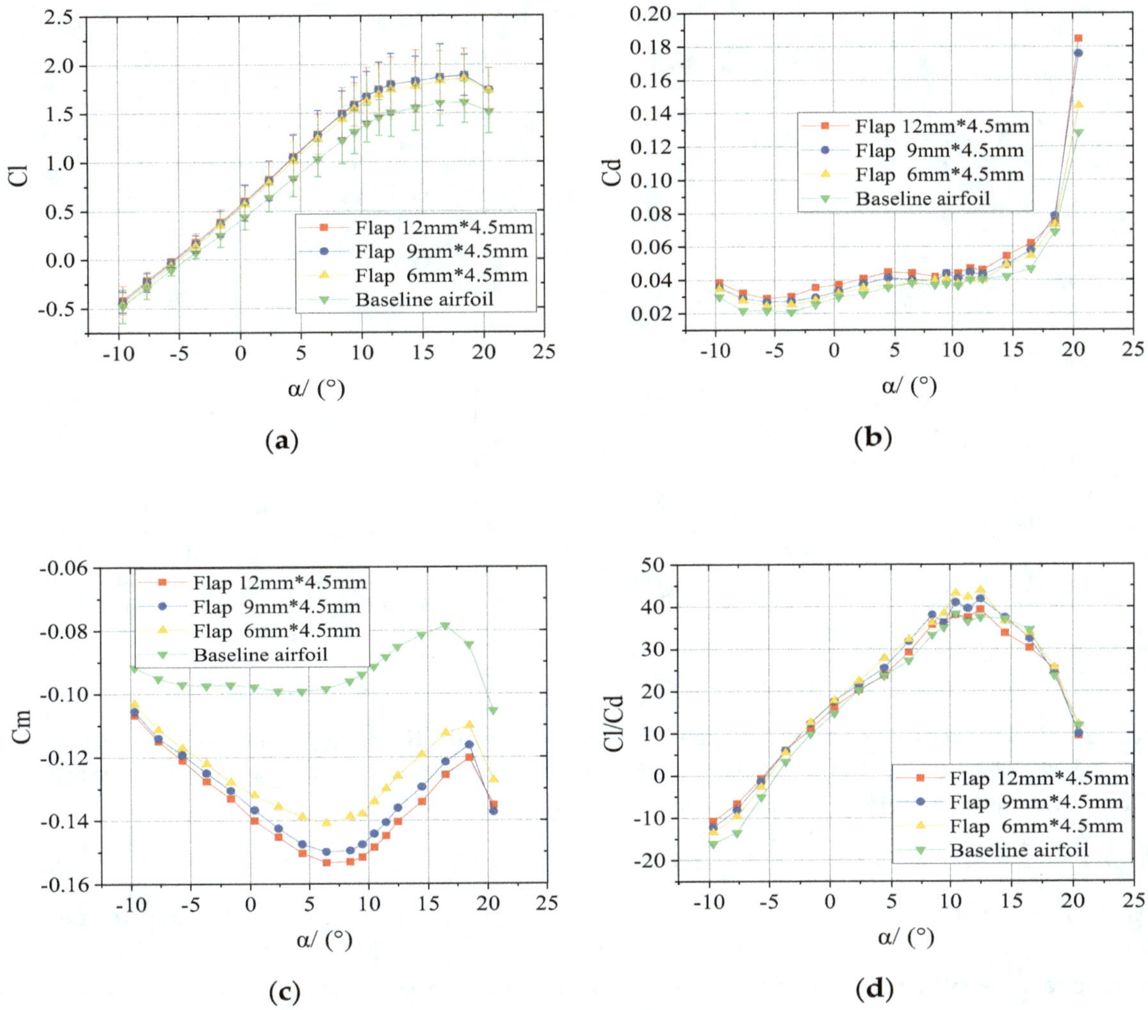

Figure 12. Variation of the aerodynamic characteristics with different heights of Gurney flaps (T.I. of 10.5%, Re: 0.8×10^6). (**a**) Lift coefficient comparison; (**b**) drag coefficient comparison; (**c**) pitching-moment coefficient comparison; (**d**) lift-to-drag ratio comparison.

In addition, the effects of various flap thicknesses were investigated. Figure 13 indicates the change of the aerodynamic characteristics of the baseline airfoil and airfoil with a height of 4.5 mm of the Gurney flaps when the turbulence intensity was 10.5%. As Figure 13 shows, like low turbulence conditions, after changing the thickness of the Gurney flap under turbulence condition of 10.5%, the aerodynamic characteristics of flapped airfoil almost show no change. The main difference in the Figure 13d was concentrated in the large lift-to-drag ratio zone, obviously because the pressure fluctuations caused by the turbulence and the flow separation at the airfoil surface result in the deviation of the measurement.

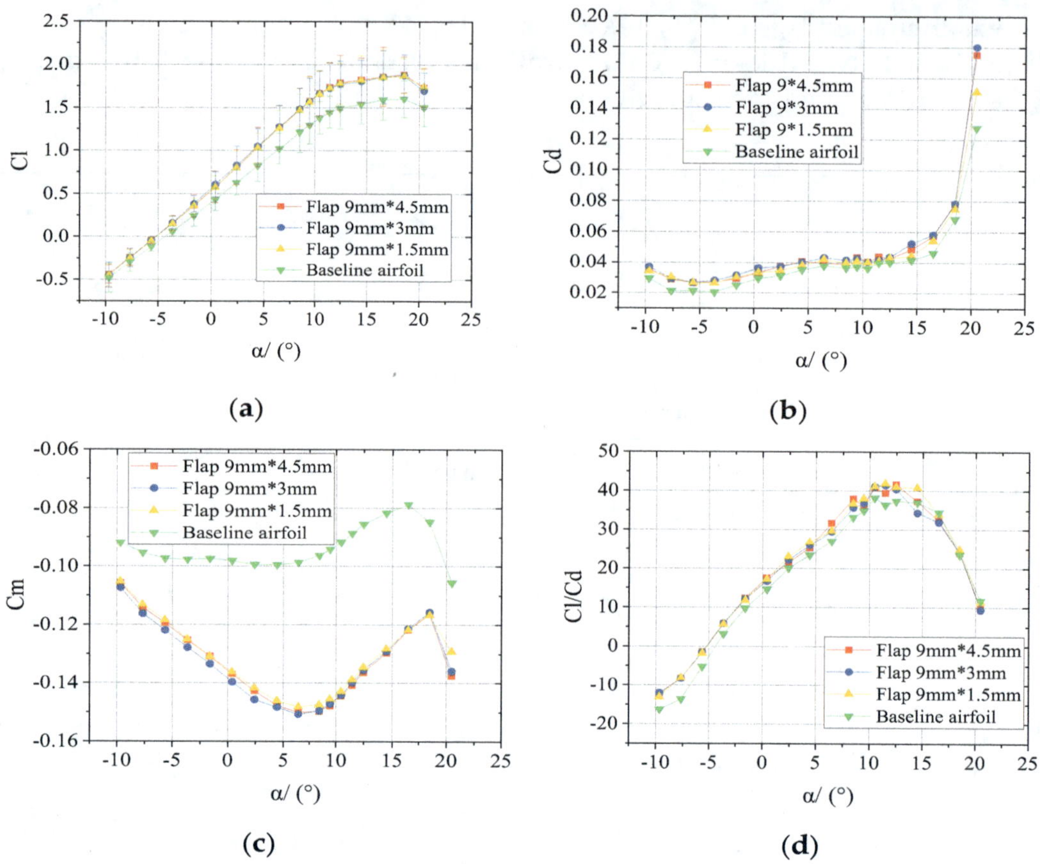

Figure 13. Variation of the aerodynamic characteristics with different thicknesses of Gurney flaps (T.I. of 10.5%, Re: 0.8×10^6). (**a**) Lift coefficient comparison; (**b**) drag coefficient comparison; (**c**) pitching-moment coefficient comparison; (**d**) lift-to-drag ratio comparison.

3.5. Experimental Results of Gurney Flap under the 19.0% Turbulence Intensity

To study the aerodynamic characteristics of the Gurney flap under high turbulence conditions, Figure 14 shows the changes of aerodynamic characteristics under the turbulence condition of 19.0%. The measured angle of attack ranged from −9.6° to 24.4°. As seen in Figure 14, the presence of turbulence enhanced the lift coefficient and drag coefficient for all cases. In Figure 14a, with the increase of turbulence intensity, the maximum lift coefficient continually increased, and the maximum lift coefficient of the baseline airfoil reaches 1.781 at 18.5 degree, with a 30.00% increase relative to that under low turbulence condition. The maximum lift coefficients increased by 2.13%, 3.37%, 5.39% at 18.5 degrees, respectively relative to the baseline airfoil, when the heights of the flaps were 6 mm, 9 mm, and 12 mm. Furthermore, since the pressure fluctuations were determined by the fluctuation's energy of the turbulent flow, the fluctuations of lift coefficient were increased relative to that under the turbulence condition of 10.5%. According to Figure 14b, the drag coefficient of the baseline airfoil changes more obvious than those under the low turbulence condition. The drag coefficients were no longer gentle within the range before stall, and it also rises sharply in the stall state. Such a trend was more noticeable after adding the Gurney flap. For example, at the angle of attack was 22.4°, the drag coefficients increased by 32.11%, 37.22% and 37.96%, when the flap heights were 6 mm, 9 mm, and 12 mm respectively. As can be seen from Figure 14c, the Gurney flap also produced a significant increase in the pitching-moment coefficient and such situation still existed in turbulent flow condition. Based on Figure 14d, unlike that under the condition of 10.5%, the lift gain obtained with Gurney flap is, unfortunately, coupled to a larger increase in drag. Adding Gurney flap with different heights at most of the angles reduces the lift-to-drag relative to the baseline airfoil, and the Gurney flap doesn't show obvious effect under the turbulence condition of 19.0%. At the same time the same situation occurred at other thicknesses of the Gurney flaps.

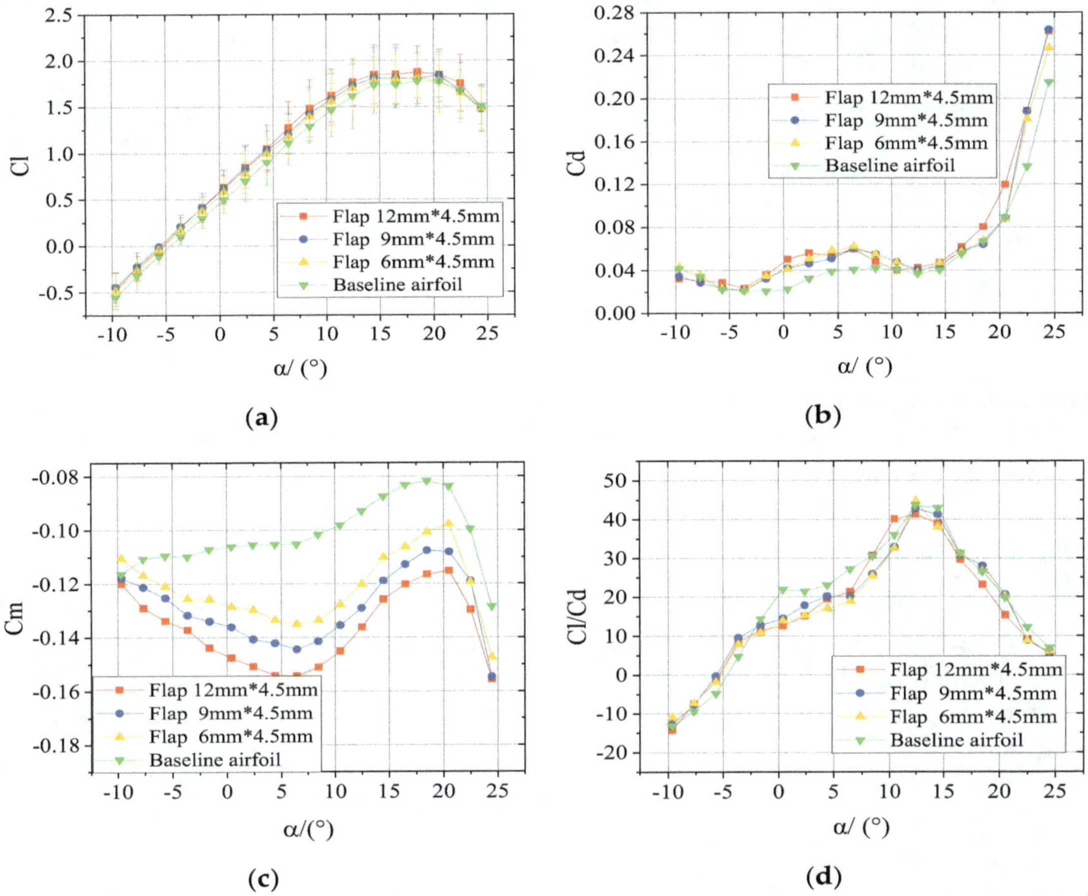

Figure 14. Variation of the aerodynamic characteristics with different heights of Gurney flaps (T.I. of 19%, Re: 0.8×10^6). **(a)** Lift coefficient comparison; **(b)** drag coefficient comparison; **(c)** pitching-moment coefficient comparison; **(d)** lift-to-drag ratio comparison.

Figure 15 exhibits the changes in the aerodynamic characteristics of the baseline airfoil and airfoil with Gurney flaps of height 4.5mm under the turbulence condition of 19.0%. As can be seen from it, with the increase of turbulence intensity, obviously, there were more pressure fluctuations. The curves of flapped airfoils with different thickness no longer overlap totally, in the majority of angles, the difference from 2% to 5%, though it may sometimes reach 20%.

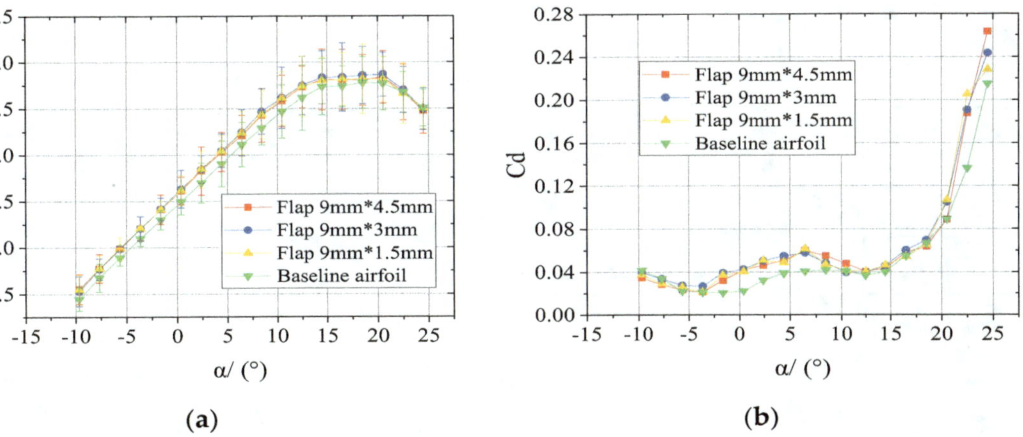

Figure 15. *Cont.*

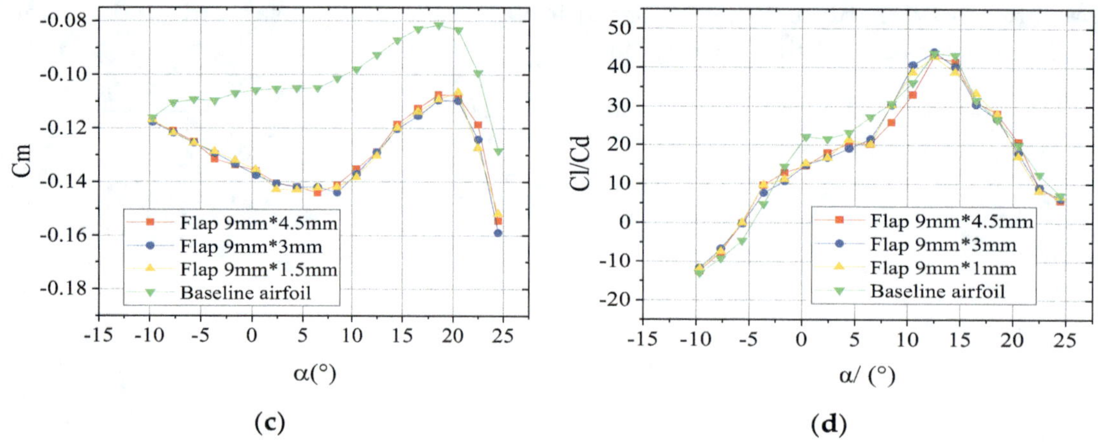

Figure 15. Variation of the aerodynamic characteristics with different thicknesses of Gurney flaps (T.I. of 19%, Re: 0.8 × 10^6). (**a**) Lift coefficient comparison; (**b**)drag coefficient comparison; (**c**) pitching-moment coefficient comparison; (**d**) lift-to-drag ratio comparison.

3.6. Surface Pressure Characteristics

To analyze the mechanism of turbulent inflow coupled with Gurney flap, Figure 16 illustrates the pressure coefficient distribution of the baseline airfoil and airfoil attached with the Gurney flap under different turbulent inflow when Reynolds number Re = 0.8 × 10^6. The x-axis shows the chord-wise position x/c and the y-axis shows the pressure coefficients C_p. The pressure coefficients C_p on the airfoil surface were given

$$C_{\mathrm{p}} = \frac{p_{\mathrm{i}} - p_0}{0.5\rho U_0^2} \tag{7}$$

where p_{i} is the pressure at the pressure tap of i (i = 1:77), p_0 is the free stream static pressure at the airfoil tested by the Pitot tube, ρ is the air density and U_0 is the free stream velocity.

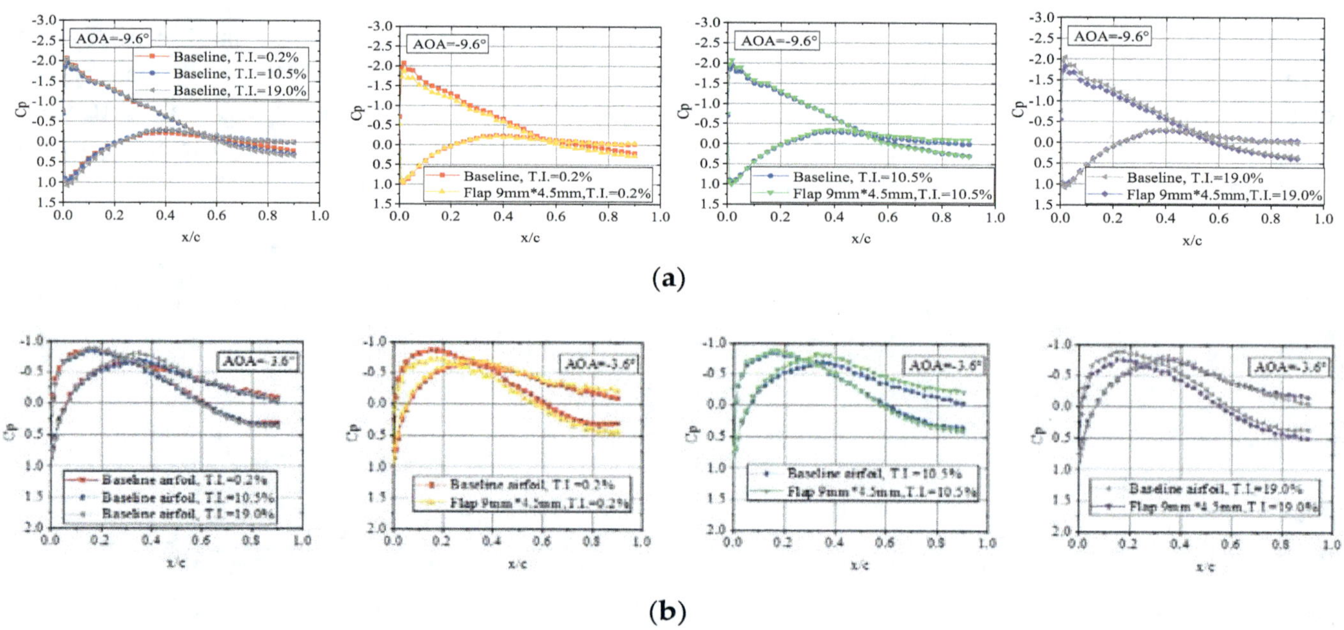

Figure 16. *Cont.*

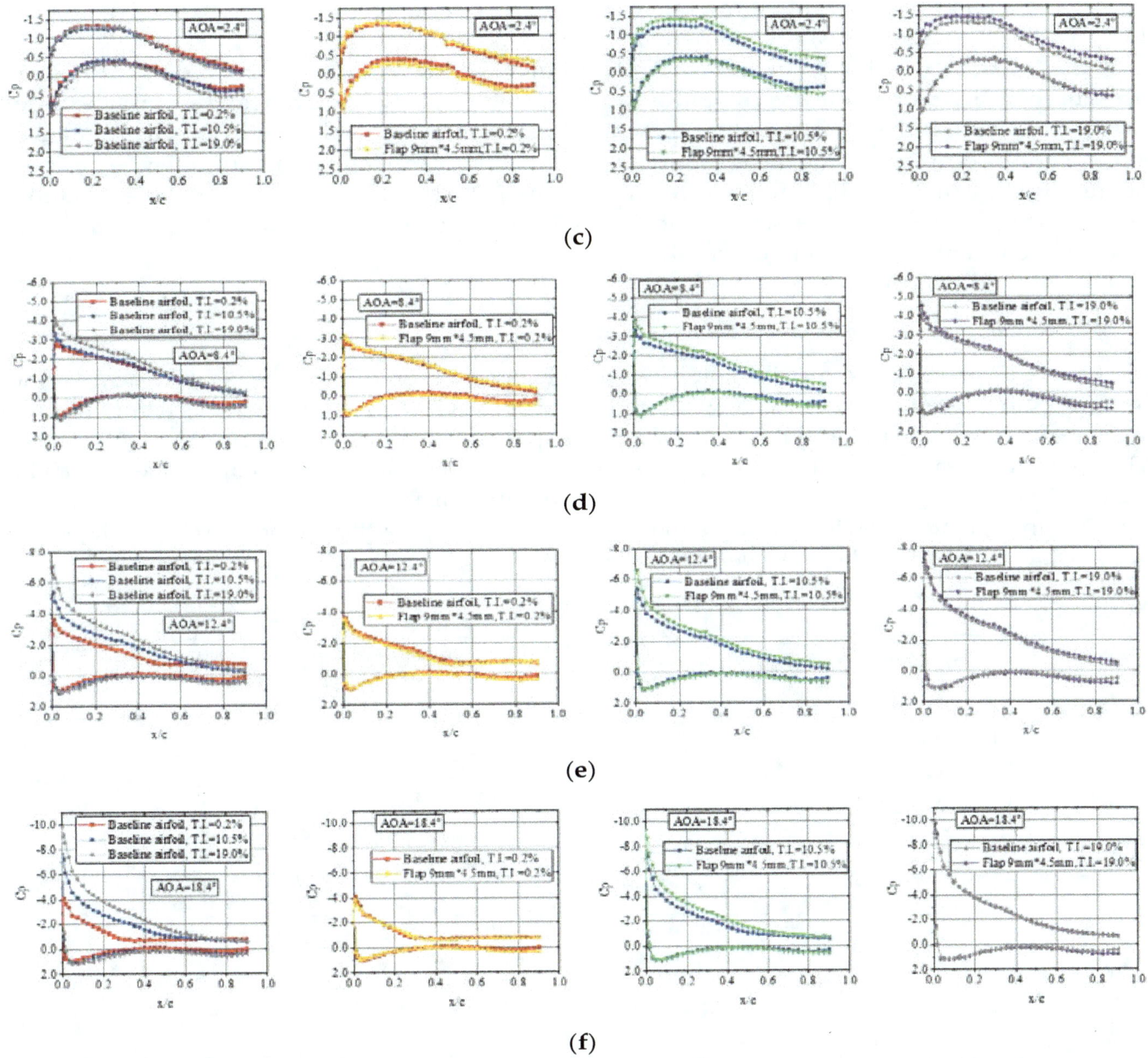

Figure 16. The surface pressure coefficients under different angle of attack (Re: 0.8×10^6). **(a)** $\alpha = -9.6°$; **(b)** $\alpha = -3.6°$; **(c)** $\alpha = 2.4°$; **(d)** $\alpha = 8.4°$; **(e)** $\alpha = 12.4°$; (f) $\alpha = 18.4°$.

Figure 16 illustrates that the pressure coefficients of the baseline airfoil under different turbulence conditions with small angles were similar to each other. However, with the increase of angle of attack, the differences become more pronounced at the leading edge. In contrast to the Gurney flaps, Figure 16a shows that at the angle = −9.6°, the differences between the flapped airfoils were small. With the increase of the angle of attack, as the angle of −3.6°, compared with the baseline airfoils, the intersection of the airfoil with Gurney flap moves to the leading edge under different conditions. Besides, the pressure coefficients (with the angle = −3.6° and 2.4°), the pressure distribution indicate a larger the pressure difference in the range of $0.8 < x/c < 0.9$. When the angle of attack was 2.4° and 8.4°, both the pressure coefficient absolute values of the baseline airfoil and the Gurney flap tendency to increase with the increase of turbulence intensity, and the flapped airfoil becomes more obvious than the baseline airfoil. At a large angle of attack, due to the condition of stalling (Figure 16e,f), there was significant flow separation both on the suction side of flapped airfoil and baseline airfoil under uniform inflow. The larger the turbulent inflow, the larger the peak value on the leading edge of the suction side, and also the larger pressure coefficient on the pressure side.

As seen from the above, the influence of turbulent inflow coupled with Gurney flap on the pressure coefficients of airfoil was complicated. Turbulent inflow changes the flow around the airfoil surface and restrains the flow separation, after attaching the Gurney flap, the pressure distribution on both suction and pressure sides changed more significantly. Especially in the turbulence condition, the trailing edge pressure distributions were farther apart for the flapped airfoil. In these situations, the increased peak values of pressure coefficients at the leading edge increased the integral area of the pressure coefficients, so the lift force was therefore increased.

3.7. Wake Profile Characteristics

Figure 17 presents the distribution of wake velocity measured by the wake rake array under uniform inflow. As can be observed, when the angle of attack was 8.4°, which was before the stall angle of attack, the wake of airfoil with the Gurney flap was inclined to the pressure side of the airfoil. When the angle of attack was 11.4°, which was after the stall angle, the Gurney flap weaken the ability of the wake position deflection. Moreover, the wake velocity deficit was significantly increased, and the Gurney flap began to have side effects on the aerodynamic characteristics of the airfoil. The higher the Gurney flap is, the larger the wake velocity deficit is, so it indicates a higher mean drag.

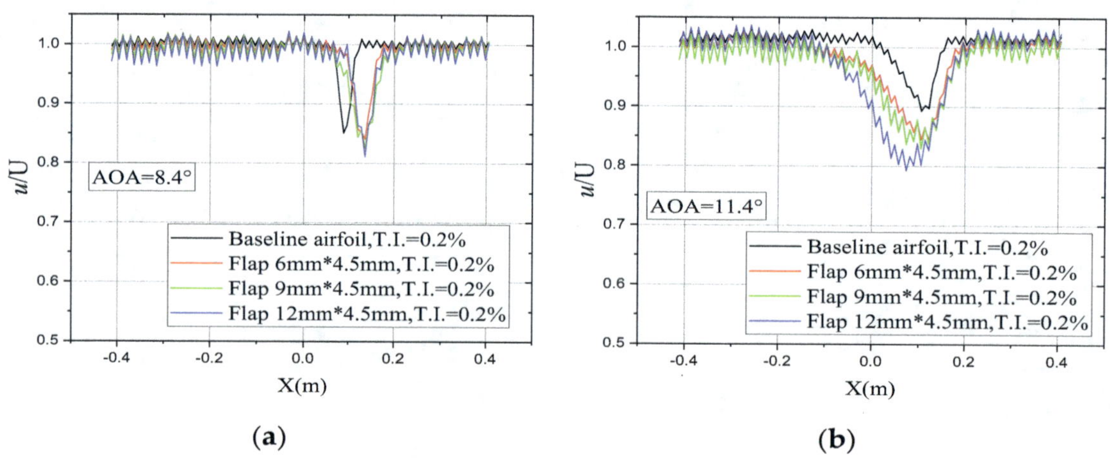

Figure 17. The wake velocity measured by the wake rake array (uniform inflow, Re: 0.8×10^6). (**a**) $\alpha = 8.4°$; (**b**) $\alpha = 11.4°$.

By comparing the flows in the near trailing edge region of the airfoil with Gurney flap. The resultant velocity distribution of the baseline airfoil and the airfoil attached with Gurney flap (9 mm in height and 4.5 mm in thickness) at the angle of attack 8.4° (before the stall angle) and 11.4° (after the stall angle) under uniform inflow were depicted in Figure 18. The vertical direction Y denoted the forward distance from the hot-wire probe to the trailing edge, and the horizontal direction X denoted the lateral distance from the hot-wire probe to the trailing edge. As shown in Figure 18, the black solid spots were the actual measurement positions by using the electronically controlled traverser. As shown, the wake flows shift to the pressure side in both the two situations. When the angle of attack was 8.4°, the airfoil with Gurney flap had a larger velocity deficit than the baseline airfoil, which suggests an increased drag. When the angle of attack was 11.4°, the velocity deficit of the baseline airfoil was also less than the Gurney flap, but there was no obvious difference between the offset direction caused by the flapped airfoil and the baseline airfoil which were being similar to the far field results measured by the wake rake array.

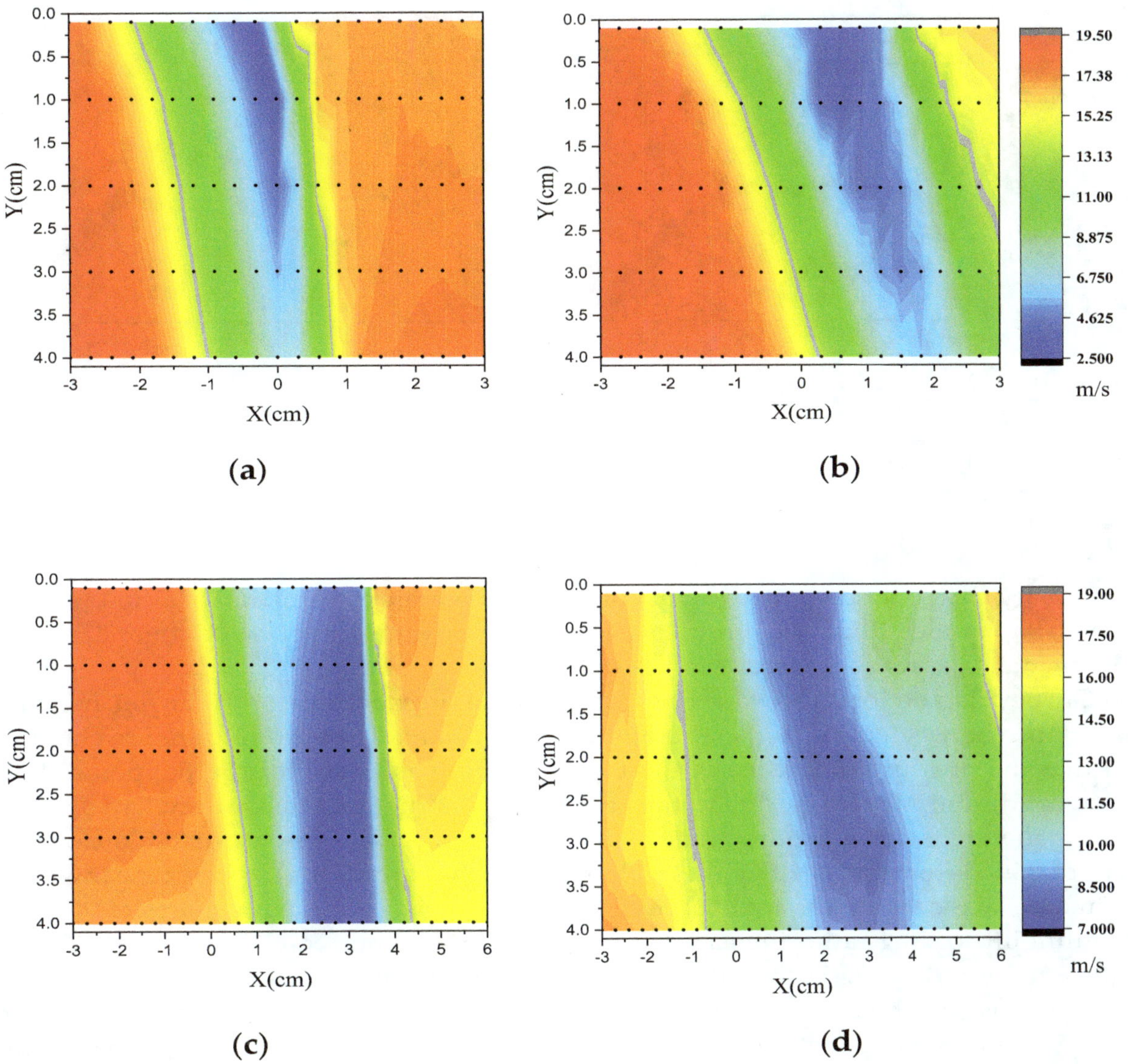

Figure 18. Velocity distribution of the region behind trailing edge (uniform inflow, Re: 0.8×10^6). (**a**) $\alpha = 8.4°$, baseline airfoil; (**b**) $\alpha = 8.4°$, Gurney flap 9 mm × 4.5 mm; (**c**) $\alpha = 11.4°$, baseline airfoil; (**d**) $\alpha = 11.4°$, Gurney flap 9 mm × 4.5 mm.

Figure 19 presents the power density spectrum measured by the region 1 cm directly behind the trailing edge under uniform inflow. As can be observed, at the angle of attack 8.4°, both along-wind and across-wind display an obvious peak in the power density spectrum, meanwhile the across-wind energy increased in the flapped airfoil. It indicates there are counter rotating vortices generated by the Gurney flap, while the wake of baseline airfoil does not show the existence of vortex shedding [33]. Gurney flaps also have a wake structure similar to the Karman vortex street behind the cylinder at bigger Reynolds number (i.e., of 0.8×10^6), the flow oscillations were caused by the separating of the shear layer on the pressure side and suction side of the trailing edge. The Karman vortex street both can be found in uniform flow and turbulent flow, so the increased lift produced by the Gurney flap seems can be explained in turbulent flow condition, either. Moreover, from the lift coefficients shown above, the size of structures was affected by turbulence level. Besides, as compared with the reference data under similar condition [33], with the increase of the Reynolds number, from Re = 1×10^5 to 0.8×10^6, the Strouhal number decreased from 0.151 (Re: 1×10^5, 2% chord length) to 0.128 (Re: 2×10^5, 2% chord length) and to 0.088 (Re: 0.8×10^6, 1.5% chord length).

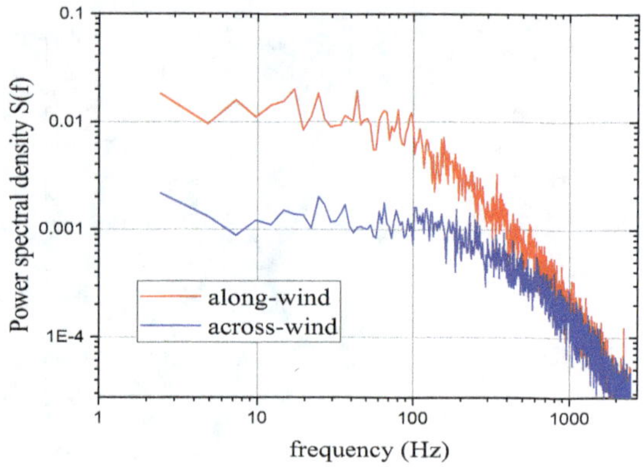

(a)

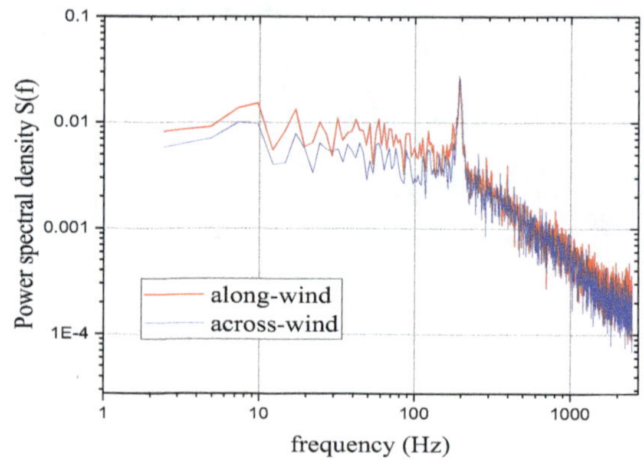

(b)

Figure 19. Comparison of the power density spectrum (uniform inflow, Re: 0.8×10^6). (**a**) $\alpha = 8.4°$, baseline airfoil; (**b**) $\alpha = 8.4°$, Gurney flap 9 mm × 4.5 mm.

4. Conclusions

In summary, this paper presented a study of the aerodynamic performance of airfoils with/without the Gurney flap under different turbulence conditions by means of wind tunnel experiments. The experimental observations were as follows:

The Gurney flap deflects the wake position from the pressure side of the airfoil, increasing the vertical distance between the airfoil chord and the middle arc. The surface pressure characteristics revealed that the pressure difference of the trailing edge was larger for the flapped airfoil in the turbulence condition. Moreover, due to the special wake structure, the Gurney flap also indicates the vortex generated at the trailing edge, which results in a greater lift of the airfoil. The lift increment obtained from Gurney flap was influenced by the turbulence intensity, meanwhile a large increase in drag was observed as well. Under a medium range of turbulence intensity (i.e., of 10.5%), the maximum lift-to-drag ratios increased from 14.35% (flap 6 mm × 4.5 mm, a = 12.4°) to 14.47% (flap 6 mm × 1.5 mm, a = 12.4°), the 1.0% Gurney flap case has the best performance, that's probably because the decrease of boundary layer thickness under turbulent conditions, so the height of 1.0% was optimal. However, under a much larger turbulence inflow (i.e., of 19.0%), all experimental data have shown that the impact was negligibly small. The drag increment was found to surpass the benefit of the increase in lift, resulting in a worse lift-to-drag ratio than the baseline airfoil.

It was of importance to note that there are two kinds of measurement uncertainties could affect the acquired results. One was stochastic uncertainties caused by the discrete pressure data which can be eliminated by multiple measurements. The other one was systematic uncertainties caused by measuring instruments which can be reduced by calibration of the instruments before the experiment.

Based on the observations from the present measurements, it was found that under very high turbulence level, e.g., with 19% of turbulence intensity, the Gurney flap has very small influence on the airfoil aerodynamic performance. The measurement results show the Gurney flap may be used in areas with slightly lower turbulence. Nevertheless, wind turbines in real life were running in the turbulent atmospheric boundary layer, as shown above, the effects of Gurney flap are highly dependent on turbulent inflow, it seems that the Gurney flap would be installed as a scalable form and preferred to have smaller size, such that it would get more potential benefits at all periods of rotation. In addition, it should be noted that the incoming flow is often a combination of rotational wake from the previous wind turbines. Research on the lift increment of Gurney flap was not isolated, it always interacted with such highly unsteady aerodynamic issues. Therefore, to analyze more precisely the effects of combining such turbulence levels with Gurney flap, further experiments should be carried out for the rotor blades in highly turbulent flows. This was in progress on a wind turbine blade composed of several DTU-LN221 wing sections.

Author Contributions: Conceptualization, J.Y.; conducted the data collection, J.Y. and Y.Y.; funding acquisition, H.Y.; supervision, N.L., H.Y. and W.Z.; formal analysis, J.Y., H.Y., W.Z.; writing—original draft preparation, J.Y.; writing—review and editing, H.Y., W.Z.; All authors have read and agreed to the published version of the manuscript.

Acknowledgments: The authors wish to express acknowledgement to the Postgraduate Research &Practice Innovation Program of Jiangsu Province under grant number KYCX19_2104 and Youth Fund Project of Jiangsu Natural Science Foundation under grant number BK20170510.

References

1. Liebeck, R.H. Design of subsonic airfoils for high lift. *J. Aircr.* **1978**, *15*, 547–561. [CrossRef]
2. Amini, Y.; Emdad, H.; Farid, M. Adjoint shape optimization of airfoils with attached Gurney flap. *Aerosp. Sci. Technol.* **2015**, *41*, 216–228. [CrossRef]
3. Xu, H.R.; Yang, H.; Liu, C. Numerical value analysis on aerodynamic performance of DU series airfoil with thickened trailing edge. *Trans. Chin. Soc. Agric. Eng.* **2014**, *30*, 101–108.
4. Yashodhar, V.; Humrutha, G.; Kaushik, M.; Khan, S.A. CFD Studies on Triangular Micro-Vortex Generators in Flow Control. In *IOP Conference Series: Materials Science and Engineering*; IOP Publishing: Bristol, UK, 2017; Volume 184, p. 012007.
5. Lee, T.; Su, Y.Y. Unsteady airfoil with a harmonically deflected trailing-edge flap. *J. Fluids Struct.* **2011**, *27*, 1411–1424. [CrossRef]
6. Seyednia, M.; Masdari, M.; Vakilipour, S. The influence of oscillating trailing-edge flap on the dynamic stall control of a pitching wind turbine airfoil. *J. Braz. Soc. Mech. Sci. Eng.* **2019**, *41*, 192. [CrossRef]
7. Lu, W.S.; Tian, Y.; Liu, P.P. Aerodynamic optimization and mechanism design of flexible variable camber trailing-edge flap. *Chin. J. Aeronaut.* **2017**, *30*, 988–1003. [CrossRef]
8. Zhang, W.G.; Bai, X.J.; Wang, Y.F.; Han, Y.; Hu, Y. Optimization of sizing parameters and multi-objective control of trailing edge flaps on a smart rotor. *Renew. Energy* **2018**, *129*, 75–91. [CrossRef]
9. Zhang, H.; Zhao, Z.D.; Zhou, G.X.; Kang, S. Experimental investigation of the effect of Gurney flap on DU93-W-210 airfoil aerodynamics performance. *Acta Energ. Solaris Sinica* **2017**, *38*, 601–606.
10. Amini, Y.; Liravi, M.; Izadpanah, E. The effects of Gurney flap on the aerodynamic performance of NACA 0012 airfoil in the rarefied gas flow. *Comput. Fluids* **2018**, *170*, 93–105. [CrossRef]
11. Lee, T.; Su, Y.Y. Lift enhancement and flow structure of airfoil with joint trailing-edge flap and Gurney flap. *Exp. Fluids* **2011**, *50*, 1671–1684. [CrossRef]
12. Chen, Z.J.; Stol, K.A.; Mace, B.R. Wind turbine blade optimization with individual pitch and trailing edge flap control. *Renew. Energy* **2016**, *103*, 750–765. [CrossRef]
13. Medina, A.; Ol, M.V.; Mancini, P.; Jones, A. Revisiting Conventional Flaps at High Deflection Rate. *AIAA J.* **2017**, *55*, 1–10. [CrossRef]
14. Elsayed, O.A.; Asrar, W.; Omar, A.A.; Kwon, K.; Jung, H.J. Experimental Investigation of Plain and Flapped-Wing Tip Vortex. *J. Aircr.* **2009**, *46*, 254–262. [CrossRef]
15. Little, J.; Nishihara, M.; Adamovich, I.; Saminy, M. High-lift airfoil trailing edge separation control using a single dielectric barrier discharge plasma actuator. *Exp. Fluids* **2010**, *48*, 521–537. [CrossRef]
16. Bergami, L.; Poulsen, N.K. A smart rotor configuration with linear quadratic control of adaptive trailing edge flaps for active load alleviation. *Wind Energy* **2015**, *18*, 625–641. [CrossRef]
17. Edward, T.; Christoph, B. Upstream shear-layer stabilization via self-oscillating trailing edge flaplets. *Exp. Fluids* **2018**, *59*, 145.
18. Straub, F.K.; Anand, V.R.; Lau, B.H.; Birchette, T.S. Wind Tunnel Test of the SMART Active Flap Rotor. *J. Am. Helicopter Soc.* **2018**, *63*, 012002. [CrossRef]
19. Traub, L.W. Prediction of Gurney-Flap Lift Enhancement for Airfoils and Wings. *AIAA J.* **2014**, *52*, 2087–2090. [CrossRef]
20. Lario, A.; Arina, R. Discontinuous Galerkin Method for the Study of Active Gurney Flaps. *J. Aircr.* **2017**, *54*, 1–11. [CrossRef]
21. Li, C.F.; Xu, Y.; Zhao, X.L.; Xu, J.Z. Analysis on Dynamic Performance Trailing Edge Flap on Wind Turbine Airfoil. *J. Eng. Thermophys.* **2014**, *35*, 883–887.

22. Zhu, W.J.; Behrens, T.; Shen, W.Z.; Sørensen, J.N. Hybrid Immersed Boundary Method for Airfoils with a Trailing-Edge Flap. *AIAA J.* **2013**, *51*, 30–41. [CrossRef]
23. Ng, B.F.; Palacios, R.; Kerrigan, E.C.; Graham, J.M.R.; Hesse, H. Aerodynamic load control in horizontal axis wind turbines with combined aeroelastic tailoring and trailing edge flaps. *Wind Energy* **2016**, *19*, 243–263. [CrossRef]
24. Colman, J.; Marañón Di Leo, J.; Delnero, J.S.; Martinez, M.; Boldes, U.; Bacchi, F. Lift and drag coefficients behaviour at low Reynolds number in an airfoil with Gurney flap submitted to a turbulent flow: Part 1. *Lat. Am. Appl. Res.* **2008**, *38*, 195–200.
25. Di Leo, J.M.; Martínez, M.A.; Delnero, J.S.; Saínz, M.G. Experimental study of the effect of the wake generated by oscillating Gurney flap. *Lat. Am. Appl. Res.* **2019**, *49*, 289–296.
26. Li, W.J.; Zhang, P.; Yang, S.F.; Fu, X.H.; Xiao, Y. An Experimental Method for Generating Shear-Free Turbulence Using Horizontal Oscillating Grids. *Water* **2020**, *12*, 591. [CrossRef]
27. Cheng, J.T.; Zhu, W.J.; Fischer, A.; García, N.R.; Madsen, J.; Chen, J.; Shen, W.Z. Design and validation of the high performance and low noise CQU-DTU-LN1 airfoils. *Wind Energy* **2013**, *17*, 1817–1833. [CrossRef]
28. ROHA. *LSWT Campaign Report on DTU-C21*; LM Internal Report: Jupitervej, Denmark, 2012.
29. Allen, H.J.; Vlncenti, W.G. Wall Interference in a Two-Dimensional Flow Wind Tunnel, with Consideration of the effect of Compressibility. *NACA Rep.* **1944**, *782*, 155–184.
30. Timmer, W.A.; van Rooij, R.P.J.O.M. Summary of the Delft University Wind Turbine Dedicated Airfoils. *J. Sol. Energy Eng.* **2003**, *125*, 11–21. [CrossRef]
31. Llorente, E.; Gorostidi, A.; Jacobs, M.; Timmer, W.A.; Munduate, A.; Pires, A. Wind Tunnel Tests of Wind Turbine Airfoils at High Reynolds Numbers. *J. Phys. Conf.* **2014**, *524*, 012012. [CrossRef]
32. Pires, O.; Munduate, X.; Ceyhan, O.; Jacobs, M.; Madsen, J.; Schepers, J.G. Analysis of the high Reynolds number 2D tests on a wind turbine airfoil performed at two different wind tunnels. *J. Phys. Conf. Ser.* **2016**, *749*, 012014. [CrossRef]
33. Troolin, D.R. A Quantitative Study of the Lift-Enhancing Flow Field Generated by an Airfoil with a Gurney Flap. Ph.D. Dissertation, University of Minnesota, Minnesota, MN, USA, 2009.

6

CFD Simulations of Flows in a Wind Farm in Complex Terrain and Comparisons to Measurements

Matias Sessarego [1], Wen Zhong Shen [1,*], Maarten Paul van der Laan [2], Kurt Schaldemose Hansen [1] and Wei Jun Zhu [3]

1 Department of Wind Energy, Technical University of Denmark, Lyngby Campus, 2800 Kgs. Lyngby, Denmark; matse@dtu.dk (M.S.); kuhan@dtu.dk (K.S.H.)
2 Department of Wind Energy, Technical University of Denmark, Risø Campus, 4000 Roskilde, Denmark; plaa@dtu.dk
3 School of Hydraulic, Energy and Power Engineering, Yangzhou University, Yangzhou 225127, China; wjzh@dtu.dk
* Correspondence: wzsh@dtu.dk

Abstract: This article describes Computational Fluid Dynamics (CFD) simulations of flows in a wind farm in complex terrain in Shaanxi, China and the comparisons of the computational results with utility scale field measurements. The CFD simulations performed in the study are using either a Reynolds-Averaged Navier–Stokes (RANS) or Large-Eddy Simulation (LES) solver. The RANS method together with an Actuator Disc (AD) approach is employed to predict the performance of the 25 wind turbines in the farm, while the LES and Actuator Line (AL) technique is used to obtain a detailed description of the flow field around a specific wind turbine #14 near two met masts. The AD-RANS simulation results are compared with the mean values of power obtained from field measurements. Furthermore, the AL-LES results are compared with the mean values of power, rotor speed, and wind speed measured from the wind turbine and its nearby two masts. Results from the simulations indicate that both AD-RANS and AL-LES methods can reasonably predict the performance of the wind farm and wind turbine #14, respectively, in complex terrain in Shaanxi. The mean percent difference obtained for power in the AD-RANS simulations was approximately 20%. Percent differences obtained for power and rotor RPM in the AL-LES varied between 0.08% and 11.6%. The mean percent differences in the AL-LES for power and rotor RPM are approximately 7% and 1%, respectively.

Keywords: computational fluid dynamics; wind farm; complex terrain; SCADA; met mast measurements

1. Introduction

Onshore wind is one of the most competitive sources of new generation capacity [1]. According to the International Renewable Energy Agency (IRENA), the global weighted average cost of electricity was USD 0.06/kWh for onshore wind power in 2017, which falls within the range of generation costs for fossil-based electricity [1]. However, the performance and lifetime of onshore wind turbines are strongly influenced by the site-specific wind resource and terrain environment. Wind flow and wind farm modeling software aims to predict and simulate the characteristics of the wind resource as well as the performance of a proposed or existing wind farm, e.g., to estimate its annual energy production. The current study focuses on high-fidelity numerical approaches, i.e., Computational Fluid Dynamics (CFD), for estimating the performance of wind turbines and wind farms in complex terrain.

A large variety of CFD methodologies have been employed to predict the flow characteristics of wind farms in offshore [2], flat [3] and complex [4–9] terrain environments. The most commonly employed high-fidelity CFD methods involve either the Reynolds-Averaged Navier–Stokes (RANS)

and Large-Eddy Simulation (LES) techniques to determine the wind flow over a farm. LES solvers are approximately three orders more computationally expensive than RANS solvers, since the large scale turbulence is resolved on a grid and only the smaller scale turbulence is modeled, e.g., using a sub-grid scale (SGS) model [3]. RANS solvers are generally performed in a steady-state manner where all scales of turbulence are modeled [3]. Since the turbulence length scale on a wind turbine rotor surface is too small in relation to the size of the wind farm, the full three-dimensional geometry of the rotor is commonly simplified using the Actuator Disc (AD) or Actuator Line (AL) techniques [10].

Makridis and Chick [4] investigated wind turbine wakes and neutral atmospheric wind flow over a moderately complex terrain using the CFD software Fluent (version 12.0, ANSYS, Inc., Canonsburg, PA, USA). RANS was used to model the wind flow and an AD implementation of Fluent was used to model the wind turbine rotor effects. Comparisons were made to measurements and WAsP (Wind Atlas Analysis and Application Program) simulations. Castellani et al. [9] performed AD-RANS analyses of a sub-cluster of four wind turbines from a wind farm sited in Italy in complex terrain. The dominant 270 degrees wind flow direction over the four wind turbines was studied using simulation results and yaw measurements. Schulz [5,6] performed Delayed-Detached-Eddy-Simulations (DES) of a 2.4 MW wind turbine in complex terrain including a turbulent atmospheric boundary layer. The wind turbine was modeled by meshing the tower, nacelle, hub, joints and blades. Tabib [7] performed LES and RANS simulations of the onshore Bessaker wind farm. Tabib [8] then analyzed wake-terrain interactions by comparing two LES simulations of a multi-wind-turbine layout with annual power data at the actual wind farm. Tabib used the AL technique to model the wind turbine.

In the current article, both AD-RANS and AL-LES techniques are employed to compute the flow field and performance of a utility scale wind farm located in Jianbian, Shaanxi, China, consisting of twenty-five 2 MW wind turbines (WTs) in complex terrain. Measurements from a supervisory control and data acquisition (SCADA) system containing sensors from the twenty-five WTs and measurements from two met masts in the farm are used to compare with the CFD simulations. The main contribution of the present work is the better understanding of the level of accuracy and capabilities of AD-RANS and AL-LES solvers when predicting the wind farm performance in complex terrain environments. The results of this study can be used by other researchers, who do not have access to measurement data, to compare their own wind farm CFD simulation results in similar complex terrain environments. The comparison will enable researchers to quantify how close their wind farm CFD simulation results might be if field measurements were available. A description of how the mesh, wind turbine model and wind farm layout are prepared, and the input parameters used in the CFD simulations can be used as reference.

The background regarding the CFD solvers, wind farm and methodology for creating the terrain grids will be described first in Section 2. Then, results from both the AD-RANS and AL-LES simulations compared with field measurements will be shown in Section 3. Finally, conclusions are given in Section 4.

2. Methodology

The RANS and LES simulations are performed using the in-house developed flow solver EllipSys3D [11,12] with the AD and AL methodologies as described in [13,14], respectively, in neutral atmospheric conditions. EllipSys3D solves the Navier–Stokes equations in general curvilinear coordinates using a block structured finite-volume approach [10]. Both the AD and AL include controllers for the rotational speed and pitch. Section 2.1 contains information about the wind turbine model used in the AD and AL techniques. Sections 2.2 and 2.3 describe the AD-RANS and AL-LES implementations in more detail, respectively. Section 2.4 summarizes the flow cases used in the present study.

2.1. Wind Turbine Model

The twenty-five 2 MW wind turbines in the wind farm are designed and manufactured in China by China Shipbuilding Industry Corporation (CSIC) Haizhuang (HZ) Windpower Equipment Co., Limited (Chongqing, China). Table 1 contains the basic information about the wind turbine. A computer-aided design (CAD) model of the blade was provided by the blade manufacturer (confidential), which was used to extract six airfoil profiles needed for the AD and AL models. The airfoil profiles were used as input to the viscous-inviscid airfoil flow solver, Q^3UIC [15], to generate the polar data containing the lift and drag coefficients as a function of the angle of attack.

The power and thrust as a function of wind speed were provided by the manufacturer, but not the rotor RPM (revolutions per minute) and pitch schedule. Therefore, the AD and AL models were developed iteratively to match as closely as possible to the manufacturer's data as well as the wind turbine measurement data for the RPM and pitch. Figure 1a,b depict the electrical power and rotor thrust as a function of the wind speed from the manufacturer's data and the developed model used in the AD and AL techniques. Figure 1b shows that the developed wind turbine model over-predicts the rotor thrust between 3 and 10 m/s compared to the manufacturer's data. However, an estimate of rotor thrust based on strain-gage measurements showed deviations from the manufacturer's data as well [16], and is in close agreement to the wind turbine model. Nevertheless, rotor thrust discrepancies will influence the wake deficit comparisons shown in Section 3.2.

Table 1. China Shipbuilding Industry Corporation (CSIC) Haizhuang (HZ) wind turbine specifications.

Name	CSIC HZ93-2.0MW
Diameter	93 m
Power rating	2.0 MW
Hub height	67 m
Power control	Variable speed, collective blade pitch control

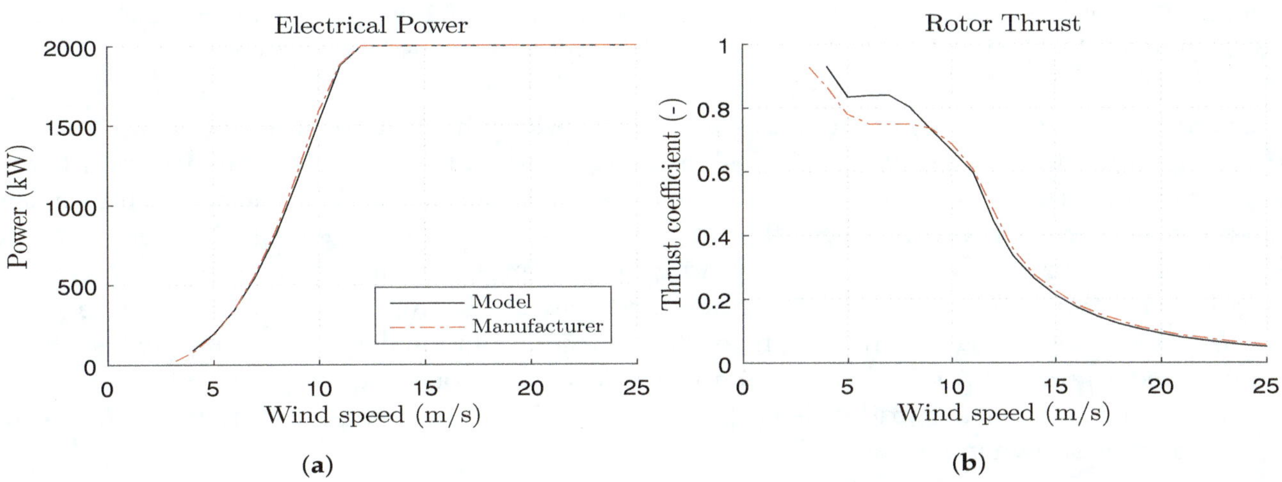

Figure 1. Comparison of the manufacturer's data and the developed model for the (**a**) electrical power and (**b**) rotor thrust as a function of the wind speed.

2.2. Actuator Disc Reynolds-Averaged Navier–Stokes Simulation

The k-ϵ-f_P eddy-viscosity model [3,17] is used to model the turbulence in RANS. A log-law is used with a surface roughness of $z_0 = 0.1$ m over the entire domain to simulate the site vegetation. Ideally, z_0 should be based on measurement data, e.g., by considering the inflow profile according to measurements in the wind farm, however a rough estimate for z_0 was used instead. Figure 2a depicts the terrain and wind turbine setup for the AD-RANS where the wind turbine rotor diameter,

D, is 93 m. In the AD-RANS setup, the mesh is cylindrical in shape with a radius of 28 km and a height of approximately 25 km. The 7.8 km^2 terrain grid is situated at the center of the mesh at WT10, since WT10 is approximately located at the center of the wind farm. The cylindrical mesh allows variable wind directions without changing the grid. Inflow/outflow boundary conditions are applied on the outer-boundary of the mesh according to the wind direction. The velocities at the top and inlet boundaries are prescribed. The mesh contains 280 blocks with each block having 48^3 grid points. The terrain grid is 10 × 10 blocks. There are approximately 6 (horizontal) × 16 (vertical) grid cells on each AD.

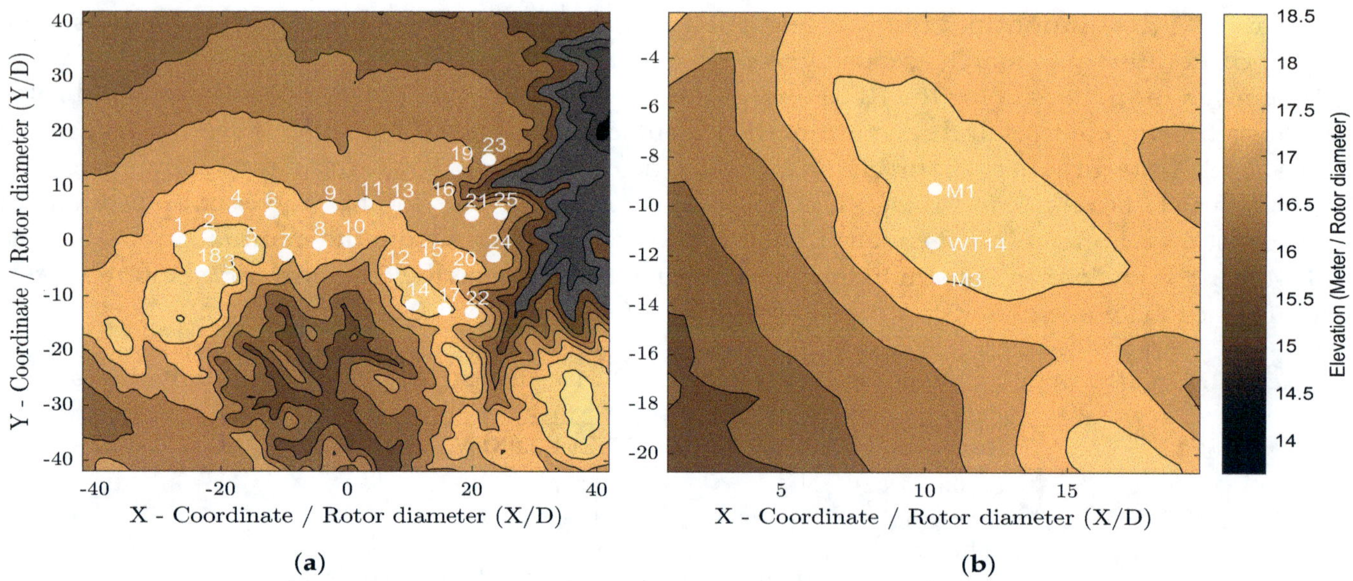

Figure 2. Computational layouts of the (**a**) Shaanxi wind farm with twenty-five wind turbines and of (**b**) wind turbine 14 with met masts M1 and M3. The wind turbines are labeled from 1 to 25.

Given an amount of computational resources to work with, since the amount of computational resources is limited, the computational mesh is designed such that there is more grid points at the center to capture terrain effects and the effect of each AD on the flow field. At the center, the velocities gradients due to complex terrain and the wind turbine presence is greatest. Further away from the center and towards the cylindrical boundary of the mesh, a much coarser grid is used over a long distance to minimize the computational cost and at the same time reduce the effect of the boundary conditions on the flow field at the center. The computational grid gave slightly less than the recommended amount of grid points on each AD, which based on experience should be a minimum of 8 horizontal and 8 vertical grid cells [17]. The effect of a refined mesh, where the terrain grid is 24 × 10 blocks, is shown in Section 3.1.

Sensors are placed at hub height for all twenty-five wind turbines as well as on the met masts in the CFD setup. The (a) outer and (b) terrain computational grids for the AD-RANS computations are shown in Figure 3. Each AD-RANS computation is performed using one-hundred-and-forty 2.8 GHz processors on a Linux cluster, where each processor is allocated for two blocks out of the 280 blocks. Each AD-RANS computation requires approximately 3 to 4 h for completion.

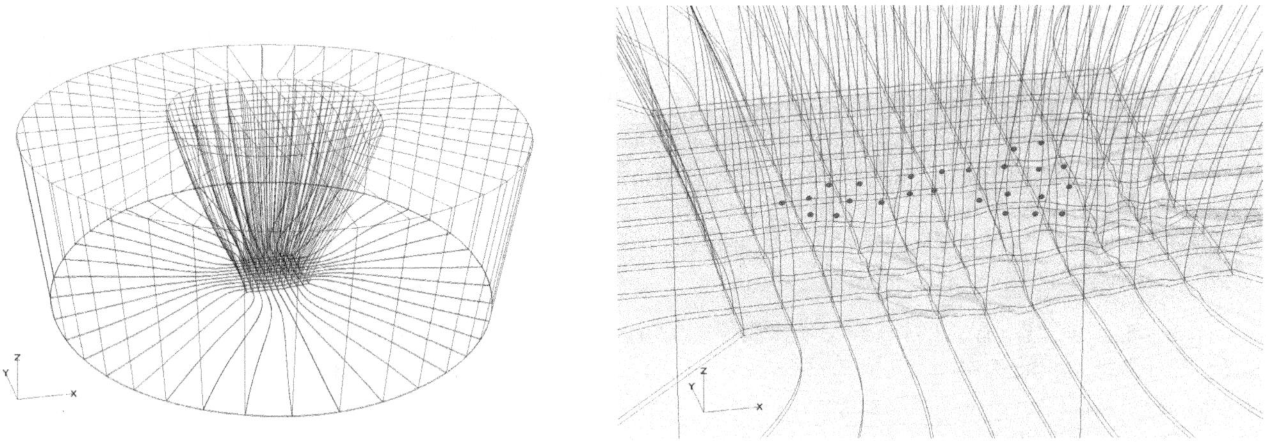

(a) (b)

Figure 3. (**a**) outer and (**b**) terrain computational grids used in the actuator disc Reynolds-averaged Navier-Stokes (AD-RANS) computations. The circles in (**b**) are the actuator discs.

2.3. *Actuator Line Large-Eddy Simulation*

Figure 2b depicts the terrain and wind turbine setup. The AL-LES provides much more detailed information about the wind flow than the AD-RANS, but at the expense of a much higher computational cost. Therefore, in contrast to the AD-RANS simulations of the entire Shaanxi wind farm, the AL-LES is applied to a much smaller area, specifically around WT14. The smaller area contains WT14 and the two met masts M1 and M3. The area around WT14 was selected because it is near the two met masts and the results from the simulations can be compared with measurements from the met masts. The terrain is centered at WT14 and is 1728 m^2 in size. The mesh is also cylindrical in shape with a radius of 6.5 km, a height of approximately 3.2 km and contains a total of 1404 blocks. The terrain grid is 18 $\times$ 18 blocks. Each block contains 48^3 grid points as well. There is approximately 47 (horizontal) $\times$ 39 (vertical) grid cells on the rotor swept area, which is similar to the number of grid cells used in Refs. [10,18]. The boundary conditions are the same as in Section 2.2. The (a) outer and (b) terrain computational grids for the AL-LES computations are shown in Figure 4. Each AL-LES computation is performed using seven-hundred-and-to 2.8 GHz processors on a Linux cluster, where each processor is allocated for two blocks out of the 1404 blocks.

To attain a flow field over the vicinity of WT14, the flow field in AL-LES was scaled using the sensor at 70 m height on the upstream met mast, since it has almost the same height as the wind turbine hub (67 m). For the 4 and 7 m/s at 180 degrees, the upstream and downstream met masts are M3 and M1, respectively. For the 4 and 7 m/s at 0 degree, the upstream and downstream met masts are M1 and M3, respectively. In addition, WT14, M1, and M3 are located at elevation levels of 1695, 1692, and 1678 m, respectively. To simulate the wind shear, a prescribed profile is imposed at the first time-step in the AL-LES using a shear exponent of 0.3, roughly estimated from the met-mast measurements.

Two sets of simulations are carried out: inflow with and without synthetic turbulence. The inflow simulations without synthetic turbulence are performed to scale the flow field more quickly, as mentioned above, since the computational cost is much less than the inflow simulations with synthetic turbulence. The AL-LES computations without and with synthetic inflow turbulence require approximately two and five days for completion, respectively. Simulations without synthetic inflow turbulence are performed until steady-state values for power, rotor RPM, and wind speed are achieved (i.e., approximately 34 s real time). Simulations with synthetic inflow turbulence are performed to obtain 600 s (10 min) of real time, excluding initial transients.

Synthetic inflow turbulence is simulated by prescribing a plane, 372 $\times$ 372 m^2 in size, 2.5 rotor diameters upstream of the wind turbine. Turbulence boxes are generated as a pre-processing step using the Mann model [19,20] and slices from the box are gradually fed into the plane during the simulation. In an attempt to represent the high turbulence magnitudes observed in the measurements, a high surface roughness of 0.4 was selected to generate the turbulence boxes. The turbulence is induced in a square 2.5 rotor diameters upstream of the wind turbine because it is scaled, using the surface roughness in the Mann model, in an attempt to match the turbulence intensity seen in the measurements in that specific area of the domain. Of course, a much larger square can be imposed upstream of the wind turbine with a constant surface roughness to cover the entire domain. However, this would imply that the turbulence intensity is uniform across the entire domain, which is likely not the case. It is assumed that creating a plane larger than $372 \times 372\,\mathrm{m}^2$ would not influence the results.

In addition, the magnitude of the turbulence released at the plane was multiplied by a scaling factor of 1.5. The scaling factor does not have a physical meaning and is used when comparing with measurements where the source of the additional turbulence observed in the measurements cannot be captured in the LES simulation alone. The time-averaged solutions from the simulations without and with synthetic inflow turbulence are compared to the mean values from SCADA.

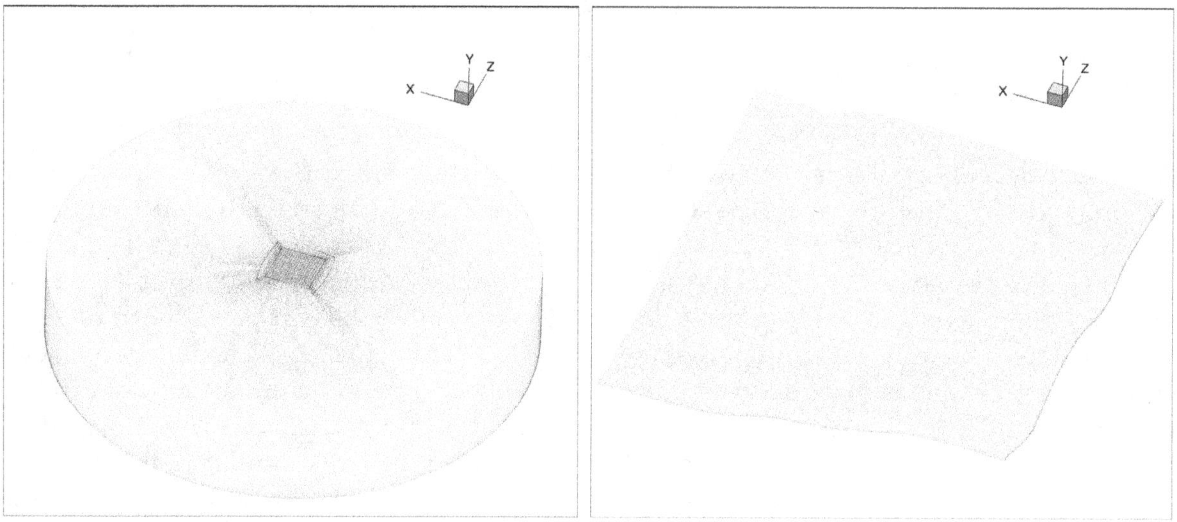

(**a**) (**b**)

Figure 4. (**a**) outer and (**b**) terrain computational grids used in the AL-LES computations.

2.4. Flow Cases

Based on twelve months of met mast and wind turbine measurements, nine wind farm flow cases (FC1) have been identified for comparing the AD-RANS simulations. The AD-RANS flow cases are shown in Table 2. The wind turbine furthest upstream and that is the least influenced by the wakes of nearby wind turbines is chosen as the wind speed reference in Table 2. The met masts were not used as references for the wind speed because they are influenced by the wakes of nearby wind turbines for the 300° and 330° wind directions. It was not possible to derive the wind direction from the wind turbine SCADA, thus the wind turbine could not be used as the wind direction reference. The met masts are used to determine the reference wind direction. Due to the large computational expense, only four cases (FC2) are investigated for the AL-LES simulations. The AL-LES flow cases are shown in Table 3. The met masts were used as reference for the wind speed and direction in FC2. The $\pm 5°$ and $\pm 1\,\mathrm{m/s}$ cases in Tables 2 and 3 represent the bin intervals used to collect the SCADA measurements. Met mast and SCADA measurements used in the comparisons are filtered and represent neutral atmospheric conditions. Section 3 contains the AD-RANS and AL-LES simulation results for the FC1 and FC2 flow cases, respectively.

Table 2. Inflow conditions for the actuator disc Reynolds-averaged Navier-Stokes (AD-RANS) flow cases (FC1) where h = height and WT = wind turbine.

FC1	Sector	Wind Speed, h = 67 m	Reference Wind Speed	Reference Wind Direction
1.1	$180 \pm 5°$	4 ± 1 m/s	WT14, h = 67 m	M3, h = 70 m
1.2	$300 \pm 5°$	4 ± 1 m/s	WT11, h = 67 m	M1, h = 70 m
1.3	$330 \pm 5°$	4 ± 1 m/s	WT11, h = 67 m	M1, h = 70 m
1.4	$180 \pm 5°$	7 ± 1 m/s	WT14, h = 67 m	M3, h = 70 m
1.5	$300 \pm 5°$	7 ± 1 m/s	WT11, h = 67 m	M1, h = 70 m
1.6	$330 \pm 5°$	7 ± 1 m/s	WT11, h = 67 m	M1, h = 70 m
1.7	$180 \pm 5°$	10 ± 1 m/s	WT14, h = 67 m	M3, h = 70 m
1.8	$300 \pm 5°$	10 ± 1 m/s	WT11, h = 67 m	M1, h = 70 m
1.9	$330 \pm 5°$	10 ± 1 m/s	WT11, h = 67 m	M1, h = 70 m

Table 3. Inflow conditions for the actuator line large-eddy simulation (AL-LES) flow cases (FC2) where h = height.

FC2	Sector	Wind Speed, h = 67 m	Reference Wind Speed	Reference Wind Direction
2.1	$180 \pm 5°$	4 ± 1 m/s	M3, h = 70 m	M3, h = 70 m
2.2	$180 \pm 5°$	7 ± 1 m/s	M3, h = 70 m	M3, h = 70 m
2.3	$0 \pm 5°$	4 ± 1 m/s	M1, h = 70 m	M1, h = 70 m
2.4	$0 \pm 5°$	7 ± 1 m/s	M1, h = 70 m	M1, h = 70 m

3. Results

The current section summarizes the results obtained from the AD-RANS and AL-LES simulations and their comparisons to the SCADA measurements. The results are divided into two subsections, actuator disc Reynolds-averaged Navier–Stokes and actuator line large-eddy simulation.

3.1. *Actuator Disc Reynolds-Averaged Navier–Stokes Results*

Prior to simulating the entire wind farm using AD-RANS, the inputs for the far upstream wind direction and speed should be modified to include the effect of the turning angle and speed-up. The wind direction and speed throughout the complex terrain will not be equal to the far upstream wind direction and speed where the terrain is assumed to be flat (see e.g., Figure 3). The flow will change direction and gain or lose speed when encountering, e.g., hills, valleys, and cliffs. For an accurate comparison with the measurements, the computed wind direction and speed at the reference met mast and wind turbine must take into account the effect of the turning angle and speed-up. The turning angle and speed-up are obtained from a pre-computation without wind turbines at the reference met mast and wind turbine locations, respectively. The turning angle and speed-up are calculated using the reference wind speed and direction as shown in Table 2. Table 4 summarizes the turning angles and speed-ups obtained for FC1. In the pre-computation without wind turbines, e.g., FC 1.1 in Table 4, the far upstream wind direction and speed were set to 180 degrees and 4 m/s, respectively. The wind direction and speed computed at M3 and WT14 was $(180 + 4.28)$ degrees and (4×1.25) m/s, respectively. The far upstream wind direction and speed are then set to $(180 - 4.38)$ degrees and $(4/1.25)$ m/s to obtain approximately 180 degrees and 4 m/s at M3 and WT14, respectively.

Table 4. Turning angles and speed-ups obtained for FC1.

FC1	1.1	1.2	1.3	1.4	1.5	1.6	1.7	1.8	1.9
Turning angle (degrees)	4.28	−4.78	1.29	4.28	−4.76	1.29	4.28	−4.75	1.28
Speed-up (-)	1.25	1.12	1.13	1.25	1.12	1.13	1.25	1.12	1.13

Figure 5 contains the results for the flow cases defined in Table 2, which consist of the AD-RANS results and the comparisons with the measurement data from SCADA. The notation *wd* and *ws* shown in the title for each figure denotes wind direction and wind speed, respectively. The wind direction corresponds to the measurements on met masts M1 and M3, while the wind speed corresponds to the measurements on the reference wind turbine (ref. WT). The figure contains the predicted, labeled as AD# and AD# $\pm$ 5° where # is the wind direction in degrees, and measured, labeled as SCADA, electrical power output for all 25 WTs. Each WT is labeled between 1 and 25, refer to Figure 2a. The AD# results represent a single run with the wind direction, #, which is also specified in the sub-figure titles. The AD# $\pm$ 5° results represent the average of five runs: −5, −2, 0, +2, and +5 degrees relative to *wd*. For example, AD# for FC 1.1 is a single run at 180 degrees (# = 180°), and AD# $\pm$ 5° is the average of 5 runs: 180 − 5, 180 − 2, 180, 180 + 2, and 180 + 5 degrees. The SCADA measurements represent averaged values in $\pm$5 degree bins. Although difficult to see, the standard error (SE) in Figure 5 is equal to SCADA $\pm$ 5° but includes the error bars (may require zoom-in).

In Figure 5, AD# $\pm$ 5° should more closely represent the SCADA measurements than AD#, since the SCADA measurements are grouped and averaged in $\pm$5 degree bins. The SCADA results cannot be determined exactly at 180, 300 or 330 degrees in Figure 5. Nevertheless, the AD-RANS results in Figure 5 show that AD# $\pm$ 5° is almost identical to AD#. In general, there is an agreement between the AD-RANS and the SCADA. However, there are significant differences in, for example, FC 1.1 for all the WTs and WT16 in FC 1.7.

A possible explanation for the former case is that for the south wind at a small wind speed of 4 m/s, the wind distribution over the terrain may be non-uniform. Large variations in wind speed is captured in the SCADA but not in the RANS. The latter discrepancy is due to the delayed wake recovery caused by low dissipation in the employed turbulence model. The percent difference for the latter discrepancy is 57.4% and 66.4% based on AD# $\pm$ 5° and AD#, respectively. The latter discrepancy is the same as observed in WT16 for FC 1.4 as well. Figure 6 shows the wind speed magnitude (a) for the entire wind farm and (b) in the vicinity of WT16 for FC 1.7. In Figure 6b, WT16 is in the wake of WT17 and possibly WT20 as well, despite being more than 10 diameters away. The higher power output for WT16 in the SCADA indicates that WT16 may not be influenced as much by the WT17 and WT20 wakes.

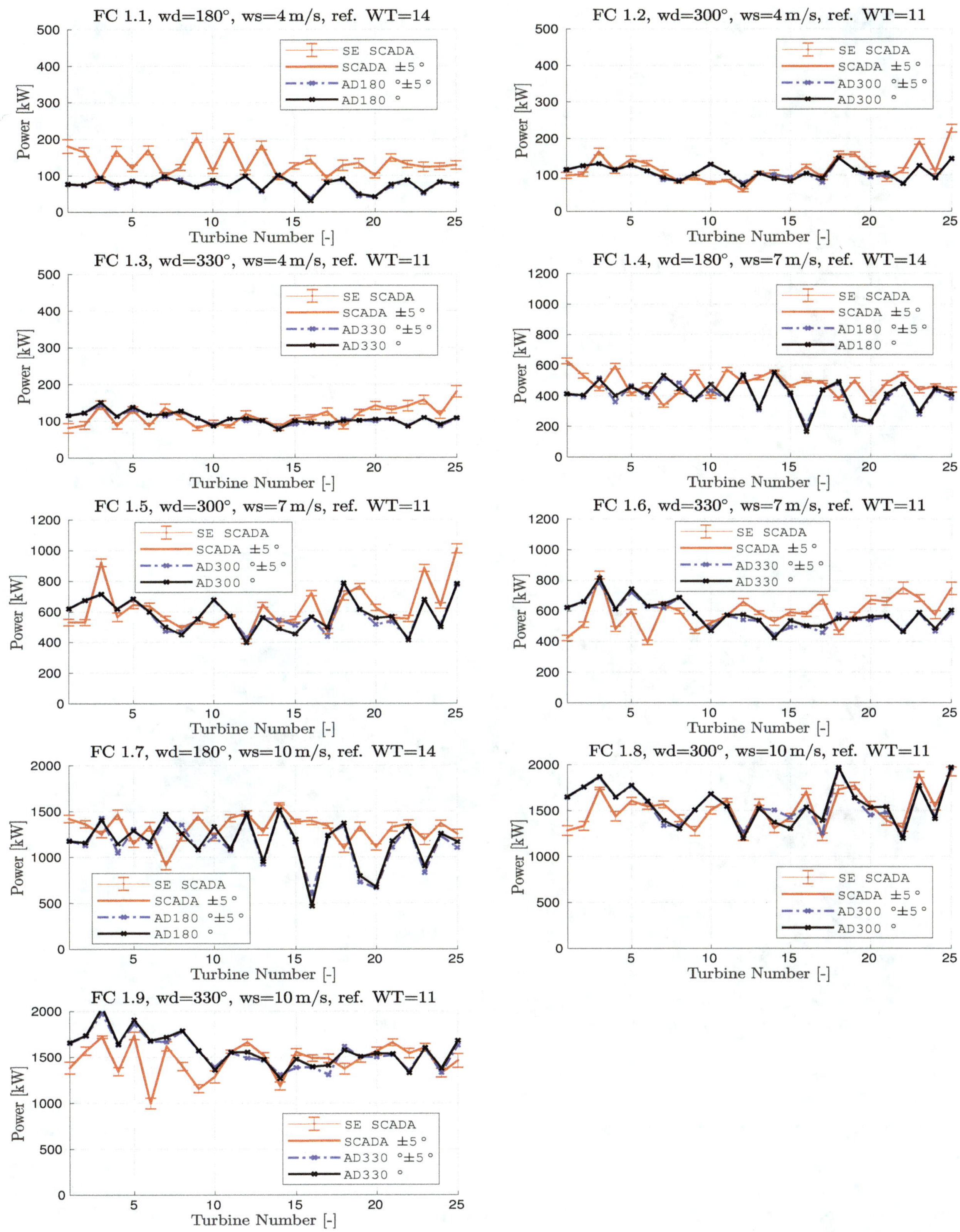

Figure 5. Predicted power from AD-RANS (AD#) for wind turbines 1 to 25 compared to supervisory control and data acquisition (SCADA) for 4, 7 and 10 m/s wind speeds (ws) at wind directions (wd) of 180, 300 and 330 degrees. The wind speed, ws, is referenced from the reference wind turbine (ref. WT). SE is the standard error and # is the wind direction in degrees.

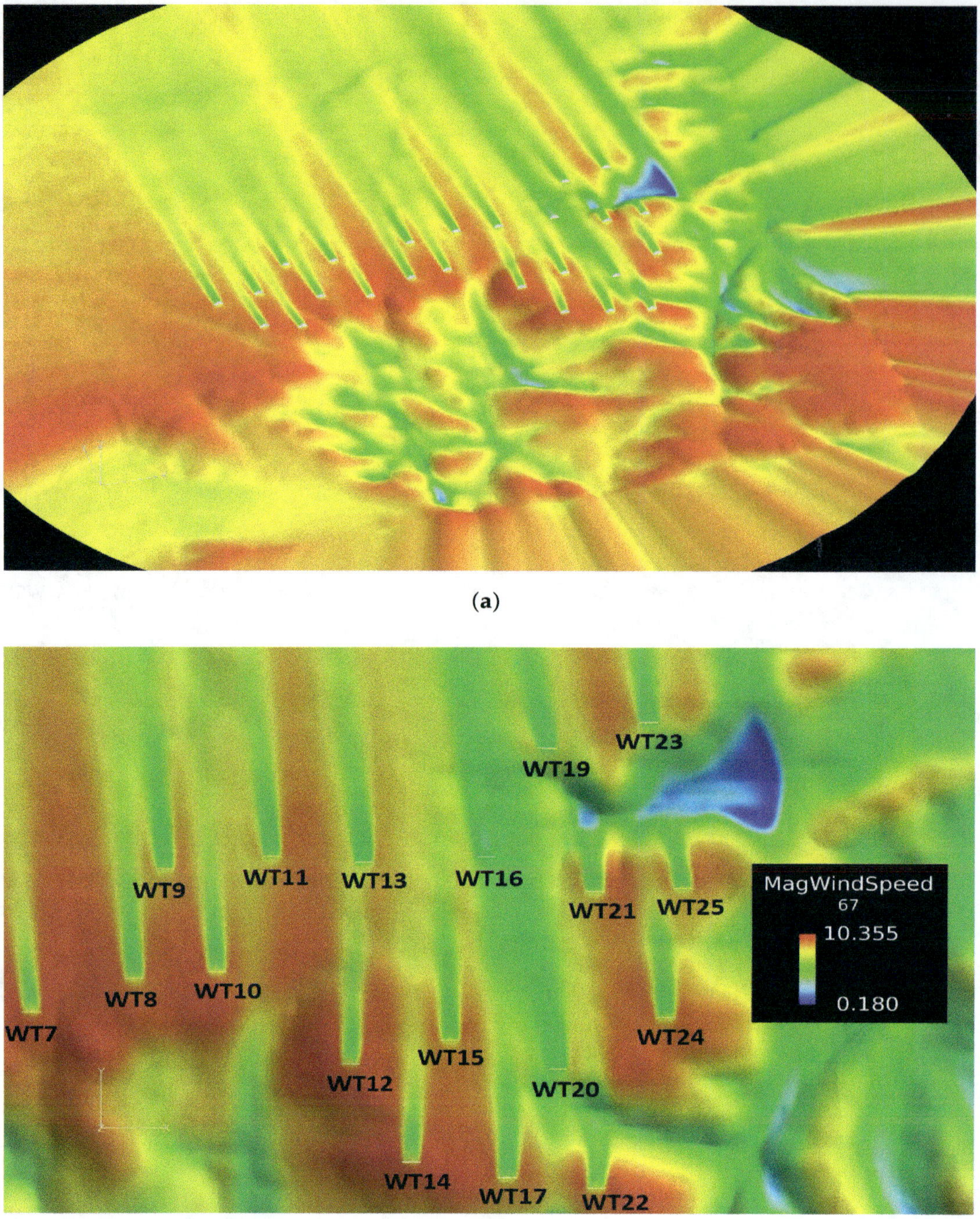

(b)

Figure 6. Wind speed magnitude (**a**) for the entire wind farm and (**b**) in the vicinity of wind turbine 16 for FC 1.7.

The predicted power from AD-RANS for two terrain mesh grids: 10 × 10 (AD Base) and 24 × 10 blocks (AD Fine), compared to SCADA for a wind direction of 180 degrees at (a) 4 m/s, (b) 7 m/s and (c) 10 m/s is shown in Figure 7. The mesh with 24 × 10 blocks is shown in Figure 7d. In comparison to the terrain mesh with 10 × 10 blocks, the refined mesh has a greater concentration of points on the left (west) and right (east) sides to capture the terrain effects in greater detail. The non-uniform mesh results in between 7 and 15 horizontal grid points per AD. Figure 7a–c shows larger differences in predicted power between AD Fine and AD Base for larger wind speeds.

A possible explanation is that the refined mesh includes more terrain detail, and the effects of terrain are pronounced at higher wind speeds. Nevertheless, refining the mesh did not result in significant improvements in comparison to SCADA measurements.

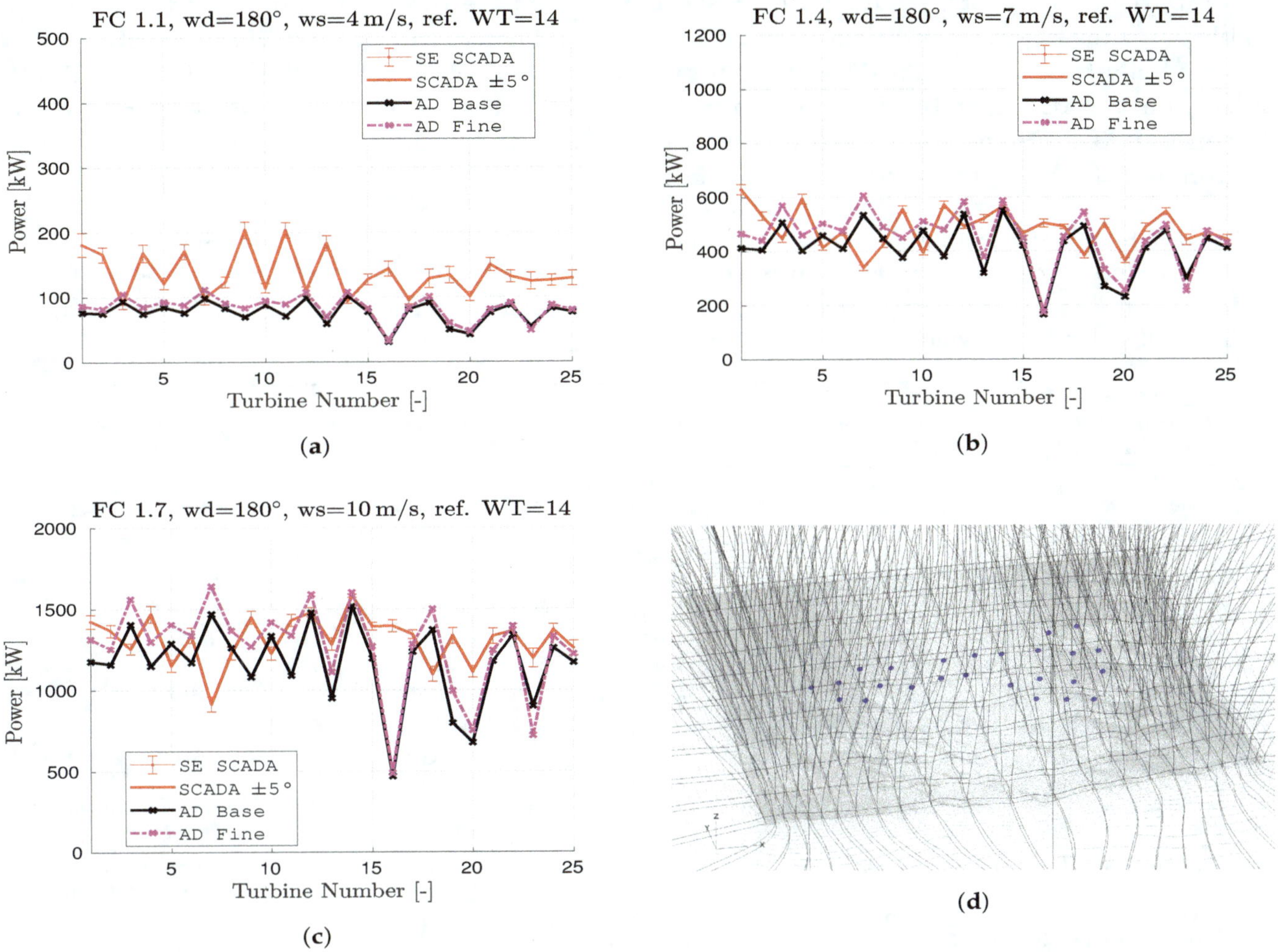

Figure 7. Predicted power from AD-RANS for two terrain mesh grids: 10 × 10 (AD Base) and 24 × 10 blocks (AD Fine), compared to SCADA for a wind direction of 180 degrees at (**a**) 4 m/s, (**b**) 7 m/s and (**c**) 10 m/s. The mesh (**d**) with 24 × 10 blocks.

In summary, the comparison of the RANS simulations with the measurements is a challenging task. The wind directions and wind speeds are always changing in the wind farm, and this cannot be measured precisely. In other words, the reference wind directions and wind speeds are difficult to determine, which may lead to a high measurement uncertainty when the data is processed into bins. The significant changes in power levels between 0 and 360 degrees is seen in the power measurements. Percent differences obtained for power in the AD-RANS varied between 0.1% (FC 1.3, WT3) and 73.4% (FC 1.1, WT16) based on AD# ± 5°. Based on AD#, minimum and maximum values are 0.1% (FC 1.7, WT12) and 78.4% (FC 1.1, WT16), respectively. The mean percent difference for power from all twenty-five wind turbines varied between 9.6% (FC 1.8) and 42.1% (FC 1.1) based on AD# ± 5°, or 10.1% (FC 1.8) and 40.9% (FC 1.1) based on AD#. The mean percent difference for power from all flow cases is 20.5% and 20.0% for AD# ± 5° and AD#, respectively. Therefore, averaging RANS simulations in the ±5 degree range may reduce extreme percent differences slightly (e.g., 78.4% down to 73.4% for FC 1.1, WT16), but does not reduce the mean percent difference.

3.2. Actuator Line Large-Eddy Simulation Results

Figure 8 depicts the iso-surface of vorticity for the AL-LES simulation without synthetic inflow turbulence of WT14 at the 7 m/s wind speed and 180 degrees wind direction, i.e., south to north. The comparison of AL-LES results with measurement data are shown in Table 5 and Figures 9–11. In Table 5, the mean power (kW), rotor speed (RPM), and the horizontal streamwise velocity at 30, 50, and 70-m heights on met masts M1 and M3 for the four flow cases are shown. The table includes the results from the AL-LES in black, the mean measurement values in blue, and the percent difference between the AL-LES and the measurements in purple. The standard deviations (STDs) from the measurements for a 1-year period are also included in the table.

Table 5. Summarized values from AL-LES without synthetic inflow turbulence in black, measurements in blue, and percent differences in purple for the FC2 flow cases. The flow field is scaled using the horizontal streamwise velocity (streamwise vel.) measured at the upstream met mast at 70 m (in bold). STD is the standard deviation of the measurements for a 1-year period.

AL-LES (Black), Measured (Blue), % diff. (Purple)	**180°, 4 m/s**	**STD**	**180°, 7 m/s**	**STD**	**0°, 4 m/s**	**STD**	**0°, 7 m/s**	**STD**
	142.0		714.3		123.1		685.1	
Power (kW)	156.6	81.2	709.0	177.4	117.8	65.8	775.0	234.3
	−9.32%		0.75%		4.50%		−11.60%	
	8.860		14.720		8.860		14.520	
Rotor speed (RPM)	8.868	3.253	14.680	0.587	8.940	1.742	14.646	0.613
	−0.08%		0.27%		−0.89%		−0.86%	
Streamwise vel. (m/s)	**4.149**		**7.027**		1.560		3.066	
@ 70 m	**4.143**	0.659	**7.036**	0.528	2.263	0.894	3.862	0.824
M3	**0.15%**		**−0.12%**		−31.06%		−20.61%	
Streamwise vel. (m/s)	3.891		6.590		2.308		4.302	
@ 50 m	3.867	0.681	6.669	0.523	2.025	0.927	3.643	0.887
M3	0.62%		−1.19%		13.98%		18.10%	
Streamwise vel. (m/s)	3.547		6.009		3.217		4.845	
@ 30 m	3.719	0.766	6.294	0.621	2.463	0.705	5.143	0.761
M3	−4.62%		−4.54%		30.57%		−5.80%	
Streamwise vel. (m/s)	1.902		3.294		**4.068**		**7.155**	
@ 70 m	2.671	0.712	3.707	0.695	**4.050**	0.406	**7.114**	0.620
M1	−28.81%		−11.13%		**0.46%**		**0.58%**	
Streamwise vel. (m/s)	2.554		4.457		3.741		6.582	
@ 50 m	2.614	0.738	3.933	0.923	3.888	0.501	6.814	0.658
M1	−2.30%		13.30%		−3.76%		−3.42%	
Streamwise vel. (m/s)	3.554		6.042		3.306		5.818	
@ 30 m	3.090	0.767	5.104	0.830	3.588	0.704	6.343	0.728
M1	15.01%		18.38%		−7.85%		−8.28%	

Larger percent differences were obtained for the horizontal streamwise velocities, particularly when comparing AL-LES and the measurements for the met mast located downstream of WT14. For example, the percent difference in the horizontal streamwise velocity at 70 m height on met mast M1 for the 180 degrees and 4 m/s flow case is 28.81%. The reason for the large difference is because M1 is located downstream of WT14 as shown in Figure 2b. The complexity and stochastic nature of the turbulence from the wake of WT14 introduces larger variations and uncertainty in the flow field. There is a reasonable agreement between the AL-LES and the measurements for the power and rotor speed on WT14, the largest percent difference being 11.60% in power for the 0 degree and 7 m/s flow case. The smallest percent difference is 0.08% in rotor speed for the 180 degrees and 4 m/s flow case. The mean percent differences from the four flow cases for power and rotor speed are 6.54% and 0.53%, respectively.

Figures 9–11 are the results in Table 5 displayed in a bar-chart format. The results for the horizontal streamwise velocity comparisons are grouped by upstream (Figure 9) and downstream (Figure 10) locations. In Figures 9 and 10, bars are paired to represent comparisons between AL-LES and measurements for each one of the sensor locations: 70 m, 50 m and 30 m heights on the met mast. The horizontal streamwise velocity comparisons for the upstream mast in Figure 9 are in better agreement than the comparisons for the downstream mast shown in Figure 10 for two reasons. First, as mentioned in Section 2.3, the flow field in the AL-LES was scaled with the measurements using the sensor at 70 m height on the upstream met mast. Second, as mentioned previously for Table 5, larger differences occur for the downstream mast because of the complexity and stochastic nature of the turbulence generated from the wake of WT14. The wake of WT14 introduces larger variations and uncertainty in the flow field. In Figure 11, the largest difference between the AL-LES and SCADA appears for the power in the 0° and 7 m/s flow case. All the other comparisons in Figure 11 generally agree.

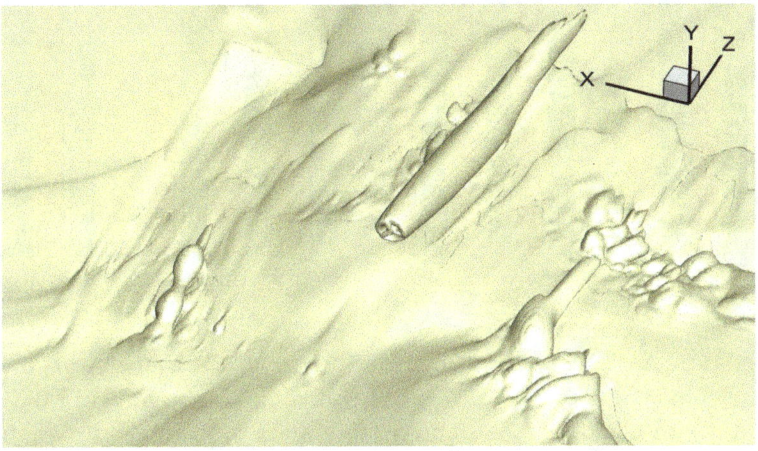

Figure 8. Iso-surface of vorticity for the actuator line large-eddy simulation of wind turbine 14 without synthetic inflow turbulence for 7 m/s wind speed and 180 degrees wind direction (south to north). The domain shown is the same as in Figure 4b.

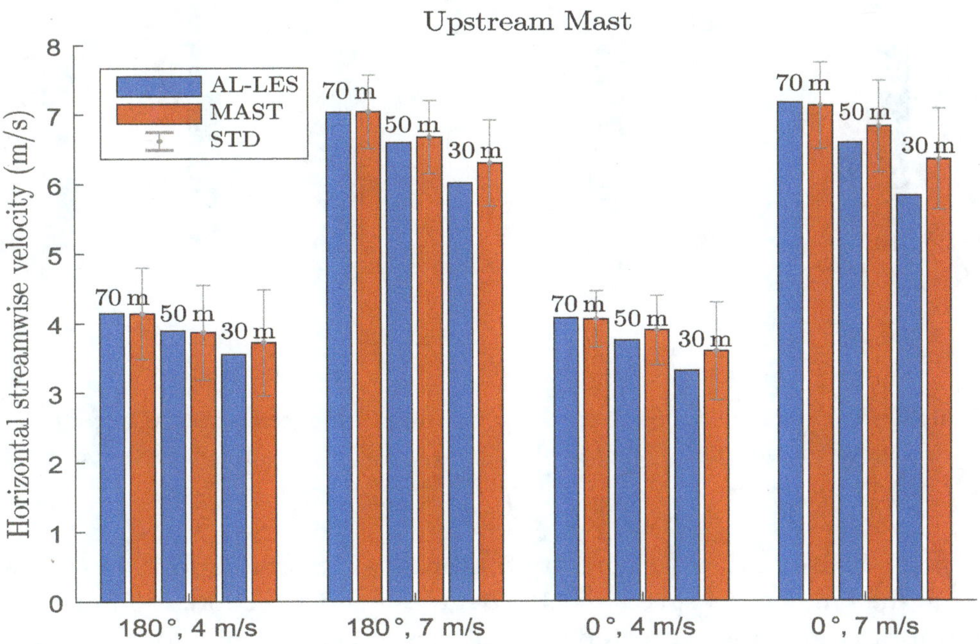

Figure 9. Comparison of AL-LES without synthetic inflow turbulence and met mast data for the horizontal streamwise velocity at 30-, 50- and 70-m heights on masts M1 and M3 when M1 and M3 are upstream of wind turbine 14. STD is the standard deviation of the measurements for a 1-year period.

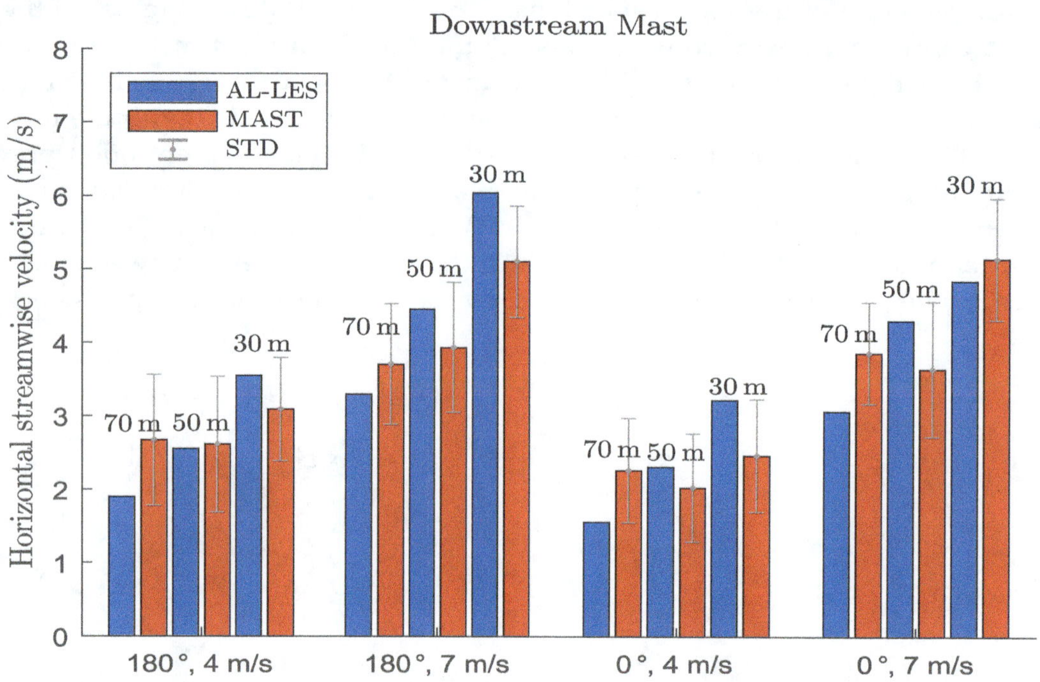

Figure 10. Comparison of AL-LES without synthetic inflow turbulence and met mast data for the horizontal streamwise velocity at 30-, 50- and 70-m heights on masts M1 and M3 when M1 and M3 are downstream of wind turbine 14. STD is the standard deviation of the measurements for a 1-year period.

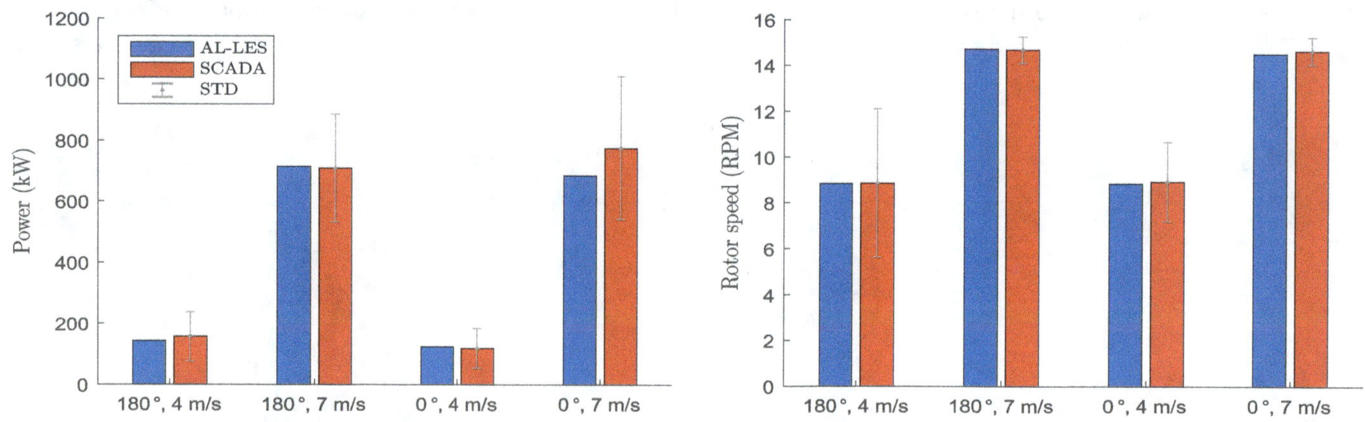

(a) (b)

Figure 11. Comparison of AL-LES without synthetic inflow turbulence and SCADA for (**a**) power and (**b**) rotor speed of wind turbine 14 for the four FC2 flow cases. STD is the standard deviation of the measurements or a 1-year period.

Figure 12 depicts the iso-surface of vorticity for the AL-LES simulation of WT14 with synthetic inflow turbulence at the 7 m/s wind speed and 180 degrees wind direction. In Figure 12, the synthetic turbulent inflow input is seen by the rectangular plane, $372 \times 372\,\text{m}^2$ in size, 2.5 rotor diameters upstream of the wind turbine. Since slices from the Mann turbulence box were gradually fed into the plane during the unsteady simulation, a rectangular box is seen encompassing the wind turbine and its wake. The comparison of AL-LES results with measurement data is shown in Table 6.

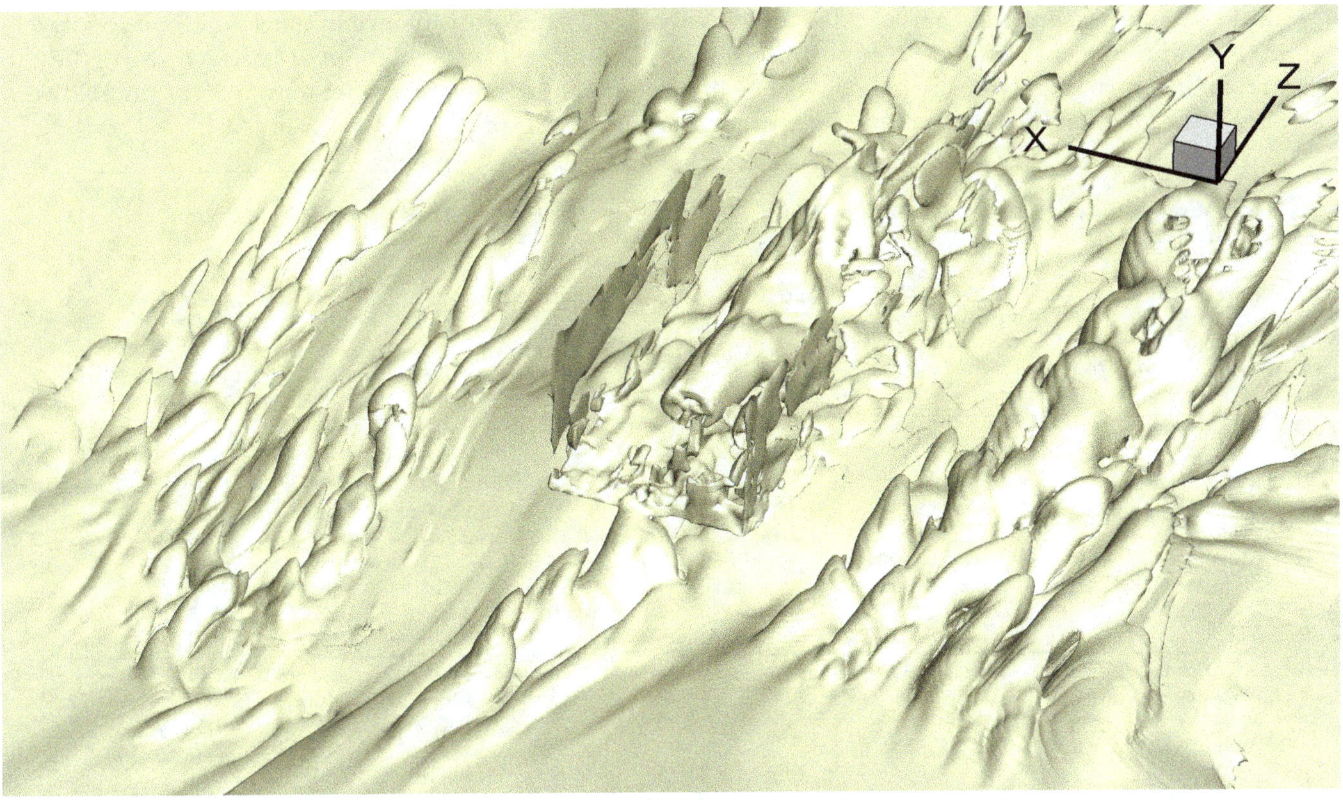

Figure 12. Iso-surface of vorticity for the actuator line large-eddy simulation of wind turbine 14 with synthetic inflow turbulence for 7 m/s wind speed and 180 degrees wind direction (south to north). The domain shown is the same as in Figure 4b. The synthetic inflow turbulence input is seen by the rectangular plane 2.5 rotor diameters upstream of the wind turbine.

The flow field in the AL-LES simulation with synthetic inflow turbulence was scaled using the sensor at 70 m height on the upstream met mast based on the simulations without synthetic inflow turbulence. The scaling is more difficult and not as accurate as in the case without synthetic inflow turbulence because of the large variation in wind speed in the AL-LES simulation with synthetic inflow turbulence. For example, for the 7 m/s wind speed and 0 degree wind direction case at 70 m on M1, the percent difference between the measurement and the AL-LES is 1.81%. For the case without synthetic inflow turbulence in Table 5, the percent difference is only 0.58%. The percent differences obtained for the horizontal streamwise velocity at 30-, 50- and 70-m heights on masts M1 and M3 are similar to the case without synthetic inflow turbulence. The percent differences obtained for the power and RPM are also similar. The percent differences obtained for power and RPM vary between 0.44% and 11.42%. The mean percent differences from the four flow cases for power and rotor speed are 6.77% and 1.47%, respectively. Figures 13–15 are the results in Table 6 displayed in a bar-chart format identical to Figures 9–11. All comparisons in Figures 13–15 are very similar to Figures 9–11.

The turbulence intensities based on the horizontal streamwise velocity in Table 6 are shown in bar-chart format in Figures 16 and 17. It was not possible to attain very good agreement for the turbulence intensities. The highest percent difference is 55.23% at 30 m on M1 for the 4 m/s and 180 degrees case. The turbulence intensities obtained from the AL-LES largely depended on the chosen input parameters for the surface roughness (=0.4) when generating the Mann turbulence box and the turbulence scaling factor (=1.5), see Section 2.3.

Table 6. Summarized values from AL-LES with synthetic inflow turbulence in black, measurements in blue, and percent differences in purple for the FC2 flow cases. The flow field is scaled using the horizontal streamwise velocity (streamwise vel.) measured at the upstream met mast at 70 m (in bold). STD is the standard deviation of the measurements for a 1-year period.

AL-LES (Black), Measured (Blue), % diff. (Purple)	**180°, 4 m/s**	**STD**	**180°, 7 m/s**	**STD**	**0°, 4 m/s**	**STD**	**0°, 7 m/s**	**STD**
	141.6		730.3		114.2		686.5	
Power (kW)	156.6	81.2	709.0	177.4	117.8	65.8	775.0	234.3
	−9.59%		3.01%		−3.05%		−11.42%	
	8.988		14.523		8.899		14.207	
Rotor speed (RPM)	8.868	3.253	14.684	0.587	8.938	1.742	14.646	0.613
	1.36%		−1.10%		−0.44%		−3.00%	
Streamwise vel. (m/s)	**4.152**		**7.156**		1.623		3.538	
@ 70 m	**4.143**	0.659	**7.036**	0.528	2.263	0.894	3.862	0.824
M3	**0.23%**		**1.70%**		−28.28%		−8.38%	
Streamwise vel. (m/s)	3.858		6.648		1.970		4.151	
@ 50 m	3.867	0.681	6.669	0.523	2.025	0.927	3.643	0.887
M3	−0.22%		−0.33%		−2.71%		13.94%	
Streamwise vel. (m/s)	3.644		6.291		2.914		5.484	
@ 30 m	3.719	0.766	6.294	0.621	2.463	0.705	5.143	0.761
M3	−2.01%		−0.06%		18.34%		6.63%	
Streamwise vel. (m/s)	2.095		3.934		**4.031**		**7.243**	
@ 70 m	2.671	0.712	3.707	0.695	**4.050**	0.406	**7.114**	0.620
M1	−21.59%		6.12%		**−0.47%**		**1.81%**	
Streamwise vel. (m/s)	2.558		4.705		3.695		6.655	
@ 50 m	2.614	0.738	3.933	0.923	3.888	0.501	6.814	0.658
M1	−2.17%		19.61%		−4.95%		−2.33%	
Streamwise vel. (m/s)	3.459		6.033		3.202		5.805	
@ 30 m	3.090	0.767	5.104	0.830	3.588	0.704	6.343	0.728
M1	11.94%		18.20%		−10.75%		−8.48%	
Turbulence intensity (-)	0.123		0.123		0.206		0.234	
@ 70 m	0.108	0.046	0.095	0.028	0.264	0.122	0.229	0.063
M3	13.92%		29.60%		−22.21%		2.16%	
Turbulence intensity (-)	0.131		0.138		0.212		0.221	
@ 50 m	0.131	0.062	0.105	0.030	0.278	0.126	0.255	0.069
M3	0.00%		31.60%		−23.52%		−13.40%	
Turbulence intensity (-)	0.064		0.084		0.185		0.185	
@ 30 m	0.134	0.068	0.115	0.025	0.271	0.096	0.242	0.041
M3	−52.06%		−27.08%		−31.72%		−23.53%	
Turbulence intensity (-)	0.145		0.161		0.105		0.097	
@ 70 m	0.258	0.083	0.231	0.056	0.111	0.076	0.110	0.026
M1	−43.99%		−30.17%		−5.17%		−11.82%	
Turbulence intensity (-)	0.206		0.188		0.107		0.107	
@ 50 m	0.274	0.092	0.262	0.054	0.128	0.080	0.115	0.028
M1	−24.75%		−28.07%		−16.01%		−6.59%	
Turbulence intensity (-)	0.104		0.098		0.144		0.161	
@ 30 m	0.232	0.076	0.215	0.038	0.153	0.094	0.132	0.029
M1	−55.23%		−54.23%		−5.51%		21.74%	

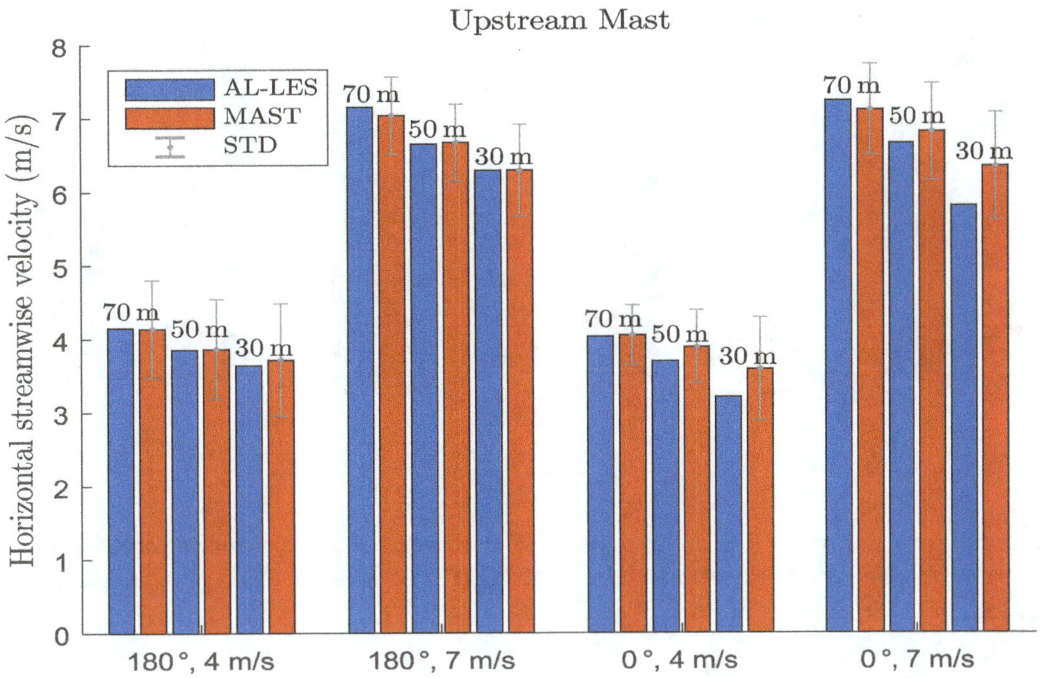

Figure 13. Comparison of mean AL-LES results with synthetic inflow turbulence and met mast data for the horizontal streamwise velocity at 30-, 50- and 70-m heights on masts M1 and M3 when M1 and M3 are upstream of wind turbine 14. STD is the standard deviation of the measurements for a 1-year period.

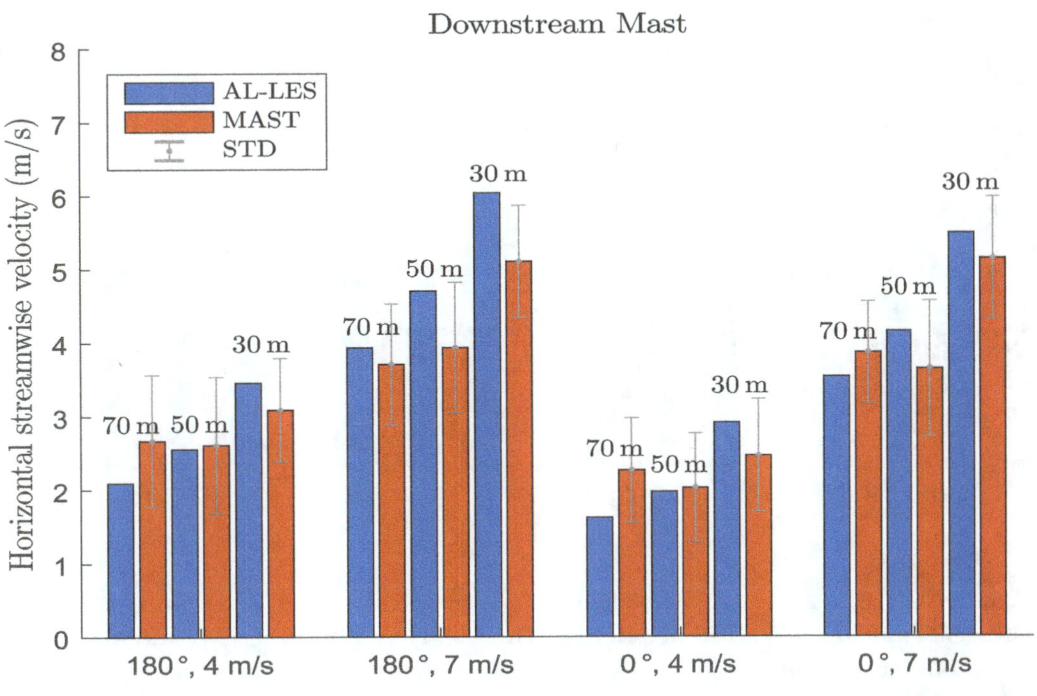

Figure 14. Comparison of mean AL-LES results with synthetic inflow turbulence and met mast data for the horizontal streamwise velocity at 30-, 50- and 70-m heights on masts M1 and M3 when M1 and M3 are downstream of wind turbine 14. STD is the standard deviation of the measurements for a 1-year period.

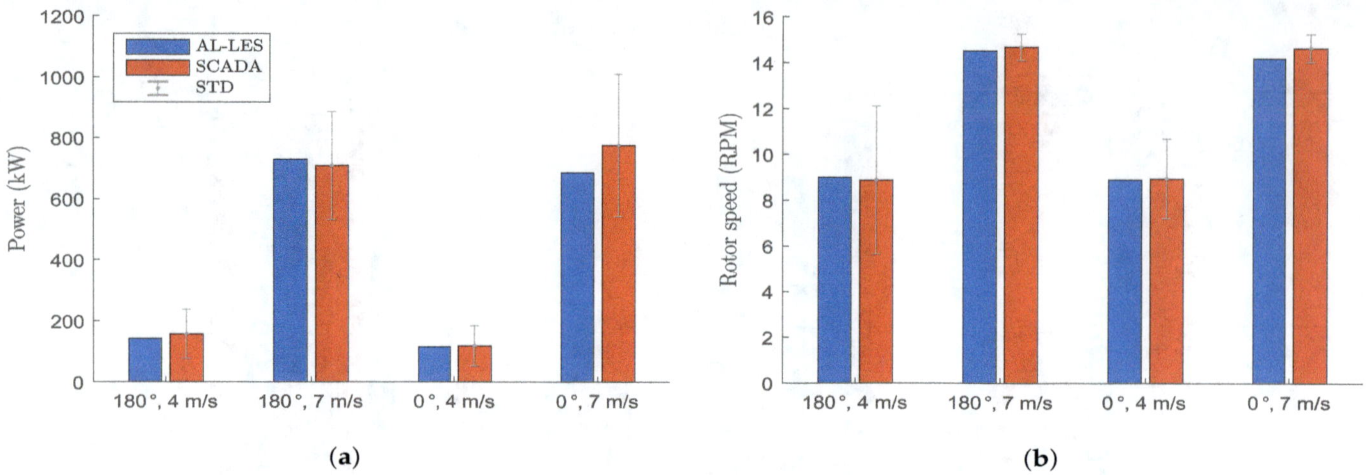

Figure 15. Comparison of mean AL-LES results with synthetic inflow turbulence and SCADA for (**a**) power and (**b**) rotor speed of wind turbine 14 for the four FC2 flow cases. STD is the standard deviation of the measurements for a 1-year period.

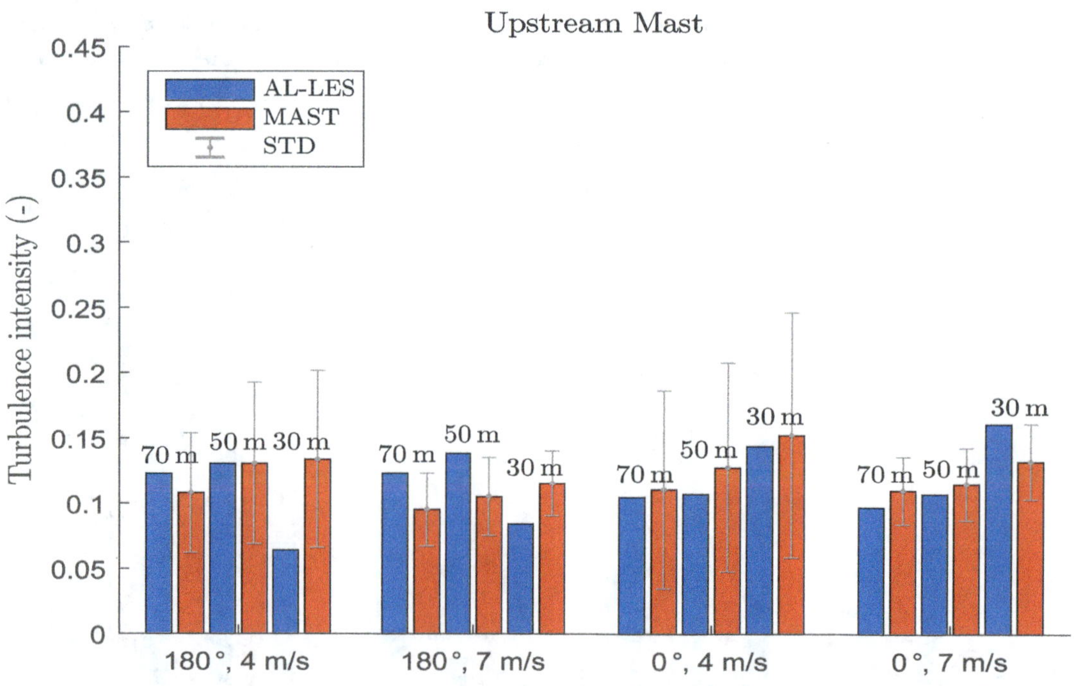

Figure 16. Comparison of mean AL-LES results with synthetic inflow turbulence and met mast data for turbulence intensities at 30-, 50- and 70-m heights on masts M1 and M3 when M1 and M3 are upstream of wind turbine 14. The turbulence intensities are based on the horizontal streamwise velocity and STD is the standard deviation of the measurements for a 1-year period.

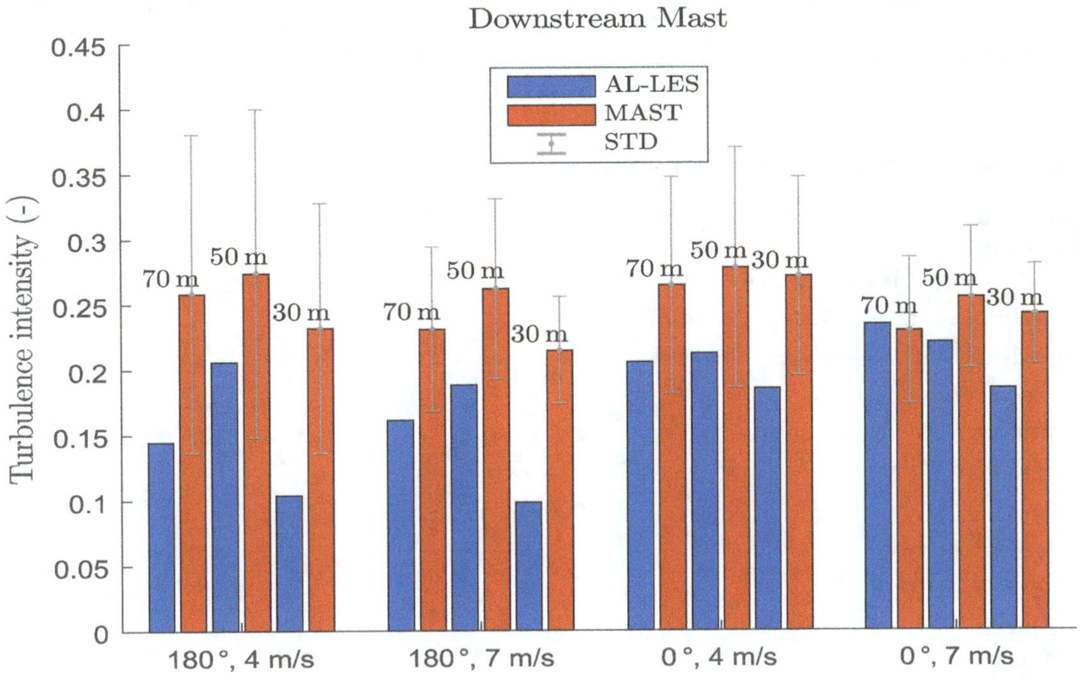

Figure 17. Comparison of mean AL-LES results with synthetic inflow turbulence and met mast data for turbulence intensities at 30-, 50- and 70-m heights on masts M1 and M3 when M1 and M3 are downstream of wind turbine 14. The turbulence intensities are based on the horizontal streamwise velocity and STD is the standard deviation of the measurements for a 1-year period.

4. Conclusions

This article described Computational Fluid Dynamics (CFD) simulations of flows in a wind farm in complex terrain in Shaanxi, China and the comparisons of the computational results with utility scale field measurements. The CFD simulations in the study were performed using Reynolds-Averaged Navier–Stokes (RANS) and Large-Eddy Simulation (LES) solvers. The RANS method together with an Actuator Disc (AD) approach was employed to predict the performance of the 25 wind turbines in the farm, while the LES method with an Actuator Line (AL) technique was used to obtain a detailed description of the flow field around one specific wind turbine near two met masts. The AD-RANS simulation results were compared with the mean values of power obtained from field measurements. Furthermore, the AL-LES results were compared with the mean values of power, rotor speed, and streamwise velocity measured from the wind turbine and its nearby two masts.

Results from both the AD-RANS and AL-LES simulations indicated that both methods can reasonably predict the flow patterns and performance of the wind farm and wind turbine #14, respectively, in complex terrain in Shaanxi. The mean percent difference obtained for power in the AD-RANS simulations was approximately 20%. Percent differences obtained for power and rotor RPM in the AL-LES varied between 0.08% and 11.6%. The mean percent differences in the AL-LES for power and rotor RPM are approximately 7% and 1%, respectively. Careful consideration of the measurement data as well as the computational setup, such as the meshing of terrain and flow domain, wind turbine model and wind farm layout, and the flow input parameters is needed to obtain reasonable or improved results. Uncertainty quantification and statistical models should be employed. For example, the uncertainty in the measurement data can be incorporated into the CFD solver to provide more reliable outputs of power, rotor rotational speed, wind velocities, etc.

Author Contributions: W.Z.S. and K.S.H. conceived and designed the project; M.S. performed the computations and analyzed the results under the guidance of M.P.v.d.L. and W.J.Z.; M.S., W.Z.S. and M.P.v.d.L. wrote the paper.

Acknowledgments: The authors wish to give special thanks to the Danish and Chinese partners (DTU Wind Energy, EMD International A/S, North West Survey and Design Institute of Hydro China Consultant Corporation, and HoHai University) for their active collaborations.

Abbreviations

The following abbreviations are used in this manuscript:

AD	actuator disc
AL	actuator line
CFD	computational fluid dynamics
CSIC	China Shipbuilding Industry Corporation
DES	delayed-detached-eddy simulation
FC1/FC2	flow case 1, flow case 2
HZ	Haizhuang
LES	large-eddy simulation
MW	megawatt
RANS	Reynolds-Averaged Navier–Stokes
RPM	revolutions per minute
SCADA	supervisory control and data acquisition
SE	standard error
SGS	sub-grid scale
STD	standard deviation
wd	wind direction
ws	wind speed
WT(s)	wind turbine(s)

References

1. International Renewable Energy Agency. *Renewable Power Generation Costs in 2017*; Technical Report; IRENA: Abu Dhabi, UAE, 2018; ISBN 978-92-9260-040-2.
2. Storey, R.C.; Norris, S.E.; Stol, K.A.; Cater, J.E. Large eddy simulation of dynamically controlled wind turbines in an offshore environment. *Wind Energy* **2013**, *16*, 845–864. [CrossRef]
3. Van der Laan, M.P.; Sørensen, N.N.; Réthoré, P.E.; Mann, J.; Kelly, M.C.; Troldborg, N. The k-ε-f_P model applied to double wind turbine wakes using different actuator disk force methods. *Wind Energy* **2015**, *18*, 2223–2240. [CrossRef]
4. Makridis, A.; Chick, J. Validation of a CFD model of wind turbine wakes with terrain effects. *J. Wind Eng. Ind. Aerodyn.* **2013**, *123*, 12–29. [CrossRef]
5. Schulz, C.; Klein, L.; Weihing, P.; Lutz, T.; Krämer, E. CFD Studies on Wind Turbines in Complex Terrain under Atmospheric Inflow Conditions. *J. Phys. Conf. Ser.* **2014**, *524*, 012134. [CrossRef]
6. Schulz, C.; Klein, L.; Weihing, P.; Lutz, T. Investigations into the Interaction of a Wind Turbine with Atmospheric Turbulence in Complex Terrain. *J. Phys. Conf. Ser.* **2016**, *753*, 032016. [CrossRef]
7. Tabib, M.; Rasheed, A.; Kvamsdal, T. LES and RANS simulation of onshore Bessaker wind farm: Analysing terrain and wake effects on wind farm performance. *J. Phys. Conf. Ser.* **2015**, *625*, 012032. [CrossRef]
8. Tabib, M.; Rasheed, A.; Fuchs, F. Analyzing complex wake-terrain interactions and its implications on wind-farm performance. *J. Phys. Conf. Ser.* **2016**, *753*, 032063. [CrossRef]
9. Castellani, F.; Astolfi, D.; Mana, M.; Piccioni, E.; Becchetti, M.; Terzi, L. Investigation of terrain and wake effects on the performance of wind farms in complex terrain using numerical and experimental data. *Wind Energy* **2017**, *20*, 1277–1289. [CrossRef]
10. Sørensen, J.N.; Mikkelsen, R.F.; Henningson, D.S.; Ivanell, S.; Sarmast, S.; Andersen, S.J. Simulation of wind turbine wakes using the actuator line technique. *Philos. Trans. Ser. A Math. Phys. Eng. Sci.* **2015**, *373*, 20140071. [CrossRef]
11. Sørensen, N.N. General Purpose Flow Solver Applied to Flow over Hills. Ph.D. Thesis, Technical University of Denmark, Roskilde, Denmark, 1995.

12. Michelsen, J.A. *Basis3D—A Platform for Development of Multiblock PDE Solvers*; Technical Report; Technical University of Denmark: Roskilde, Denmark, 1992.
13. Réthoré, P.E.; van der Laan, P.; Troldborg, N.; Zahle, F.; Sørensen, N.N. Verification and validation of an actuator disc model. *Wind Energy* **2014**, *17*, 919–937. [CrossRef]
14. Sørensen, J.N.; Shen, W.Z. Numerical Modeling of Wind Turbine Wakes. *J. Fluids Eng.* **2002**, *124*, 393–399. [CrossRef]
15. Ramos-García, N.; Sørensen, J.N.; Shen, W.Z. A strong viscous-inviscid interaction model for rotating airfoils. *Wind Energy* **2014**, *17*, 1957–1984. [CrossRef]
16. Han, X.; Liu, D.; Xu, C.; Shen, W.Z. Atmospheric stability and topography effects on wind turbine performance and wake properties in complex terrain. *Renew. Energy* **2018**, *126*, 640–651. [CrossRef]
17. Van der Laan, M.P.; Sørensen, N.N.; Réthoré, P.E.; Mann, J.; Kelly, M.C.; Troldborg, N.; Schepers, J.G.; Machefaux, E. An improved k-ε model applied to a wind turbine wake in atmospheric turbulence. *Wind Energy* **2015**, *18*, 889–907. [CrossRef]
18. Andersen, S.J.; Sørensen, J.N.; Mikkelsen, R. Simulation of the inherent turbulence and wake interaction inside an infinitely long row of wind turbines. *J. Turbul.* **2013**, *14*, 1–24, [CrossRef]
19. Mann, J. The spatial structure of neutral atmospheric surface-layer turbulence. *J. Fluid Mech.* **1994**, *273*, 141–168. [CrossRef]
20. Mann, J. Wind field simulation. *Probab. Eng. Mech.* **1998**, *13*, 269–282. [CrossRef]

7

Variable Pitch Approach for Performance Improving of Straight-Bladed VAWT at Rated Tip Speed Ratio

Zhenzhou Zhao [1,*], Ruixin Wang [1], Wenzhong Shen [2], Tongguang Wang [3], Bofeng Xu [1], Yuan Zheng [1] and Siyue Qian [1]

[1] College of Energy and Electrical Engineering, Hohai University, Nanjing 210098, Jiangsu, China; wrx123wy@163.com (R.W.); bfxu1985@hhu.edu.cn (B.X.); zhengyuan@hhu.edu.cn (Y.Z.); 15950553095@163.com (S.Q.)

[2] Department of Wind Energy, Technical University of Denmark, 2800 Lyngby, Denmark; wzsh@dtu.dk

[3] Jiangsu Key Laboratory of Hi-Tech Research for Wind Turbine Design, Nanjing University of Aeronautics & Astronautics, Nanjing 210016, Jiangsu, China; tgwang@nuaa.edu.cn

* Correspondence: joephy@163.com

Featured Application: This research provides a new pitching approach to increase the maximum power efficiency of vertical axis type wind turbines and water turbines used to absorb marine current energy. This approach did not deny the validity of the traditional methods. On the contrary, the new approach combined with traditional pitching method will improve power efficiency at all tip speed ratios, and well solve the problems of low efficiency and poor self-starting ability of the turbines.

Abstract: This paper presents a new variable pitch (VP) approach to increase the peak power coefficient of the straight-bladed vertical-axis wind turbine (VAWT), by widening the azimuthal angle band of the blade with the highest aerodynamic torque, instead of increasing the highest torque. The new VP-approach provides a curve of pitch angle designed for the blade operating at the rated tip speed ratio (TSR) corresponding to the peak power coefficient of the fixed pitch (FP)-VAWT. The effects of the new approach are exploited by using the double multiple stream tubes (DMST) model and Prandtl's mathematics to evaluate the blade tip loss. The research describes the effects from six aspects, including the lift, drag, angle of attack (AoA), resultant velocity, torque, and power output, through a comparison between VP-VAWTs and FP-VAWTs working at four TSRs: 4, 4.5, 5, and 5.5. Compared with the FP-blade, the VP-blade has a wider azimuthal zone with the maximum AoA, lift, drag, and torque in the upwind half-cycle, and yields the two new larger maximum values in the downwind half-cycle. The power distribution in the swept area of the turbine changes from an arched shape of the FP-VAWT into the rectangular shape of the VP-VAWT. The new VP-approach markedly widens the highest-performance zone of the blade in a revolution, and ultimately achieves an 18.9% growth of the peak power coefficient of the VAWT at the optimum TSR. Besides achieving this growth, the new pitching method will enhance the performance at TSRs that are higher than current optimal values, and an increase of torque is also generated.

Keywords: variable pitch; H-type VAWT; straight blade; DMST model; NACA0012; wind energy; power coefficient; tip speed ratio

---

1. Introduction

Wind turbines absorb wind energy and convert it into mechanical energy, and are classified according to the orientation of their axis of rotation into horizontal-axis wind turbines (HAWTs) and vertical-axis wind turbines (VAWTs). The HAWTs, represented by three-bladed propeller turbines, are the most common wind turbines because of the highest performance and easy manufacturing as a result of the great advance of aerodynamics and material engineering [1]. The VAWTs are further grouped into lift-type and drag-type. The lift-type VAWTs, which employ airfoil section blades to generate lift force as HAWTs, have higher rotation speed and better performance than the drag-type. The lift-type VAWTs, represented by the Φ-type and the straight-bladed type of Darrieus turbine, are also well-known [2]. On the other hand, most of them are used in small-scale applications, although they have several advantages over HAWTs, such as the ability to accept wind from any direction without yawing and the ability to provide direct rotary drive to a fixed load [3]. Some of the disadvantages of the VAWTs include an inability to self-start and relatively lower efficiency [4,5].

Darrieus turbine blades use airfoil sections designed as aircraft wing profiles. The NACA0012, NACA0015, and NACA0018 profiles are commonly used as blade sections [6]. Typically, these are designed to operate at small angles of attack (AoAs), lower than $\pm 10°$. At angles higher than this, the airfoil stalls and the flow separates on the upper surface of the blade, causing a loss of lift and an increase in drag [7]. At the startup of the turbine, a zero-tangential flow speed contributes to a large AoA, which then makes the airfoil stall. In this case, the pressure drag of the blade is so high that the lift cannot overcome the drag to self-start [8], not to mention any additional load from the electricity generator. Previous studies proposed a number of methods to make a self-starting lift-type VAWT. Kirke [9] reviewed those studies, and stated that high solidity turbines have the potential to increase the self-starting torque, but a higher solidity lowers the maximum power coefficient and narrows the operating range. Beri et al. [10] maintained that the Darrieus type VAWTs with asymmetrical blades have better self-starting behavior, but unfortunately, this also causes a reduction in peak efficiency. Zamani et al. [11,12] numerically investigated a 3-kW straight-bladed Darrieus type VAWT with a designed J-shaped profile. Their results showed that the J-shaped profile could suppress the vortices and improve the self-starting of the turbine. The hybrid wind turbine, comprising two vertical co-axial rotors, is normally constructed as a Darrieus with drag-type blades in the middle [13]. The primary purpose of the hybrid design is to promote the lack of torque when self-starting and at low tip speed ratios (TSRs) [4]. They can self-start normally, but after starting, they are less efficient than a turbine with normal lift due to the negative drag created at the middle [14]. It has been found that the above approaches hardly provide a way to promote both the self-starting ability and the peak power coefficient of the VAWTs at same time.

A particular method to address those issues is to develop a variable pitch (VP)-straight-bladed VAWT to control the pitch angles of the blades on each azimuthal angle in order to maximize turbine torque and start up easily [15]. A reduction of stall is the main mechanism whereby VP-technology produces significant improvement of the torque at start up and low TSRs [16]. Besides offering the greatest potential for achieving significantly increased torque at low and intermediate TSRs, VP-technology is also used to promote peak efficiency [17–19]. VP-VAWTs consist of active and passive-type turbines. The active designs are defined as those systems that produce blade pitch change through means other than the direct action of the aerodynamic forces acting on them in passive designs [16]. Staelens et al. [7] introduced three modifications for the increased power output of a straight-bladed VAWT by varying its pitch. The third modification proposed a continuous variation in the local AoA correction during the rotation cycle using a sinusoidal function. Although the power output obtained by using such a modification is less than the other two, it has the inherent advantage of being practically feasible. Also, Kiwata et al. [4] described a micro VP-VAWT that varies the pitch angle according to the azimuthal angle. The performance of the VP-VAWT was measured in an open-circuit wind tunnel, and the results were better than those of the fixed pitch (FP)-VAWT. Erickson et al. [20] tested the effects of cyclic blade pitch actuation on the efficiency and operability of a high-solidity VAWT with a cam and control rod mechanism to prescribe the pitch dynamics in the wind tunnel over a wide range of design and operational variables. The results revealed that a tuned first-order sinusoidal actuation system could achieve a maximum absolute power coefficient of 0.436, which is an increase of 35% over the optimal fixed-blade configuration, with self-starting capabilities and drastically improved performance at a wide range of suboptimal operating conditions. Hwang et al. [21] investigated the performance of a VP-turbine controlled by a cycloidal strategy. Compared with the FP-turbine, the performance at the upwind half-cycle of the VAWT with the optimized blade pitch angle was greatly improved, particularly at the azimuth region from 90° to 180°; the effective region was expanded, and was nearly twice as large as that of the FP-case. Chougule et al. [17] designed a 500-W VAWT to implement the pitch control mechanism using the cyclic VP-technology. Its aerodynamics were predicted using the double multiple stream tube (DMST) model, and their results showed that only a 5° increase of the pitch angle amplitude increased the power coefficient of the VAWT by 12%.

Despite the improvement of the self-starting ability and power coefficient, the above VP-approaches have the same limitation, which is mainly about the performance enhancement of azimuths with high AoAs in the case of the FP. In FP-VAWTs, each blade is constantly subjected to a variation of the AoA. Since the AoA constantly changes, the blade will generate large torque in some azimuthal angles, and low torque in other azimuthal angles. The overall performance is determined by the average torque of the blade in a cycle path line. Evidently, the low torques will produce a lower average value, and then degrade the power efficiency. The azimuthal position with the biggest geometric AoA in FP-VAWTs will achieve the biggest increase in the performance of the blade in VP-VAWTs. Assuming that the biggest geometric AoA is close to and below the stall value at the rated TSR, the magnitude of the maximum increase of the AoA would be much smaller if it still used the above-mentioned VP-approaches; thus, the increase of the AoA in other azimuths and the peak efficiency of the turbine would be much smaller, too.

To change this, we proposed a new VP-approach that was intended to mainly promote the performance of the blade working in the azimuths with smaller AoAs in FP-VAWTs. Ultimately, the aim is to largely enhance the peak efficiency of the VAWT in the rated or larger TSR. In this paper, the effect of the new approach on the performance of the blade or the turbine are investigated using a method implementing the double multiple stream tube (DMST) model.

2. Explanation of Variable Pitch Concept

2.1. Oscillating AoA

The Darrieus-type VAWT is a cross-flow rotor whose axis of rotation meets the flow of the working fluid at right angles. The blades perform the work twice in one cycle: i.e., the upwind half-cycle (azimuthal angle from 0° to 180°, where the 0°, 90°, 180°, and 270° positions are defined in Figure 1) and the downwind half-cycle (ranging from 180° to 360°). The azimuth angle of 0° is defined as a position where the directions of the local wind speed 'V_W' and the tangential flow speed '$U = r\omega$' are the same; the azimuthal angle of 180° is the position where the directions of the two speeds are opposite; as its anticlockwise rotation, the azimuthal angle of 90° is the position moving anticlockwise on a quarter circle from 0°, which is the fixed reference position. The resultant flow, which is the vector sum of the local wind speed 'V_W' and the tangential flow speed '$U = \omega r$', always comes from the upwind side of the blade: i.e., the outer side on the upwind half-cycle and the inner side on the downwind half-cycle [22]. Accordingly, as illustrated in Figure 1, the rotating blade experiences a changing resultant flow, including its magnitude and direction. The changing direction of the resultant velocity and chord line of the airfoil causes the oscillating AoA between the negative and positive values following a sinusoidal function of the azimuthal angle.

As shown in Figure 2, the AoA is determined by:

$$\alpha = \tan^{-1}\left[\frac{a \sin\theta}{\lambda_0 + a\cos\theta}\right] \tag{1}$$

where α is the AoA, a is the axial induction factor defined as $a = V/V_W$ (V is the velocity of the flow in the stream tube, V_W is the velocity of the free wind), θ is the azimuthal angle, and $\lambda_0 = R\omega/V_W$ represents the local TSR, where ω is the rotational speed, R is the rotational radius. The 0° and

180° positions are transformation points of the AoA between the negative and positive values, which determine their zero AoAs. Affected by the oscillating AoA, the blade of the VAWT inevitably produces fluctuating torque, even in steady conditions. For a given airfoil section, the non-dimensional normal and tangential forces are written as:

$$C_N = C_L \cos\alpha + C_D \sin\alpha \tag{2}$$

$$C_T = C_L \sin\alpha - C_D \cos\alpha \tag{3}$$

where C_N and C_T are the normal force and tangential force coefficients, respectively, and C_L and C_D are the lift and drag coefficients, respectively. According to Equations (2) and (3), the blade in the 0° and 180° azimuths will generate a zero-normal force and a negative tangential force, which cause a negative torque, according to Equation (4):

$$M = \frac{1}{2}\rho c R(2H) C_T V_r^2 \tag{4}$$

where M presents the torque of the blade, ρ is the air density, c is the chord length, $2H$ is the height of the blade, V_r is the local resultant velocity, and R is the rotation diameter. Due to the continuity of the blade torque variation, produced by the 0° and 180° azimuths, the torque in a zone around the 0° or 180° azimuth remains negative, and a wider zone may have a positive torque, but is still not big enough. In contrast, the 0° and 180°, and 90° and 270° angles produce the maximum AoAs. This is because the directions of the upwind flow and tangential flow are vertical to each other (Figure 1), and furthermore, the 90° azimuth having the larger upwind flow speed normally shows better aerodynamics than the 270° azimuth.

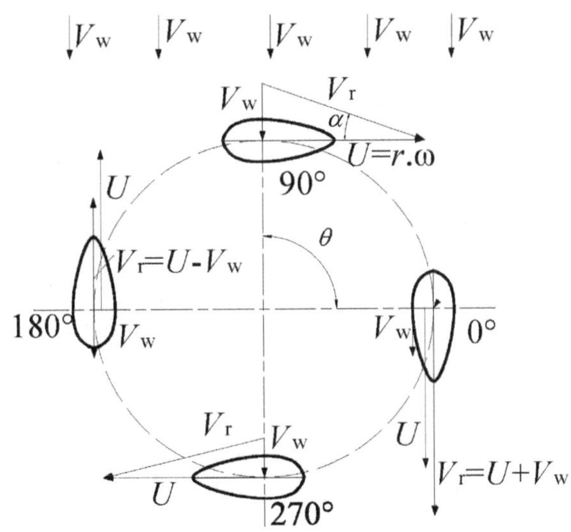

Figure 1. Illustration of Darrieus rotor.

2.2. *Aerodynamics in Traditional VP-Technology*

The VAWT blade is subject to cyclic variations in wind speed and the AoA, which are efficiency improvements that can result from corresponding variations of the pitch angle of the blade to optimize the AoA at each point in its cycle [20]. Any improvement made to the AoA can be of a cyclic nature, as shown in Figure 3 [23]. Thus, a typical curve of the pitch angle in traditional VP-technology is expressed by Equations (5) and (6):

$$\gamma = A_c \sin(\theta) \tag{5}$$

$$A_c = A_s - \left(A_s \frac{\lambda}{X_0}\right)\sin(\theta) \quad (6)$$

where γ is the variable value of the pitch angle changed with the azimuth, A_c is the maximum pitch angle at a certain TSR, A_s is the maximum pitch angle at all of the TSRs, λ is the TSR, X_0 is the TSR of max C_P at zero A_c, and θ is the azimuthal angle.

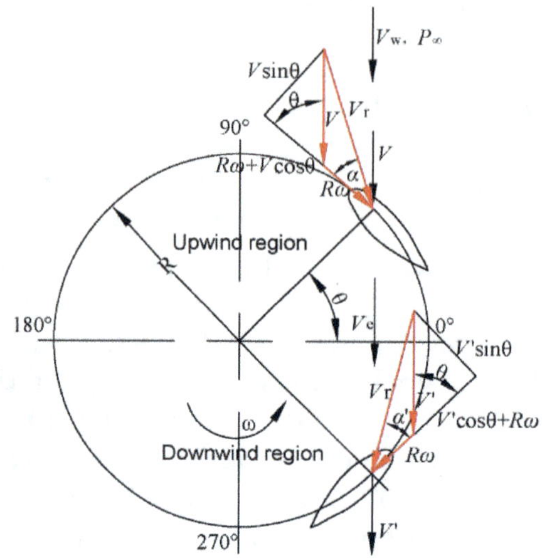

Figure 2. Relative wind speed vectors in the upstream and downstream section.

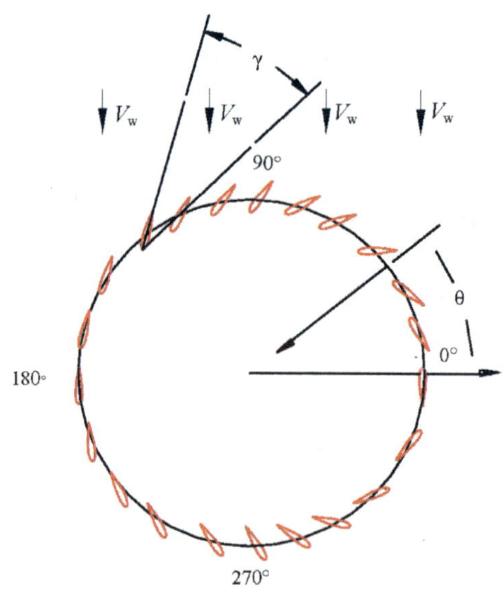

Figure 3. Illustration of cyclical variable pitch.

In another conventional VP-technology of VAWT, the blades have an oscillatory motion around their own axis, which was superimposed on the uniform rotation motion. Considering a cross-section of the turbine, the lines normal to the profile chord for every blade position intersect a given point "P" during a complete revolution (Figure 4) [24]. The variation of the pitch angle with respect to the azimuthal angle is expressed as:

$$\gamma = A_c \sin\theta - \arctan\left(\frac{\sin\theta}{\cos\theta + \lambda}\right) \quad (7)$$

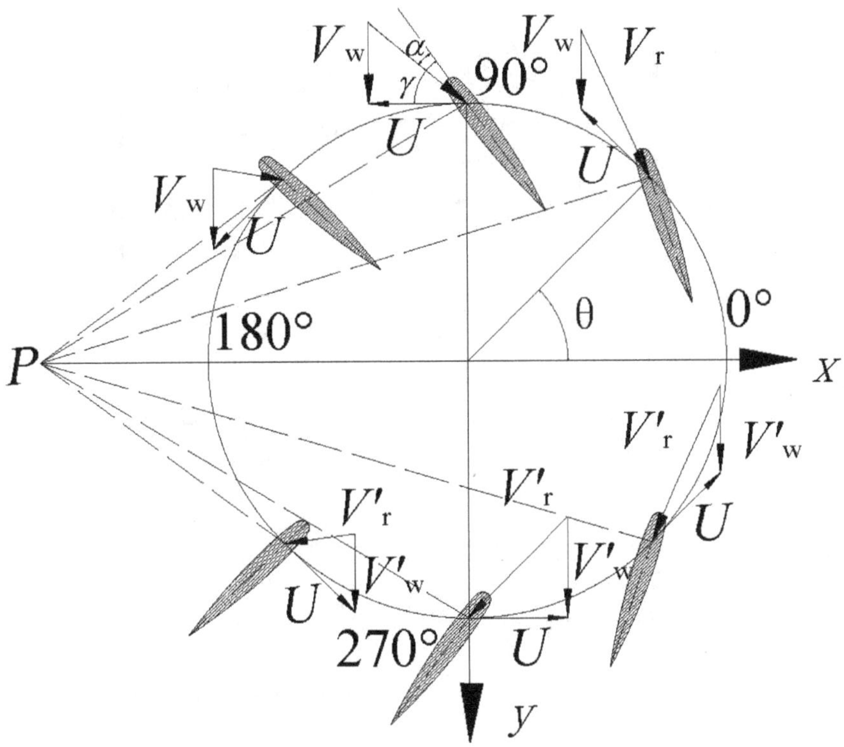

Figure 4. Illustration cycloidal variable pitch.

Two conventional VP-approaches have one point in common: the azimuthal angle with a larger geometric AoA will be given a larger pitch angle. Affected by the performance, as expressed by the lift coefficient ($C_L - \alpha$) of the airfoil, the blade has a stall static AoA of about 10°, as shown in Figure 5a [18]. At the start-up and small TSRs, the maximum AoA is much greater than the stall angle, the lift of the blade will decrease, and the drag will greatly increase after the stall occurs. In such cases, the VP-technology will change the pitch angle in the direction of the decreased AoA to avoid the stall. Two methods were used to achieve this goal. In the first method, the pitch angle changes only in the positions where the local geometric AoA exceeds the stall value, and the final local effective AoA is kept below the static stall angle [7,25], as illustrated in Figure 5b. The advantage of this method is that it makes the blade rotate with the optimum AoA in a much wider azimuthal zone. Its disadvantage is that the change of the pitch angle (represented by a dotted line in Figure 5b) is discontinuous at some points, which may not be practical to implement, and also causes early fatigue due to the abrupt dynamical loads [7]. The other method continuously changes the pitch angle following a sinusoidal function (Figure 5c). Under this control, the maximum amplitude of the sinusoidal function for the pitch angle is set to be equal to the maximum difference between the geometric AoA in the FP-VAWT and the effective AoA in the VP-VAWT. The advantage of this method is that the pitch angle as well as the forces changes smoothly and continuously, and then the method is physically and mechanically feasible to implement; the implementation procedures are presented in Figures 2 and 3. However, the power efficiency, although significantly larger than that of the FP-VAWT, is much lower than that of the other method illustrated in Figure 5b, because the final local effective AoA curve also follows a sinusoidal function in which the blade experiences the maximum performance only at two inflection points. At an optimum or larger TSR, the pitch angle should change toward the direction of the increased AoA for an improved performance in terms of the increased peak power coefficient of the VAWT. Meanwhile, if the variation of the pitch angle follows a sinusoidal function, as shown in Figure 5d, the space for performance enhancement would be exceedingly limited, because the maximum amplitude of the AoA change is much smaller (Figure 5d). A new approach is explored next to solve this problem.

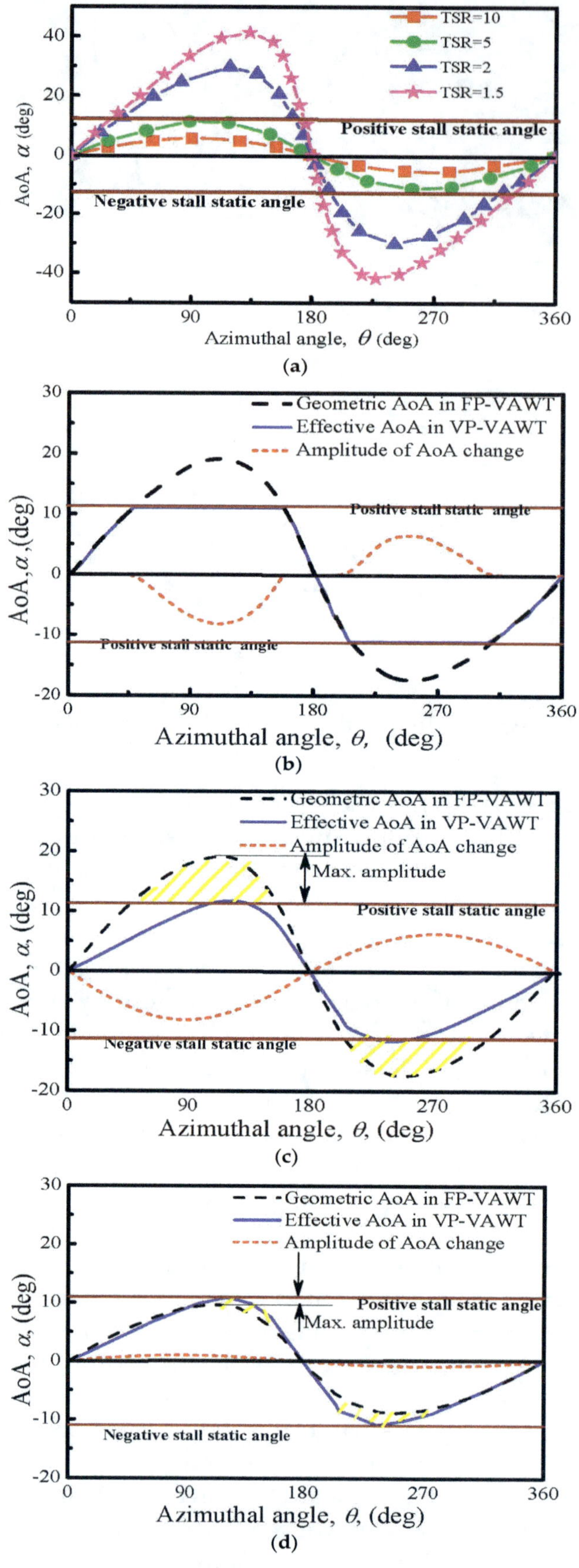

Figure 5. Local angle of attack as a function of the azimuthal angle. (**a**) AoA variation at different TSRs; (**b**) one pitching method at low TSR; (**c**) another pitching method at low TSR; (**d**) pitching method at rated TSR or higher TSR.

At an optimum or larger TSR, the pitch angle should change in the direction of the increased AoA for an improved performance in terms of the increased peak power coefficient of the VAWT. Alternatively, if the variation of the pitch angle follows a sinusoidal function, as shown in Figure 5d, the space for performance enhancement would be exceedingly limited, because the maximum amplitude of the AoA change is much smaller (Figure 5d). A new approach is explored next to solve this problem.

2.3. New VP-Approach

The new VP-strategy proposed in this paper focuses mainly on improving the performance by widening the range of the largest local effective AoA, instead of enlarging the maximum value. In the new VP-approach, the local effective AoA is kept below the stall angle and maintains a constant value in a wide zone during a rotation cycle; the curve is shown as a solid line in Figure 6. During the design of the new approach, four key points are adequately considered.

(1) How to deal with the zero AoA at the azimuthal angles of 0° and 180°: Since the positions of the 0° and 180° angles are the transformation points between the positive and negative AoAs, the design of the new VP-approach cannot avoid the zero value AoAs. The pitch angles in these two positions remain at the zero value.

(2) How to set the value of the constant effective AoA in Figure 6: At the rated TSR, the largest local AoA in a cycle should be designed as the optimum AoA corresponding to the largest lift–drag ratio, rather than the static stall AoA. Thus, if the pitch curve of the new approach is designed for a blade operating at the rated TSR, the constant value in Figure 6 should be equal to the maximum geometric AoA in a FP-VAWT, since both a larger and a smaller value would contribute to a smaller torque. Consequently, the pitch angles are also kept at zero at the two inflection-point locations, namely the azimuthal angles of 90° and 270°.

(3) Is the approach physically and mechanically feasible to implement? The amplitude of the change of the pitch angle is shown in Figure 6 as a dotted curve. In this curve, the pitch angle changes smoothly and continuously, and eliminates the drawback of the design in Figure 5b. The whole variation process of the pitch is illustrated in Figure 7, which is easily accomplished mechanically, such as a specially designed cam. As shown in Figure 7, the pitch angles varies in each 1/4 circle (0°–90°, 90°–180°, 180°–270°, and 270°–360°), and returns to zero at four positions (0°, 90°, 180°, and 270°). Here, we defined the presence of a positive pitch angle when the blade rotates counterclockwise around its own axis, and the leading edge of the blade points toward the inner side of the circular path. In contrast, the angle is negative when the blade rotates and the leading edge points in the opposite direction.

(4) What is the effect of the new VP-approach on the performance of the blade? After the mechanical feasibility of the approach is validated in theory, the following study focuses on the investigation of the effect of the new VP-approach using the DMST model.

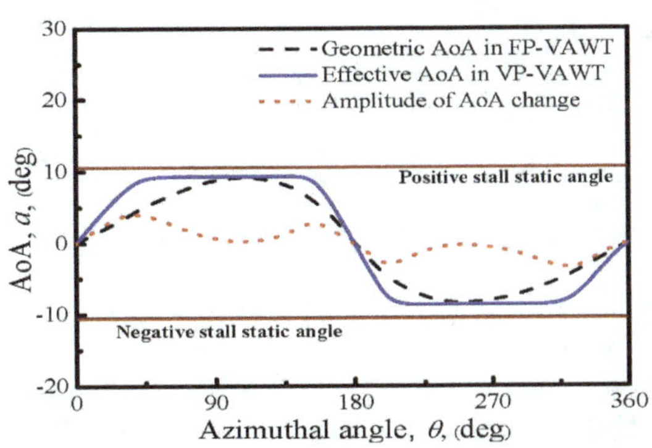

Figure 6. Curve of pitch angle in new approach.

Figure 7. Illustration of new variable pitch.

3. Computational Model

3.1. Geometric Characteristics of VATW

The geometric characteristics of the straight-bladed VAWT studied in this paper are as follows. The NACA0012 blade profile: a height of 2 m (H = 1 m, Figure 8), a rotation radius (R) of 1 m, a chord length of the blade of 0.09 m, with two blades. Two straight-bladed VAWTs are also referred to as H-type VAWTs, since their configurations are much like an "H" shape. The operating conditions of the H-type VAWT aerodynamic parameter chosen for this study are as follows: an air density of 1.225 kg/m^3, a wind speed of 9 m/s, a kinematic viscosity of 1.48×10^{-5} m^2/s. The curve of the pitch angle with respect to the azimuthal angle was designed with the condition of a TSR = 4.5, in which the FP-turbine generates the peak power coefficient. The performance at four TSRs values (i.e., TSR = 4, 4.5, 5.0, and 5.5) was predicted, and their results were compared with each other. In addition, as shown in Figure 8, the effect of the radial arm was not considered, even though it is also a source of drag that reduces the power performance, due to the use of the DMST algorithm in the subsequent research.

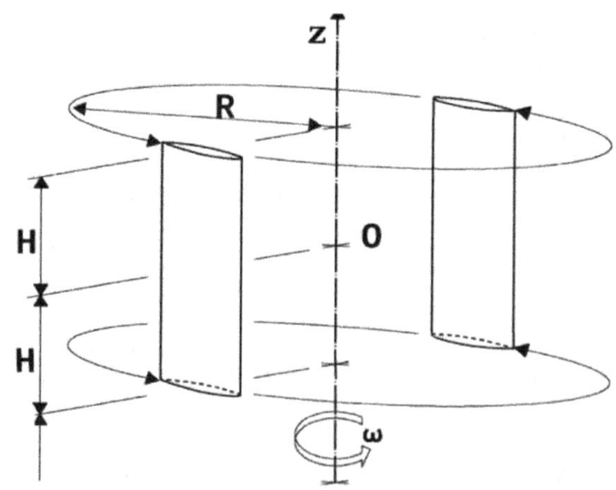

Figure 8. Configuration of an H-type vertical-axis wind turbine (VAWT).

3.2. Aerodynamic Model of Double Multi Stream Tube (DMST) Model

The DMST model makes use of the blade element momentum (BEM) theory, whereby the performance of the rotor is calculated by coupling the momentum equation in the mainstream direction of the wind and the aerodynamics analysis of the interactions between the airfoils in motion and the oncoming flow on the rotor [15,26]. The first approach using a momentum model to analyze the flow field around a VAWT was developed by Templin [27], who considered the rotor as an actuator disk enclosed in a simple stream tube, where the induced velocity through the swept volume of the turbine was assumed to be constant. An extension of this method to the multiple stream tubes model was developed by Strickland [28], who considered the swept volume of the turbine as a series of adjacent stream tubes. Additionally, Paraschivoiu [29] developed an analytical model, the DMST model, which considers a multiple-stream tube system divided into two parts, where the upwind and downwind components of the induced velocities at each level of the rotor are calculated by using the principle of two actuator disks in tandem. Subsequently, many predicted the performance of the Darrieus turbine using the DMST model, and all of the results showed that the DMST model has a high prediction accuracy for predicting the aerodynamics of the lift-type VAWTs [24,30–33]. The Darrieus-type wind turbine at low TSRs normally experiences a large variation within its AoA, and the blades are dominated by dynamic stall. In such conditions, the prediction must consider the effect of the dynamic stall on the accuracy. However, in the paper, the detailed investigation is performed at a high TSR, which is higher than four. Meanwhile, the variation of the AoA is getting much smaller and no stall is occurring; thus it has little effect on the prediction accuracy [22]. Accordingly, the dynamic stall was neglected in the paper. The model relies on pre-existing lift and drag data to calculate the blade aerodynamics. In a cross-flow wind turbine, blade–vortex interactions may occur, but its effects are not clear. There is not a well-developed algorithm for the modification of the lift and drag data due to the impingement of the vortex on the blades [34]. Thus, the interaction was also neglected in our work.

Since this investigation is focused on an H-type VAWT, a simplified version of the DMST model applicable to a straight-bladed rotor is briefly outlined below [35–37]. For an H-type VAWT, the resultant velocity, V_r, during the upwind half-cycle can be determined from the velocity vector triangle shown in Figure 2, which then gives:

$$V_r = V_w \sqrt{(\lambda_0 + a\cos\theta)^2 + (a\sin\theta)^2} \tag{8}$$

Applying the blade element theory and equating the change in momentum to the drag on the rotor for each stream tube, it is found that:

$$fa = \pi\eta(1-a) \tag{9}$$

where $\eta = r/R$ is the dimensionless parameter of the radius, and f is the unwind function that characterizes the upstream half-cycle of the rotor on the blade element rotating in this zone. The upwind function is given by the following equation:

$$f = \frac{Nc}{8\pi R}\int_0^{\pi}\left(C_N\frac{\cos\theta}{|\cos\theta|} - C_T\frac{\sin\theta}{|\cos\theta|}\right)\left(\frac{V_r}{V_w}\right)\mathrm{d}\theta \tag{10}$$

where N is the number of the blades, and C_N and C_T can be calculated from Equations (2) and (3), respectively. The lift and drag coefficients, C_L and C_D, respectively, are obtained by interpolating the known experimental data using both the local blade section Reynolds number $Re_b = V_r c/V_w$ and the local AoA. The local AoA can be calculated from Equation (1). Also, using Equation (8), Re_b is given as a function of the induced upwind velocity for each blade element in rotation.

3.3. Tip Loss Consideration

For a finite aspect ratio of the blade, tip loss is unavoidable. Thus, the effect of the blade tip loss was considered in the modeling to achieve an accurate performance prediction. The tip loss characterizes the tendency for trailing the vorticity from the blade tip, or from any other point where the circulation distribution changes, thereby reducing the blade's effectiveness. Prandtl's tip loss factor is commonly employed for VAWTs, and Soraghan et al. [30] described the mathematics in detail.

3.4. Computational Procedure

For a given H-Type rotor geometry and rotational speed ω, a value of the local TSR λ is chosen by assuming that the interference factor a is unity. Thus, Re_b and α are evaluated as first approximations, and the airfoil C_L and C_D characteristics are interpolated from the test data reported by Sheldahl and Klimas [38]. Then, using Equation (7), the normal and tangential force coefficients of the blade section are estimated, while Equation (10) allows the upwind function to be evaluated. With the first value of f, Equation (9) is used to calculate another value for the interference factor, and the iterations continue until successive sets of a are reasonably close. Convergence will be achieved, as the error is less than 0.0001. Once the final value of the upwind zone 'a' has been calculated, the local relative velocity V_r is determined from Equation (7) and the local AoA from Equation (1). A similar procedure is repeated for the downwind half-cycle. Additionally, it should be noted that here, the upwind velocity is an equivalent velocity expressed as $V_e = (2a - 1)V_W$.

4. Results Analysis

Due to the consideration of the tip loss of the blade, the results distributed along the spanwise direction of the finite length blade are not constant; their distribution shows symmetric characteristics on both sides of the middle height, namely $H = 0$, as shown in Figure 8. However, the differences between any two heights, except the position that is closest to the tip of the blade, are small for a large aspect ratio. Thus, the data from $H = 0$ are chosen to be analyzed and compared in the following sections.

4.1. AoA Variation

The final variation of the AoA in a full cycle is shown in Figure 9. As required by the original specifications of the design, the new approach increases the AoA in most azimuthal angles, which is much obvious in the upwind region. The downward parabolic AoA curve with respect to the azimuthal angle is more open, and more azimuthal angles have a higher AoA, but the largest value is not changed in the upwind section. In the downwind region, the increase of the AoA is obvious when the TSR = 4, 4.5, and 5, but it decreases as the TSR increases, and even shows a negative tendency when the TSR = 5.5. As the rotation speed increases, more wind energy is absorbed into the stream tubes of the upwind part, and the steam flows though the stream tubes of the downwind part with lower velocity, which causes a lower AoA in the condition of a constant rotation speed. It is also observed in all of the sub-figurations that the real curve of the AoA is not as smooth as the designed one, especially in the downwind region. This is because the AoA strongly depends on the considerably changed axial induction factor.

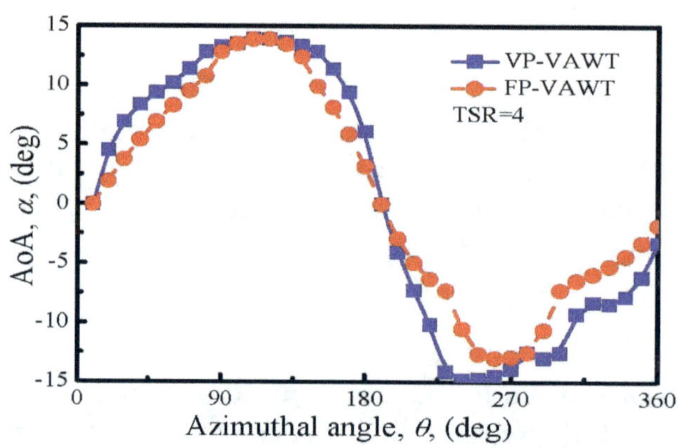

(a)

Figure 9. *Cont.*

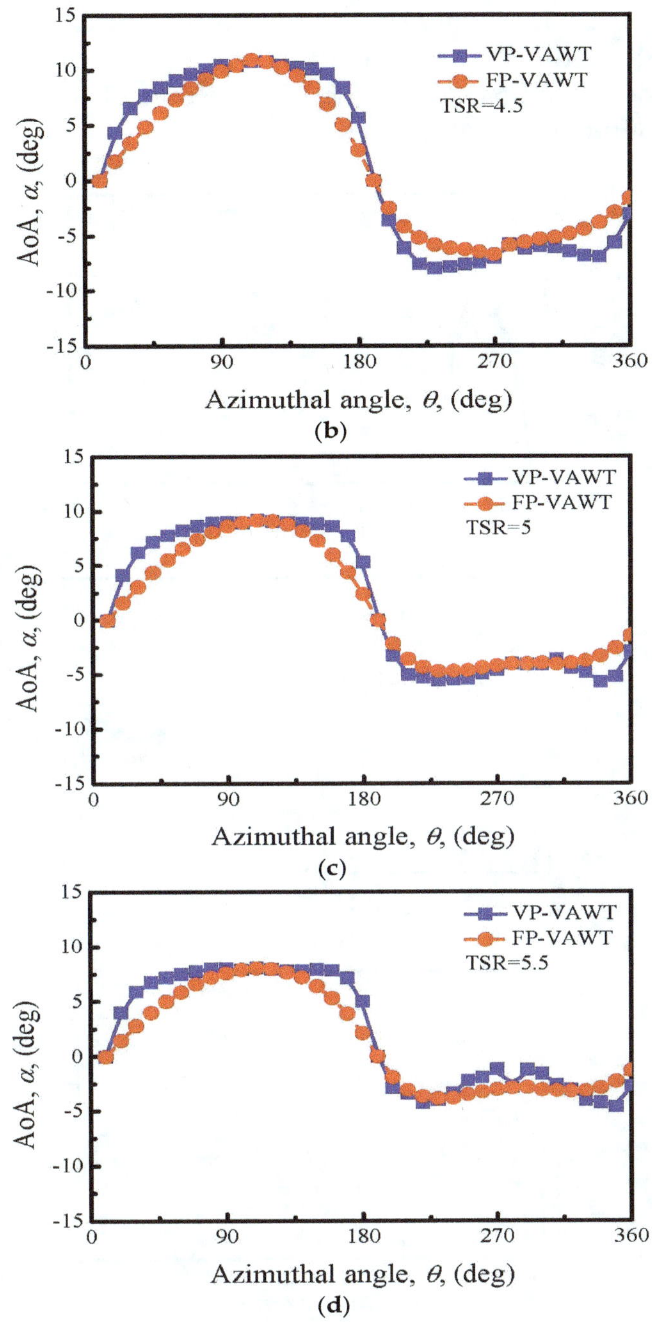

Figure 9. Angle of attack (AoA) as a function of azimuthal angle in variable pitch (VP) and fixed pitch (FP) VAWTs. (**a**) TSR = 4; (**b**) TSR = 4.5; (**c**) TSR = 5; (**d**) TSR = 5.5.

4.2. Lift Force Variation

The lift generated by the blade is a key parameter of a lift-type VAWT, because it determines the magnitude of the tangential force that causes the turbine rotation. One of the effects of the new VP-control changes the lift of the blade. The variation of the local lift coefficient during a complete rotation in four TSRs is illustrated in Figure 10, which reveals that, when the TSR = 4, the lift coefficient fluctuates intensively during a whole range of azimuthal angles, whether in the VP-VAWT or FP-VAWT, and a much lower value is observed at around 90° and 270°. This is because the AoA values in the large region around 90° and 270° are still larger than the stall static angles, which leads to the occurrence of the stall and then a decreased lift coefficient. As can be observed in Figure 10, the VP-blade does not create a larger maximum lift coefficient than the FP-blade, but it does have a wider stall zone.

The reason is that the new VP-approach increases the AoA and creates a wider azimuthal zone having a larger AoA than the stall value, but the maximum effective AoA is not changed. All of these phenomena indicate that the VP-approach designed for the rated TSR is not suitable to be applied to improve the performance of the VAWT at much lower TSRs.

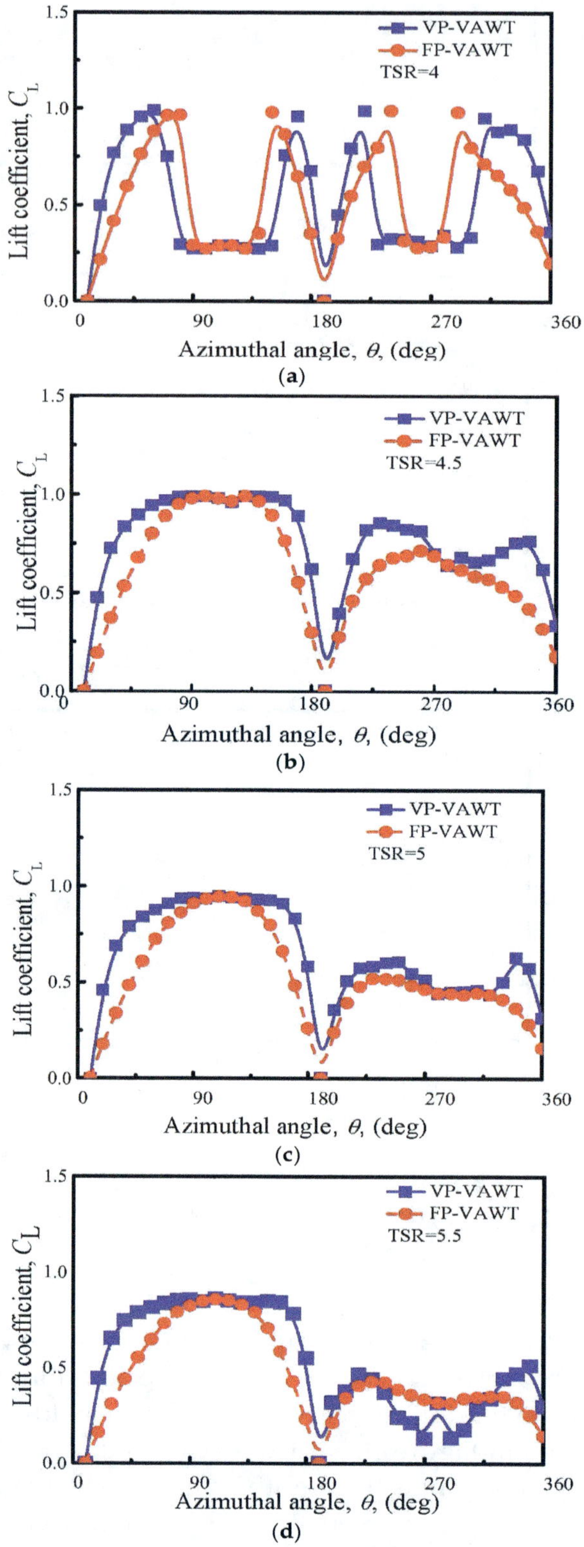

Figure 10. Lift coefficient as a function of azimuthal angle in VP-VAWTs and FP-VAWTs. (**a**) TSR = 4; (**b**) TSR = 4.5; (**c**) TSR = 5; (**d**) TSR = 5.5.

At TSRs larger than 4.5, stall does not occur; the curve of the lift coefficient changes smoothly, and shows an arched shape in the upwind or downwind half-cycles in the FP-VAWT. In the same conditions, the stall does not occur in the VP-VAWT either; larger lift coefficients are obtained almost in the whole rotation zone except four at positions, namely: 0°, 90°, 180°, and 270°. Finally, the novel approach achieves a larger zone with the largest lift coefficients in the upwind half-cycle and two new larger maximum coefficients in the downwind half-cycle. A comparison of the increase among the TSRs reveals that the increase of the lift is in the downstream region as the TSR increases. Even when the TSR = 5.5, the lift is affected by the AoA; as shown in Figure 9, the VP-blade produces lower lift than the FP-blade in an extensive zone of the downwind region.

4.3. Drag Force Variation

The drag coefficients of the blades of the FP-VAWT and VP-VAWT are illustrated in Figure 11. First, when the TSR = 4, the blade that stalls generates a large peak drag in the upwind region, which is almost five times as large as that when the TSR = 4.5, and another larger peak drag in the downwind region of the VP-VAWT. A comparison of the distribution between the VP-VAWTs and FP-VAWTs indicates that the new VP control also increases the drag of the blade because of the increased AoA. The trends of the drag changes between the VP-VAWTs and FP-VAWTs are much like those of the lift changes. However, the increased drag cannot negate the advantage of the new VP-approach. According to the aerodynamics of the airfoil, the AoA that has the largest lift coefficient normally results in a large drag coefficient. The final effect is a promotion of the blade performance or not, which is determined by whether the component of the increase in the tangential direction of the lift is larger or smaller than that of the drag coefficient.

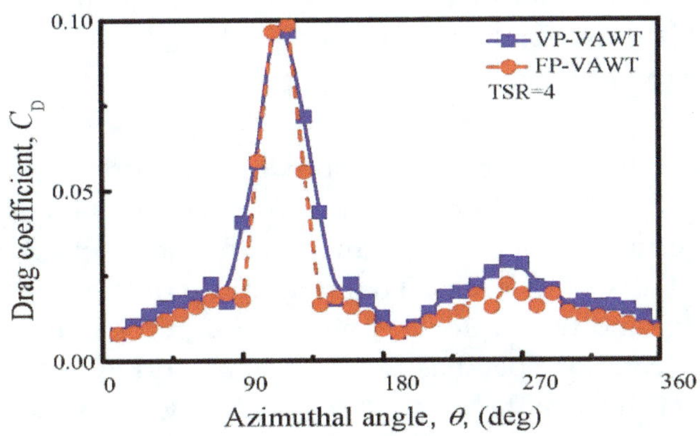

(a)

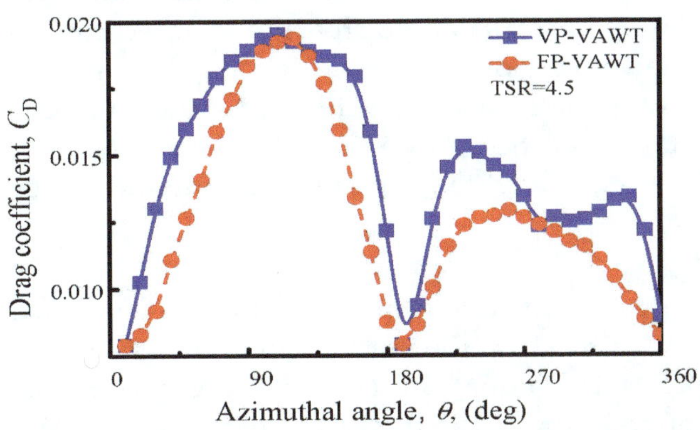

(b)

Figure 11. *Cont.*

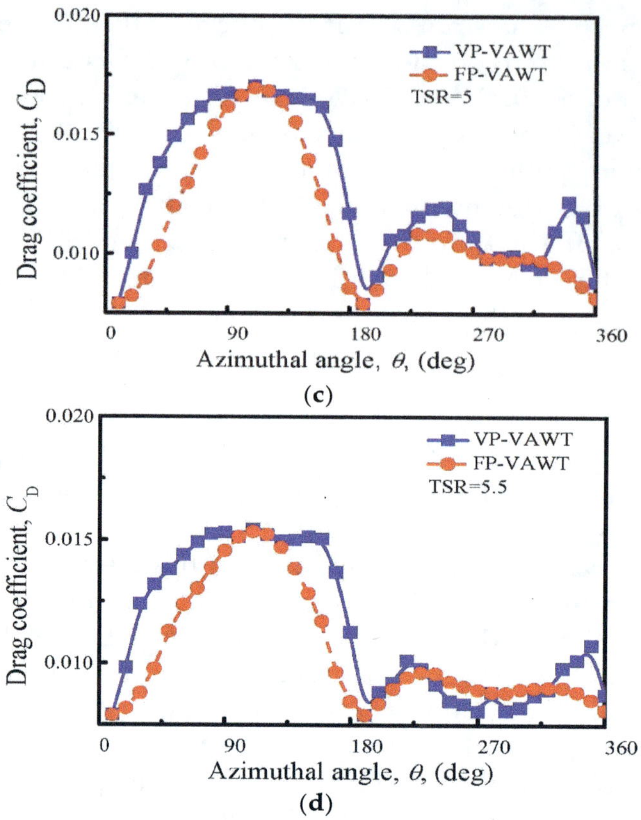

Figure 11. Drag coefficient as a function of the azimuthal angle in VP-VAWTs and FP-VAWTs. (**a**) TSR = 4; (**b**) TSR = 4.5; (**c**) TSR = 5; (**d**) TSR = 5.5.

4.4. Resultant Velocity Variation

The resultant velocity, which is the vector sum of V_W and U, is depicted as a function of the azimuthal angle in Figure 12. Other than the free wind velocity, the resultant velocity is the effective flow velocity acting on the blade, which is another factor, together with the AoA, that determines the magnitude of all of the forces. The magnitude of the resultant velocity is determined by the tangential flow velocity and flow velocity going through the stream tube according to the local velocity triangle. At the same rotation speed as the VP-VAWT and FP-VAWT, in the upwind half-cycle, the pitch angle does not affect the local tangential flow velocity and wind velocity. This causes slight changes in the resultant velocity in the upwind half-cycle, as shown in Figure 12. The slight changes are also observed when the TSR = 4 and 4.5 in the downstream half-cycle, and large changes occur when the TSR = 5 and 5.5 (Figure 12). The VP-approach decreases the wake velocities of the upwind stream tubes, and changes the velocity of the stream flowing through the stream tubes and the axial induction factor of the downwind half-cycle. The changed stream velocity and axial induction factor determine the different resultant velocities. As depicted in Figure 12, when the TSR = 5.5, the resultant velocity has a tendency to be even in a wide zone of the downwind region.

4.5. Torque of Blade Variation

The torque of a single blade as a function of the azimuthal angle is shown in Figure 13, which reflects well the effects of the new VP-approach on the improvement of the aerodynamics of the blade in every azimuthal position. As expressed by Equations (3) and (4), the torque is extensively affected by lift and drag. For the increase of both forces, the final torque in the VP-VAWT significantly differs from that in the FP-VAWT. At a TSR = 4 and the downwind half-cycle, the torque is increased in some azimuthal angles, but it is also reduced in some zones; thus, it is not easy to directly judge whether the overall torque is increased or decreased (Figure 13). In comparison, when the TSR = 4.5, 5, and 5.5, a different extent of increase is obtained on both sides of the 90° azimuthal angle, where one peak torque is generated in the FP-VAWT, and then the low torque zone gets correspondingly narrower. In the downwind region, the large increase is shown when the TSR = 4.5 and 5; the VP-blade creates two new maximum values on the two sides of the azimuthal angle of 270°, where another peak torque of the FP-blade is generated. It should be noted that the former two new maximum values are larger than the latter ones; thus, the curve of the downwind section exhibits an "M" shape. At a TSR = 5.5, the torque is increased in some positions, and decreases in the zone adjacent to the 270° azimuthal angle.

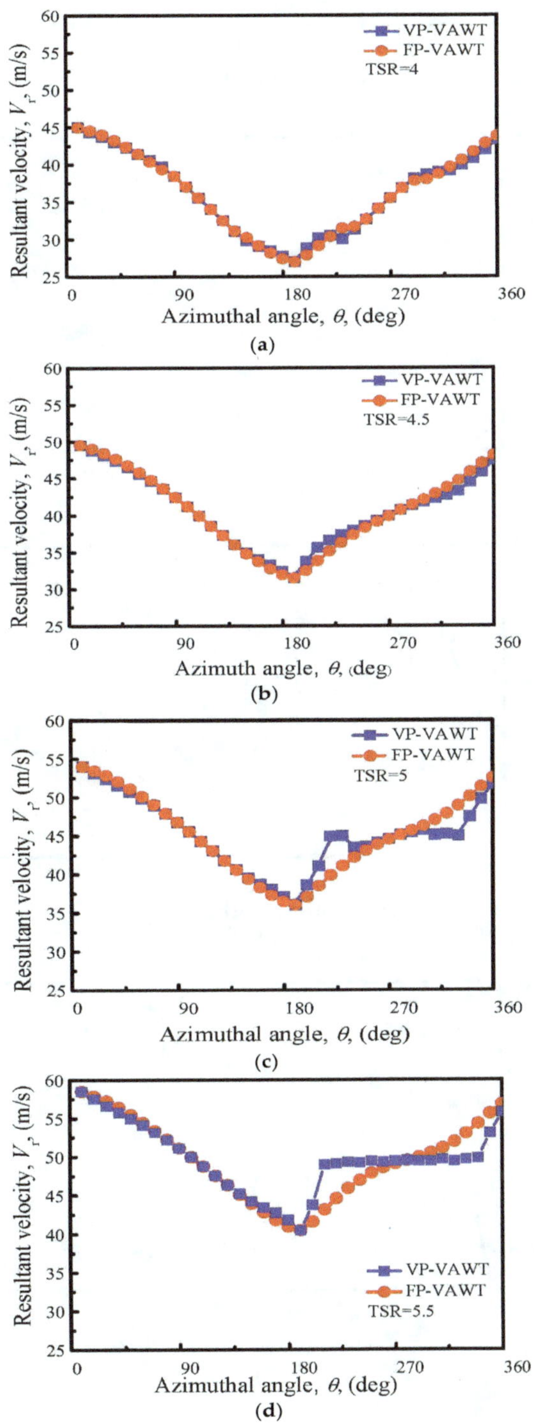

Figure 12. Resultant velocity as a function of the azimuthal angles in VP-VAWTs and FP-VAWTs. (**a**) TSR = 4; (**b**) TSR = 4.5; (**c**) TSR = 5; (**d**) TSR = 5.5.

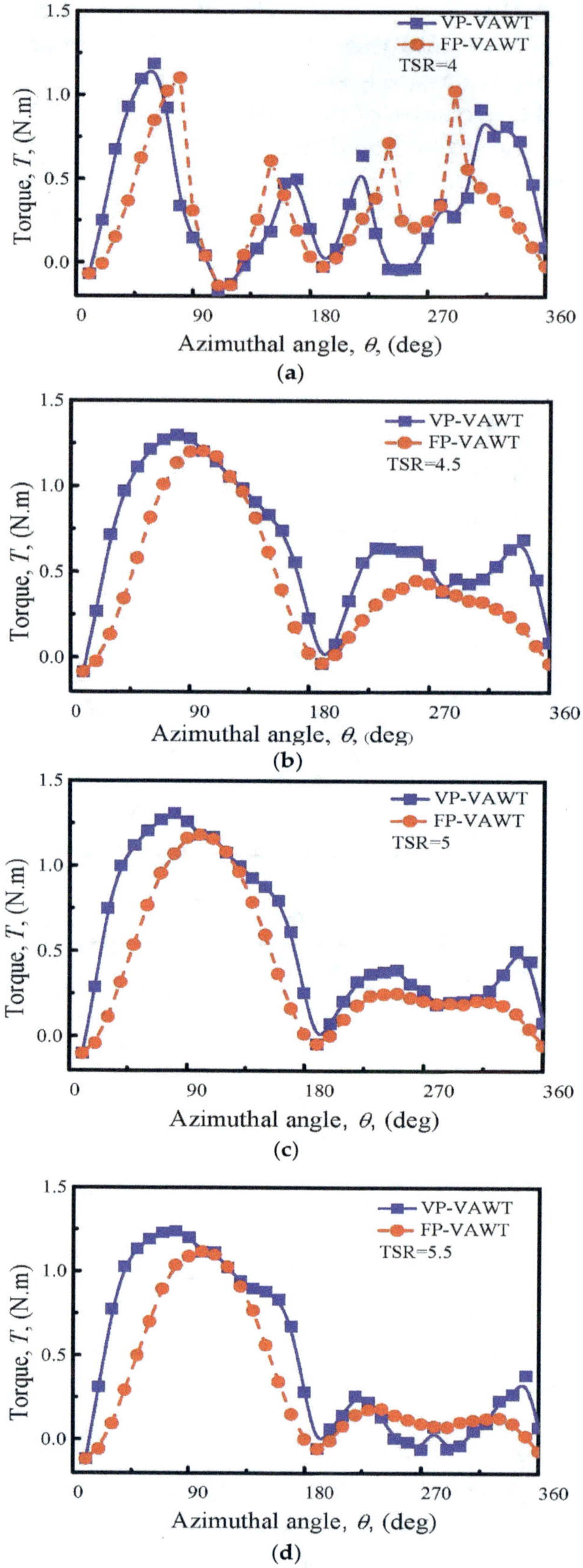

Figure 13. Torque as a function of the azimuthal angles in VP-VAWTs and FP-VAWTs. (**a**) TSR = 4; (**b**) TSR = 4.5; (**c**) TSR = 5; (**d**) TSR = 5.5.

4.6. Power Output

The power output along the radial direction is presented in Figure 14; its value is equal to the sum of the power output of a stream tube in the upwind part and the corresponding downwind part, which can be written as Equation (11). Adding the power outputs of the stream tubes in all of the radial and heights positions, and subsequently dividing the power of the free wind, the total power coefficient is obtained through Equation (12):

$$P_i = P_{up.i} + P_{dw.i} \tag{11}$$

$$C_{P.t} = \sum_{-H}^{H} \sum_{i=-R}^{R} \frac{Pi}{\frac{1}{2} S \rho V_w^3} \tag{12}$$

where C_P denotes the power coefficient; the subscript i is the radial position of the stream tube; *up* and *dw* represent the upwind and downwind region, respectively; t means total; and S is the swept area. As shown in Figure 14, the power output of the FP-blade shows the peak value in the radial position of $r = 0$; toward the two sides, the power is gradually reduced, and reaches zero or even a negative power in the $-R$ or R positions. The whole curve of the FP-blade is a downward parabolic line. In comparison, the new VP-approach extends the position generated at the peak power from one point in the FP-VAWT to a wide radial zone. Therefore, the final distribution of the power output is similar to a trapezoid when the TSR = 4.5 and 5. For a TSR = 5.5, although a drop of the torque occurs downstream, as shown in Figure 13, the VP-blade still experiences an increase in the power output in most radial positions, because of a larger torque increase in the upwind region. The results presented in Figure 14 also show a small drop in the position of $r = 0$, but a larger increase occurs in other positions. Thus, the total power output will be absolutely increased.

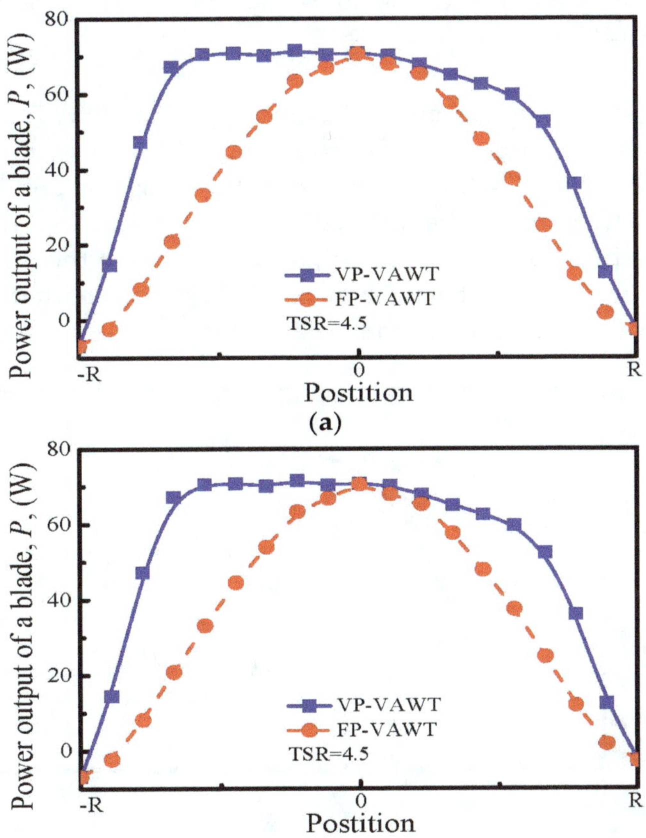

Figure 14. *Cont.*

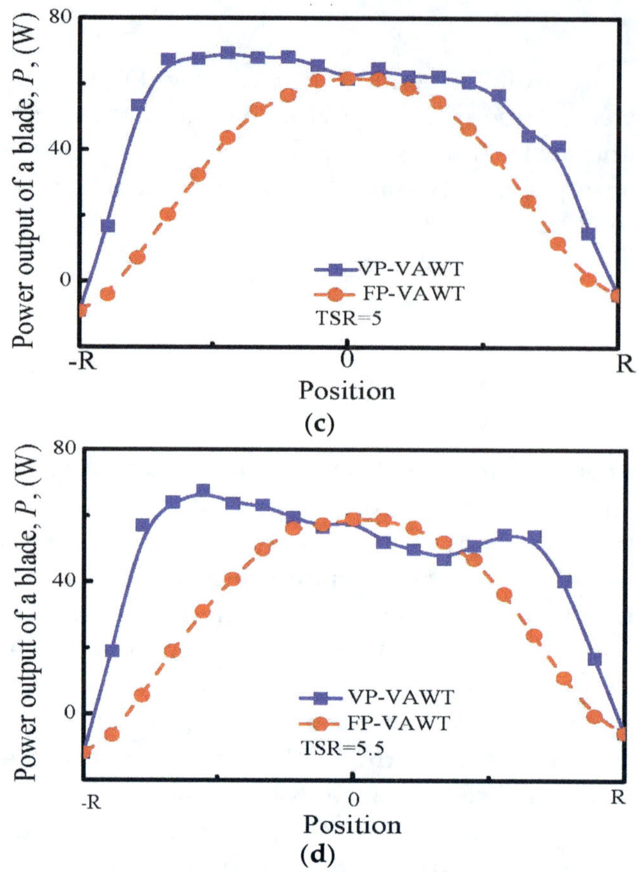

Figure 14. Power output of a blade as a function of position in VP-VAWTs and FP-VAWTs. (**a**) TSR = 4; (**b**) TSR = 4.5; (**c**) TSR = 5; (**d**) TSR = 5.5.

A comparison of the power coefficient of the two types of VAWTs when the TSR = 4 to 5.5 is shown in Figure 15, which reveals the obvious promotion of the power coefficient at a TSR = 4.5, 5, and 5.5. The increase is 16% at a TSR = 4.5, and becomes larger as the TSR increases. Finally, an 18.9% increase of the peak power efficiency of the VP-VAWT is achieved when the TSR = 5. The new curve has a wider TSR zone with a high power coefficient, and the drop in the power coefficient starts when the TSR = 5.5, but at a low rate. The results in Figure 15 clearly indicate that the new approach is highly suitable for improving the peak efficiency of the VAWT at the optimum or larger TSR.

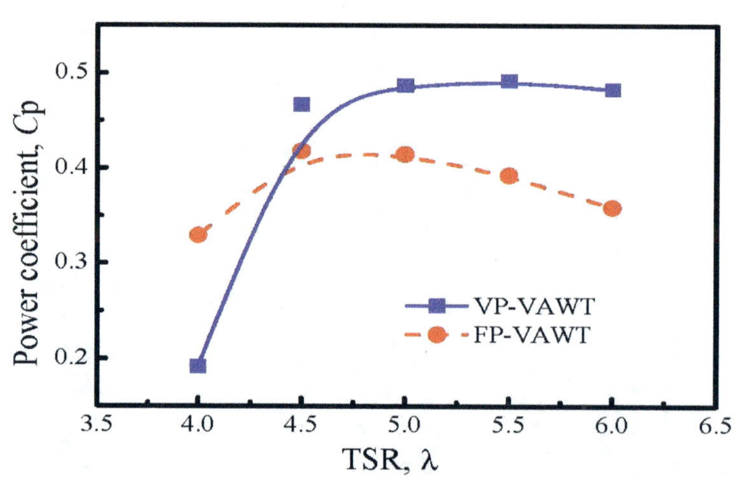

Figure 15. Power coefficient curves of VP-VAWTs and FP-VAWTs.

Since the parasitic drag of the radial arm was not considered in the investigation, the prediction value may be higher than the real one in Figure 15. While the structure of the arm can be well-designed, such as using an airfoil profile and being installed at a zero AoA relative to the free wind, several methods such as this can greatly reduce their parasitic drag. Therefore, the new pitching approach does effectively promote the overall performance of the turbine, even after considering the parasitic drag of the arm.

5. Conclusions

This paper proposed a new VP-approach to increase the peak power coefficient of a straight-blade VAWT. The new approach was designed for the blade at the optimum TSR, and focuses mainly on widening the azimuthal zone of the largest AoA, instead of increasing it. A two-blade H-type VAWT was studied using the DMST model to validate the effect of the new approach, in which the tip loss was evaluated by Prandtl's mathematics. The data of the middle height of the VAWT was adopted to analyze and compare the FP-blade and the VP-blade from six different characteristics, namely the AoA, lift, drag, resultant velocity, torque, and power output. The conclusions drawn from the study are as follows.

(1) Through the application of the new approach, a large increase in the AoA is produced in the upwind half-cycle of the VP-blade. A large increase in the AoA in the downwind half-cycle is also achieved when the TSR = 4, 4.5, and 5, but the increase becomes smaller as the TSR increases, so that a negative growth is shown when the TSR = 5.5.

(2) The new approach greatly enhances the lift of the blade and distributes the maximum lift in a wider zone in the upwind region. In the downwind region, two new and larger maximum lifts are created when the TSR = 4.5 and 5, but a smaller lift is obtained when the TSR = 5.5.

(3) The increased AoA leads to an increase of the drag of the blade in most of the positions. The trend of the distribution of the drag is similar to that of the lift.

(4) The resultant velocity experiences little changes in the upwind region, while in the downwind region, the changes are obvious, and become more obvious as the TSR increases.

(5) Influenced by the lift and drag, the ultimate effect of the torque shows that a large increase is obtained in the upwind region and downwind region at most of the TSRs. The distribution trend is also much like that of the lift.

(6) The new VP-approach also enlarges the azimuthal zone of the blade with the highest power output. Consequently, an 18.9% increase of the peak power efficiency of the VAWT is achieved when the TSR = 4.5. Additionally, the turbines are capable of working with high efficiency in a wider TSR zone. The new VP-approach designed at the rated TSR is suitable to enhance the peak efficiency of VAWTs.

Author Contributions: Z.Z. and T.W. conceived and designed the research; R.W. performed the calculations; R.W. and S.Q. analyzed the data; Y.Z. contributed analysis tools; Z.Z. and B.X. wrote the paper. W.S. revised the paper.

Acknowledgments: This paper was supported by the project of National Natural Science Foundation of China, Project No. 11502070 and 51607058; the Fundamental Research Funds for the Central Universities, Project No. 2018B24914; National Basic Research Program of China ("973" Program), Project No. 2014CB046200.

Nomenclature

Roman Letters

a	induction factor
A_c	amplitude of the pitch angle, [deg]
A_s	Max amplitude of the pitch angle, [deg]
c	length of chord line, [m]

C_D	drag coefficient
C_L	lift coefficient
C_N	normal force coefficient
C_T	tangential force coefficient
C_P	power coefficient
N	number of blade
H	1/2 height of rotor, [m]
R	equatorial radius of rotor, [m]
S	swept area of wind rotor, [m^2]
U	Tangential velocity, [m/s]
V_r	resultant velocity, [m/s]
V_w	free wind velocity, [m/s]
X_0	TSR of max C_P at zero A_c
AoA	angle of attack, [deg]
CFD	computational fluid dynamics
DMST	double multi stream tube (model)
FP	Fixed pitch
HAWT	horizontal axis wind turbine
VAWT	vertical axis wind turbine
TSR	tip speed ratio
VP	variable pitch

Greek Letters

α	angle of attack, [deg]
ρ	density of air, [kg·m^{-3}]
θ	azimuthal angle, [deg]
ω	rotor angular velocity, [rad·s^{-1}]
λ	tip speed ratio (TSR)
γ	pitch angle changed with the azimuth, [deg]
ξ	dimensionless parameter of height

References

1. Kosaku, T.; Sano, M.; Nakatani, K. Optimum pitch control for variable-pitch vertical-axis wind turbines by a single stage model on the momentum theory. In Proceedings of the IEEE Conference on Systems, Man and Cybernetics, Yasmine Hammamet, Tunisia, 6–9 October 2002. [CrossRef]
2. Bhutta, M.M.A.; Hayat, N.; Farooq, A.U.; Ali, Z.; Jamil, S.R.; Hussain, Z. Vertical axis wind turbine–A review of various configurations and design techniques. *Renew. Sustain. Energy Rev.* **2012**, *16*, 1926–1939. [CrossRef]
3. Li, Y.; Zheng, Y.F.; Zhao, S.Y.; Feng, F.; Li, J.Y.; Wang, N.X.; Bai, R.B. A review on aerodynamic characteristics ofstragith-bladed vertical axis wind turbine. *Acta Aerodyn. Sin.* **2017**, *35*, 368–385.
4. Kiwata, T.; Yamada, T.; Kita, T.; Takata, S.; Komatsu, N.; Kimura, S. Performance of a vertical axis wind turbine with variable-pitch straight blades utilizing a linkage mechanism. *J. Environ. Eng.* **2010**, *5*, 213–225. [CrossRef]
5. Aggarwal, A.; Mishra, D.; Chandramaouli, V. Study on optimization of control mechanism in vertical axis wind turbine. *Indian J. Sci. Technol.* **2017**, *10*, 1–10. [CrossRef]
6. Claessens, M.C. The Design and Testing of Airfoils in Small Vertical Axis Wind Turbines. Master's Thesis, Delft University of Technology, Delft, The Netherlands, 2006.
7. Staelens, Y.; Saeed, F.; Paraschivoiu, I. A straight-bladed variable-pitch VAWT concept for improved power generation. AIAA-2003-0524. In Proceedings of the ASME 2003 Wind Energy Symposium, Reno, NV, USA, 6–9 January 2003.
8. Kirke, B.K. Evaluation of Self-Starting Vertical Axis wind Turbines for Stand-Alone Applications. Ph.D. Thesis, Griffith University Gold Coast, Brisbane, Australia, 1998.
9. Cooper, P.; Kennedy, O.C. Development and analysis of a new novel Vertical Axis Wind Turbine. In Proceedings of the 42nd Annual Conference of the Australian and New Zealand Solar Society, Perth, Australia, 30 November–3 December 2004.

10. Beri, H.; Yao, Y. Effect of camber airfoil on self-starting of vertical axis wind turbine. *J. Environ. Sci. Technol.* **2011**, *4*, 302–312. [CrossRef]
11. Zamani, M.; Maghrebi, M.J.; Moshizi, S.A. Numerical study of airfoil thickness effects on the performance of J-shaped straight blade vertical axis wind turbine. *Wind Struct.* **2016**, *22*, 595–616. [CrossRef]
12. Zamani, M.; Maghrebi, M.J.; Varedi, S.R. Starting torque improvement using J-shaped straight-bladed Darrieus vertical axis wind turbine by means of numerical simulation. *Renew. Energy* **2016**, *95*, 109–126. [CrossRef]
13. Sun, X.J.; Chen, Y.J.; Cao, Y.; Wu, G.Q.; Zheng, Z.Q.; Huang, D.G. Research on the aerodynamic characteristics of a lift drag hybrid vertical axis wind turbine. *Adv. Mech. Eng.* **2016**, *8*, 1–11. [CrossRef]
14. Dwiyantoro, B.A.; Suphandani, V. The system design and performance test of hybrid vertical axis wind turbine. *AIP Conf. Proc.* **2017**, *1831*, 020030. [CrossRef]
15. Bianchini, A.; Ferrara, G.; Ferrari, L. Pitch Optimization in small-size Darrieus wind turbines. *Energy* **2015**, *81*, 122–132. [CrossRef]
16. Pawsey, N.C.K. Development and Evaluation of Passive Variable-Pitch Vertical Axis Wind Turbines. Ph.D. Thesis, The University of New South Wales, Sydney, NSW, Australia, 2002.
17. Chougule, P.; Nielsen, S. Overview and Design of self-acting pitch control mechanism for vertical axis wind turbine using multi body simulation approach. *J. Phys.* **2014**, *524*, 012055. [CrossRef]
18. Sagharichi, A.; Maghrebi, M.J.; ArabGolarcheh, A. Variable pitch blades: An approach for improving performance of Darrieuswind turbine. *J. Renew. Sustain. Energy* **2016**, *8*, 053305. [CrossRef]
19. Xisto, C.M.; Páscoa, J.C.; Leger, J.A.; Trancossi, M. Wind energy production using an optimized variable pitch vertical axis rotor. In Proceedings of the Mechanical Engineering Congress and Exposition, Montreal, QC, Canada, 14–20 November 2014.
20. Erickson, D.W.; Wallace, J.J.; Peraire, J. Performance Characterization of Cyclic Blade Pitch Variation on a Vertical Axis Wind Turbine. In Proceedings of the 49th AIAA Aerospace Sciences Meeting including the New Horizons Forum and Aerospace Exposition, Orlando, FL, USA, 4–7 January 2011.
21. Hwang, I.S.; Yun, H.L.; Kim, S.J. Optimization of cycloidal water turbine and the performance improvement by individual blade control. *Appl. Energy* **2009**, *86*, 1532–1540. [CrossRef]
22. Islam, M.; Ting, D.S.K.; Fartaj, A. Aerodynamic models for Darrieus-type straight-bladed vertical axis wind turbines. *Renew. Sustain. Energy Rev.* **2008**, *12*, 1087–1109. [CrossRef]
23. Schonborn, A.; Chantzidakis, M. Development of a hydraulic control mechanism for cyclic pitch marine current turbines. *Renew. Energy* **2007**, *32*, 662–679. [CrossRef]
24. Camporealea, S.M.; Magi, V. Streamtube model for analysis of vertical axis variable pitch turbine for marine currents energy conversion. *Energy Convers. Manag.* **2000**, *41*, 1811–1827. [CrossRef]
25. Zuo, W.; Kang, S. Numerical simulation of aerodynamic performance of H-type wind turbine with pitch angle changing. *J. Eng. Thermophys.* **2015**, *36*, 501–504.
26. Paraschivoiu, I. *Wind Turbine Desigh with Emphasis on Darrieus Concept*; Polytechnic International Press: Montreal, QC, Canada, 2002.
27. Templin, R.J. *Aerodynamic Performance Theory for the NRC Vertical-Axis Wind Turbine*; LTR-LA-160; National Research Council of Canada: Ottawa, ON, Canada, 1974.
28. Strickland, J.H. *Darrieus Turbine: A Performance Prediction Model Using Multiple Streamtubes*; SAND75-0431; Sandia National Laboratories: Livermore, CA, USA, 1975.
29. Paraschivoiu, I.; Delclaux, F. Double multiple streamtube model with recent improvements (for predicting aerodynamic loads and performance of Darrieus vertical axis wind turbines). *J. Energy* **1983**, *7*, 250–255. [CrossRef]
30. Paraschivoiu, I.; Trifu, O.; Saeed, F. H-Darrieus Wind Turbine with Blade Pitch Control. *Int. J. Rotat. Mach.* **2009**, *2009*, 505343. [CrossRef]
31. Soraghan, C.E.; Leithead, W.E.; Yue, H.; Feuchtwang, J. *Double Multiple Streamtube Model for Variable Pitch Vertical Axis Wind Turbines*; American Institute of Aeronautics and Astronautics: Reston, VA, USA, 2013.
32. Saeidi, D.; Sedaghat, A.; Alamdari, P.; Alemrajabi, A.A. Aerodynamic design and economical evaluation of site specific small vertical axis wind turbines. *Appl. Energy* **2013**, *101*, 765–775. [CrossRef]
33. Zhao, Z.Z.; Yan, C.; Wang, T.G.; Xu, B.F.; Zheng, Y. Study on approach of performance improvement of VAWT employing double multiple stream tubes model. *J. Sustain. Renew. Energy* **2017**, *9*, 023305. [CrossRef]

34. Ferrer, E.; Willden, R. Blade-wake interactions in cross-flow turbines. *Int. J. Mar. Energy* **2015**, *11*, 71–83. [CrossRef]
35. Elkhoury, M.; Kiwata, T.; Aoun, E. Experimental and numerical investigation of a three-dimensional vertical-axis wind turbine with variable-pitch. *J. Wind Eng. Ind. Aerodyn.* **2015**, *139*, 111–123. [CrossRef]
36. Firdaus, R.; Kiwata, T.; Kono, T.; Nagao, K. Numerical and experimental studies of a small vertical-axis wind turbine with variable-pitch straight blades. *J. Fluid Sci. Technol.* **2015**, *10*, 14–29. [CrossRef]
37. Sengupta, A.; Biswas, A.; Gupta, R. Studies of some high solidity symmetrical and unsymmetrical blade H-Darrieus rotors with respect to starting characteristics, dynamic performances and flow physics in low wind streams. *Renew. Energy* **2016**, *93*, 536–547. [CrossRef]
38. Sheldahl, R.E.; Klimas, P.C. *Aerodynamic Characteristics of Seven Symmetrical Airfoil Sections through 180-Degree Angle of Attack for Use in Aerodynamic Analysis of Vertical Axis Wind Turbines*; SAND80-2114; Sandia National Laboratories: Livermore, CA, USA, 1981.

8

Wind Turbine Optimization for Minimum Cost of Energy in Low Wind Speed Areas Considering Blade Length and Hub Height

Han Yang, Jin Chen * and Xiaoping Pang

State Key Laboratory of Mechanical Transmissions, Chongqing University, Chongqing 400044, China; yh098@126.com (H.Y.); pangxp@cqu.edu.cn (X.P.)
* Correspondence: chenjin413@cqu.edu.cn

Abstract: In recent years, sites with low annual average wind speeds have begun to be considered for the development of new wind farms. The majority of design methods for a wind turbine operating at low wind speed is to increase the blade length or hub height compared to a wind turbine operating in high wind speed sites. The cost of the rotor and the tower is a considerable portion of the overall wind turbine cost. This study investigates a method to trade-off the blade length and hub height during the wind turbine optimization at low wind speeds. A cost and scaling model is implemented to evaluate the cost of energy. The procedure optimizes the blades' aero-structural performance considering blade length and the hub height simultaneously. The blade element momentum (BEM) code is used to evaluate blade aerodynamic performance and classical laminate theory (CLT) is applied to estimate the stiffness and mass per unit length of each blade section. The particle swarm optimization (PSO) algorithm is applied to determine the optimal wind turbine with the minimum cost of energy (COE). The results show that increasing rotor diameter is less efficient than increasing the hub height for a low wind speed turbine and the COE reduces 16.14% and 17.54% under two design schemes through the optimization.

Keywords: wind turbine optimization; low wind speed areas; cost of energy; particle swarm optimization

---

1. Introduction

Wind energy is renewable and clean, which can help mitigate global climate change. Wind farms with high quality wind resources are limited. The wind farms with low quality wind resources are far more plentiful than high-quality ones and have some advantages such as being closer to the existing electrical grid. The design and development of wind turbines in low wind speed areas faces several technical and financial challenges related to maximizing energy conversion efficiency and minimizing cost of energy (COE). The classical wind turbine literature mainly deals with acquiring ideal blade geometry or structural design and improving structural properties of the tower. The research objective mainly aims to maximize annual energy production (AEP), minimize COE, minimize blades mass, or a combination of these.

Each wind turbine is designed for specific wind conditions. The IEC 61400-1 standard [1] defines wind classes according to wind speed. Currently, the Classes III, II and I of wind turbines correspond to low, medium and high wind speed locations, respectively. In the global wind energy market, wind turbines installed in high wind speed sites (Class I) have progressively lost market share for the past few years in favor of wind turbines in low wind speed locations (Class III). The Asian wind energy market has been dominated by low wind speed wind turbines during the last decade mainly due to the lower quality wind resources in most places in China and India [2].

There has been a moderate amount of literature regarding low wind speed wind turbines, mainly focusing on: low wind resources [3], blade aerodynamic performance [4,5] or structural performance [6,7], tower structural design or analysis [8,9]. In general, the rotor of a wind turbine designed for low wind speed areas has a higher aspect ratio with longer blades, which aims to acquire more wind power and reduce the cost. The costs of larger rotor diameter and higher tower height have a strong impact on the total cost of the wind turbine system. With the additional loads acting on a tower's top owing to the increasing thrust when increasing the blade length and hub height to attract more wind power, the initial capital cost of a wind turbine will also increase. Thus, the wind turbine optimization design should trade-off the hub height and rotor diameter, especially in low wind speed sites. As previously described, because of a prominent increase in the low wind speed wind turbine market, wind turbine manufacturers have invested considerable effort to design and develop wind turbine blades operating in low wind speed sites. Hence, with the wind turbine scale increased, reducing the cost of energy is important. The objective of the optimization model by Xudong et al. [4] is the minimum cost of energy. The design variables in their study are the blade geometric parameters including chord, twist and thickness distribution with fixed rotor diameter. In the present work, the entire system of a wind turbine is considered.

The literature about wind turbines for low wind speed areas is very limited. Some studies work on the shape optimization of low wind speed wind turbine blades, but the object models are always the micro wind turbine [10–13]. Some low wind speed blade designs in the literature [5,6] focus on large scale wind turbines. The work in [5] focuses on aerodynamic performance analysis and AEP enhancement of 3 MW wind turbine blade for low wind speed areas. The optimization objective is to improve AEP and to minimize the thrust based on blade element momentum (BEM) theory. The design variables such as blade length and rated wind speed are selected by existing wind turbine blade groups. The work in [6] proposes a method to quantitatively compare wind turbine operating in wind Class I and wind Class III. Their results show that the traditional design method for low wind speed blade structures is not efficient, and the research is used to improve the design process for low wind speed blades.

The minimized cost of energy is acquired by altering the wind turbine system parameters in the study [14]. They use a multi-level optimization approach to minimize the cost of energy by maintaining AEP and reducing blade root loads. However, their optimization only considers the geometric shape and structure of the blade, and the costs of other components are estimated by the cost and scaling model in reference [15]. Because the design model selects a 1 MW wind turbine for Class I wind resources, the variations of rotor diameter and hub height have less influence on the total system. The optimization process only considers the hub height to evaluate the tower mass by scaling with the product of the swept area and hub height. This cannot ensure accuracy when the tower has a height much greater than 80 m and the wind turbine operates in low wind speed areas [15]. Bortolotti [16] is concerned with the holistic optimization of wind turbines. A multi-disciplinary optimization procedure considering rotor diameter and tower height is presented. The methodology is applied to a commercial 2.2 MW onshore and a conceptual 10 MW offshore wind turbine used in high wind speed locations.

To our knowledge, no studies have focused on a wind turbine model for low wind speed areas with a full scale system taken into account. In this study, the main work is to reduce the COE with rotor and tower integrated optimization. An aero-structural design optimization for a commercial 2.1 MW wind turbine for low wind speed locations is performed. The BEM theory is adopted to evaluate the aerodynamic performance and the classical laminate heory (CLT) is used to estimate the structural performance for the wind turbine blade. The aerodynamic loads acting on blade and tower would change according to the rotor diameter and hub height. The COE model comprises the overall wind turbine system and is a function of rotor diameter and tower height. Most researchers employ the model proposed in reference [15], but this reference indicates that when the tower is much greater than 80 m, the scaling relationships should be used carefully because the cost of the tower will have a major impact on design. In addition, in order to estimate an accurate cost of the tower with much greater height, the matched tower of the commercial wind turbine is used as a reference.

2. Wind Turbine Model

A commercial 2.1 MW HAWT for low wind speed areas is selected as a case study to verify the efficacy of the design method. The length of the blade is 59.8 m and its mass is 13,240 kg. The corresponding tower is 95 m in height and the hub is considered as a concentrated mass. The blade is made of glass fiber reinforced plastic (GFRP). The core materials consist of balsa wood and polyvinyl chloride (PVC). The blade baseline is described by the chord distributions, twist distributions and selected airfoils distributions. Details of each parameter are shown in Tables 1 and 2. The geometric and material properties of the wind turbine tower are presented in Table 3. The material of the tower consists of Q345 steel, and the steel density is 7820 kg/m^3.

Table 1. Baseline blade geometry.

Span Location (m)	Chord (m)	Twist (°)	Airfoil	Thickness (%)
0	2.40	18.0	circle	100
1.0	2.40	18.0	circle	100
13.4	3.42	9.6	DU00-W2-401	40
16.2	3.17	6.9	DU00-W2-350	35
24.9	2.36	2.7	DU97-W-300	30
49.8	1.02	−1.56	DU91-W2-250	25
59.8	0.02	−1.6	DU93-W-210	21

Table 2. Wind turbine characteristics.

Rated power	2.1 MW
Rated wind speed	9 m/s
Cut in wind speed	3 m/s
Cut out wind speed	20 m/s
Number of blades	3
Design tip speed ratio	10.6
Rotor diameter	122.3m
Control type	Variable speed—variable pitch
Maximum power coefficient	0.483

Table 3. Geometric and material properties of the tower.

Tower height (m)	95
Tower top outer diameter (m)	3.15
Tower base outer diameter (m)	5.72
Tower top wall thickness (mm)	24
Tower base wall thickness (mm)	40
Density (kg/m^3)	7820
Young's modulus (GPa)	200
Poisson ratio	0.3
Yield strength (MPa)	345

2.1. Materials

The main material of the blade is GFRP, consisting of leading/trailing edge reinforcement, panel sections, two shear webs and a spar cap. The spar cap is made of unidirectional glass fiber composite materials to withstand the aerodynamic loads flap-wise. The shear webs are composed of sandwich panels to withstand the shear force. The core materials comprise of glass fiber composite materials with balsa wood and PVC. The material properties (E_1 is the longitudinal modulus, E_2 is the transverse modulus, G_{12} is the shear modulus, v_{12} is the Poisson's ratio and the thickness is the laminate thickness of each ply) are listed in Table 4.

Table 4. The material properties.

Stack ID	Materials	E1 (GPa)	E2 (GPa)	G12 (GPa)	v12	ρ (kg/m^3)	Thickness (mm)
1	Unidirectional	39.18	11.69	3.5	0.3	1940	0.892
2	Bi-axial	11.5	11.5	10.8	0.48	1970	0.555
3	Triax	25.8	13.59	7.36	0.36	1842	0.906
4	Gelcoat	3.44	3.44	1.38	0.3	1235	0.6
5	Balsa	1.321	0.03	0.006	0.39	151	25.4
6	Polyvinyl chloride (PVC)	0.035	0.035	0.022	0.3	60	5

2.2. Original Blade Layup Schedule

Figure 1 shows the distributions of the layup layers along the blade span-wise. The first two meters of span is circumferential reinforcement for the blade root. The main material for this section is triaxial composite materials with large layers. The first 13 m of span is inner and outer surface reinforcement for the blade root. Sections outboard of 5 m have thick caps and trailing edge layup by unidirectional materials. The airfoil with thickness of 40% is set at the blade span location of 13.4 m. The root region of the blade experiences the highest bending moments, and cylindrical reinforcement is used to withstand the moments. Thus, the thickness of cylindrical reinforcement is also considered as a design variable in the optimization process. The structural design variables are set as layup layers for spar caps and trailing edge reinforcement. The configuration of the blade section is also considered. The details of the design variables are described in Section 4.1.

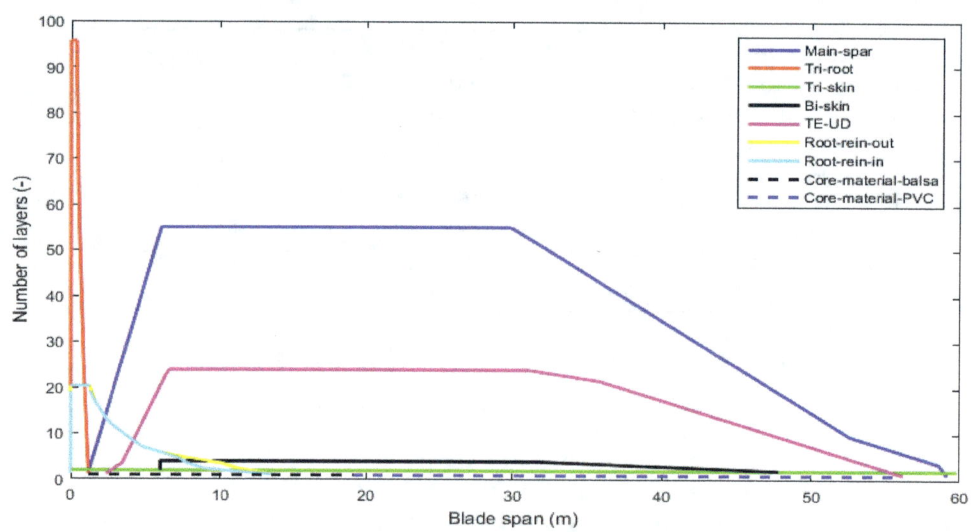

Figure 1. Distribution of section's layup layers along the blade radial coordinate.

2.3. Cost of Energy (COE) Model

The study [17] has shown the design objective of maximizing annual energy production or using sequential aerodynamic and structural optimization to be significantly suboptimal compared to the aero-structural integrated methods. For variable rotor diameter and hub height, the optimal rotor diameter may be misleading because the tower mass dominates the total mass of turbine. Thus, minimizing COE is a much better metric for the full-scale wind turbine optimization objective than minimizing the ratio of turbine mass to annual energy production. The main objective is to minimize the COE of a wind turbine used in low wind speed areas. Following the study [15], COE is calculated as:

$$\text{COE} = \frac{\text{FCR} \times \text{ICC} + \text{AOE}}{\text{AEP}} \tag{1}$$

where FCR is the fixed charge rate, ICC is the initial capital cost, AOE is the annual operating expenses, and AEP is annual energy production.

As wind turbines become more sophisticated in terms of aerodynamic shape and structural configuration with increasing size, the design of each wind turbine component is not always clear. The COE model is used to assess the true impact of the technical change and various designs. The FCR, ICC and AOE are obtained from the National Renewable Energy Laboratory (NREL) wind turbine design cost and scaling model [15]. The BEM code and the Weibull probability distribution with annual average wind speed of 6 m/s at hub height are used to calculate AEP. The blade stiffness and mass distributions are computed by the classical laminate theory code named PreComp [18]. The total cost of a wind turbine is defined by adding the value of the initial capital cost multiplied by the fixed charge rate to the annual operating expenses. The fixed charge rate is the annual amount per dollar of initial capital cost used to cover the capital cost and other fixed charges. The FCR value is set to 0.1158 per year as suggested in reference [15].

3. Theoretical Models and Verifications

3.1. Aerodynamic Model

In this work, the combination of blade element theory and momentum theory is used to estimate the aerodynamic performance and calculate the AEP. The blade element theory aims to calculate the aerodynamic force acting on each blade element. The blade momentum theory refers to analysis of both axial and tangential induced velocity by introducing axial and angular induction factors. The induced velocity in each blade section influences the angle of attack of the airfoil and therefore affects the loads acting on the blade sections. With an iterative procedure, BEM theory provides a method to calculate the aerodynamic loads on each blade element. For a rotor with a finite number of blades, the vortex system in the wake is different from that of a rotor with an infinite number of blades [19]. By introducing the Prandtl's tip loss factor, the BEM theory can simulate a wind turbine rotor with a finite blade number.

The aerodynamic performance of the total wind turbine blade can be obtained from each blade section. Thus, it is necessary to acquire the lift and drag coefficient of the adopted airfoil. However, the experimental polar of airfoils are not often available. In this work, RFoil [20] is used to estimate the aerodynamic characteristics of the airfoil. The RFoil code is an improved version of XFoil [21] and has many advantages such as better corrections after stall.

The annual average wind speed is the main source of wind power. Due to the friction of the plants and constructions, better wind resources are located at higher heights. The friction coefficient n could affect the average wind speed at hub height, and the different average wind speed at hub height could have an impact on the AEP of wind turbines. In this work, the friction coefficient is set as 0.23 which corresponds to the roughness of the countryside. The wind speed is varied depending on the hub height according to the following equation:

$$V_{hub} = V_{ref} \left(\frac{H_{hub}}{H_{ref}} \right)^{n} \tag{2}$$

where V_{hub} is the wind speed at the hub; H_{hub} is the hub height; V_{ref} is the reference wind speed at the reference height; H_{ref} is the reference height; and n is the friction coefficient and set as 0.23 for low wind speed condition.

3.2. Blade Element Momentum (BEM) Code Validation

In this work, a commercial wind turbine for Class III with blade length 59.8 m and 2.1 MW of rated power is used as a reference. The aerodynamic performance of the blade is calculated by considering a tip speed ratio of 10.6 with a range of wind velocities between 3 m/s and 20 m/s. Air density is set to 1.225 kg/m^3 and the cone and tilt angle of the wind turbine are set to zero degrees. The RFoil is used to calculate the aerodynamic characteristics of Delft University (DU) series airfoils for a specific Reynolds number according to their locations. The power coefficient and the power curves calculated by BEM code are implemented in MATLAB software (MATLAB 2014b; The MathWorks Inc.: Natick, MA, USA, 2014). In order to validate the BEM code, the aerodynamic performance data from the wind turbine manufacturer, based on simulations by WINDnovation Engineering Solutions GmbH, are compared with the BEM code. The numerical simulations and the manufacturer's data show an excellent agreement as shown in Figure 2. The power curve calculated by BEM code shows that the rated power is achieved with wind velocity of 9 m/s. Therefore, the aerodynamic performance prediction of the wind turbine blade calculated by BEM code for this study can be considered acceptable.

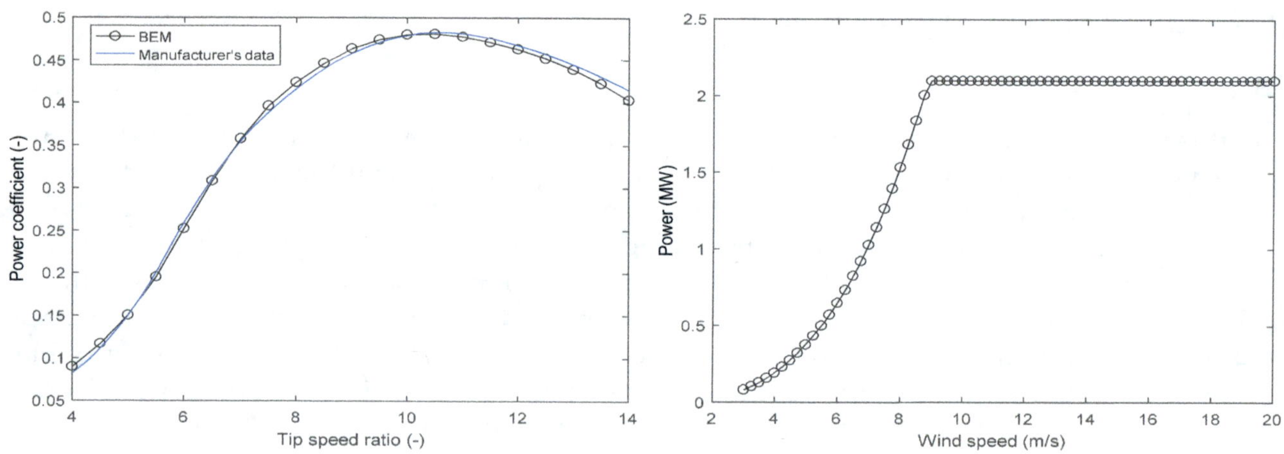

Figure 2. Comparison between the power coefficient and power curves obtained by blade element momentum (BEM) code and wind turbine manufacturer's data.

3.3. Structural Model

The structural performance can be represented by the cross-sectional composite properties of each blade section. In order to evaluate the cross-sectional composite properties of the blade section, classical lamination theory [22] is applied during the optimization procedure. For the laminated plates, the CLT model is an extension of classical plate theory [23] which has improvements in reasonable validation and computational efficiency. Thus, the CLT model can confirm the overall effective performance of composite laminated structure of wind turbine blade. Bir et al. [18] developed a FORTRAN code named PreComp (pre-processor for computing composite blade properties) to calculate the blade structural properties based on CLT. PreComp requires input files of the blade geometric shape, interior structural layout and materials properties. In addition, PreComp allows various internal geometry configurations and a general layup of composite laminates. The CLT model can transfer a complex three-dimensional problem into calculating two-dimensional blade section properties. In this paper, PreComp is applied to evaluate the structural properties of wind turbine blade sections. The structural parameters like blade geometry and layup schedule are set the same as the actual blade. After that, the stiffness and the mass per unit length distributions can be calculated. Figure 3 shows that the calculated stiffness values are in good agreement with the manufacturer's data based on simulations by WINDnovation Engineering Solutions GmbH.

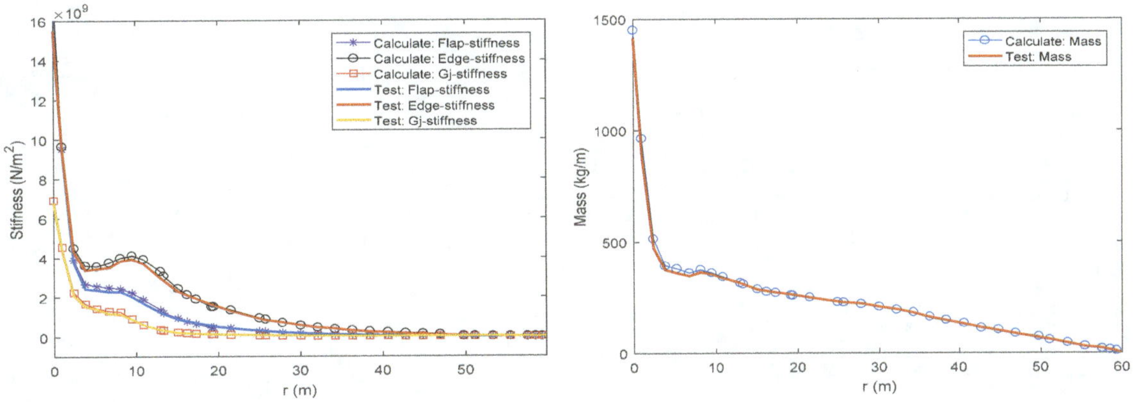

Figure 3. Comparison between the stiffness and mass per unit length distributions obtained by CLT code and actual blade.

4. The Optimization Model

In general, according to the wind turbine blades used in Class I, the same turbine family used in Class III has a length of 130~140% [6]. When the length of the blades increases, the blade mass and tip deflections will also increase. For structural and aerodynamic performance, the fitness function f in this paper is to minimize the cost of energy:

$$\text{maximize} \quad f = 1/\text{COE} \tag{3}$$

4.1. Design Variables

The geometric shape of the blade can be generated by chord distribution, twist distribution and the airfoil series. Eight control points are used for the thickness distribution, and the upper and lower bounds of the control points are set with adequate space, as shown in Figure 4. Nine control points are used for the chord and twist distributions, as shown in Figure 5. The first control point of chord distributions is at the root, the third control point is at the maximum chord station, and the ninth control point is at the blade tip. The thickness distribution is defined with a 9th order Bezier curve, and the chord and twist distributions are defined with a 10th order Bezier curve. For large scale wind turbine, a lower tip speed ratio would result in less efficient aerodynamic performance but lower cost of fabrication because of less aerodynamic loads by lower blade centrifugal stresses. The lower tip speed ratio can also reduce the tower cost because of the lower aerodynamic loads from the rotor. Thus, the tip speed ratio is set as a design variable in the design process.

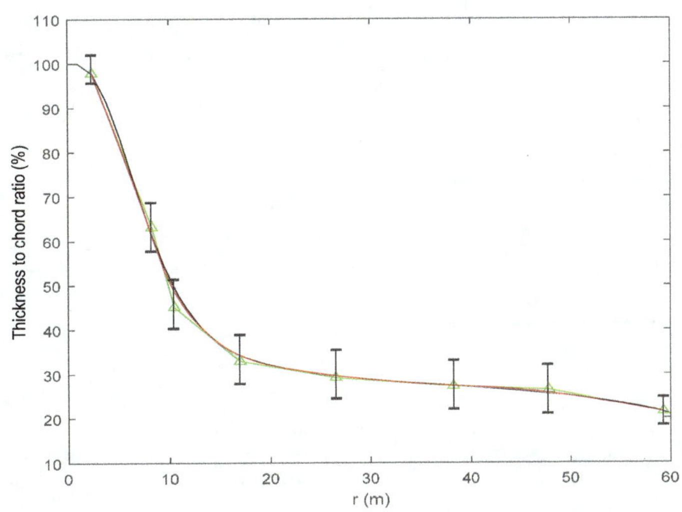

Figure 4. Thickness distribution of the blade.

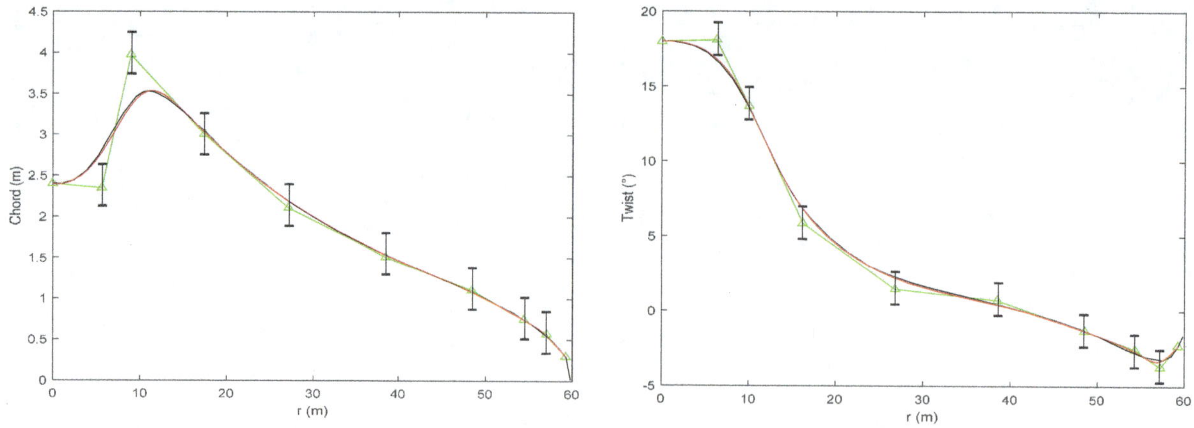

Figure 5. Chord and twist distributions of the blade.

The typical airfoil and its internal structural configuration of a typical modern wind turbine blade are shown in Figure 6 (c is the chord length). The internal structure mainly consists of spar cap and shear webs. The spar caps at the maximum thickness resist the flap-wise loads and the two shear webs resist the torsional loads. In this work, the maximum thicknesses of the spar cap and trailing edge reinforcement are the design variables for the blade layup schedule which are shown in Figure 7. The shear webs distributions are also considered in the structural optimization design. The locations of each web at blade root and tip are set as design variables.

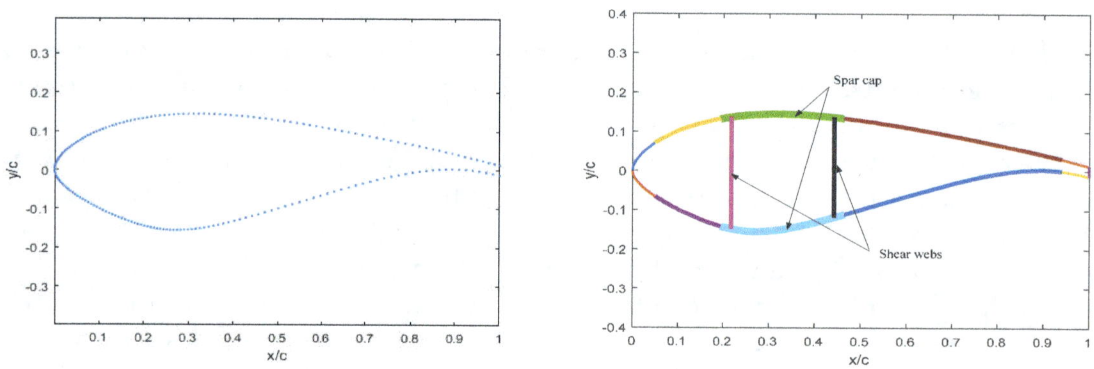

Figure 6. Typical structural parameters for section of the blade.

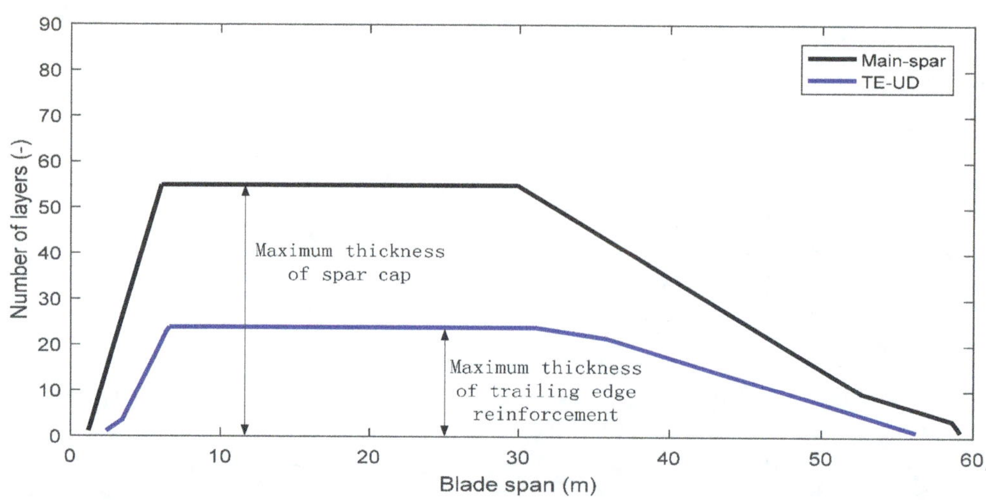

Figure 7. Design variables for spar cap and trailing edge reinforcement.

Figure 8 presents the schematic of the tower structure. With increasing tower height, the optimization design needs to trade off between the energy production and the cost of the structure. The wind turbines used in low wind speed areas always have higher hub height to obtain better wind sources than those used in high wind speed areas. The tower heights are always much greater than 80 m in low wind speed areas. Thus, the relationship of the tower mass and cost scaling are not efficient [15]. Therefore, an actual 2.1 MW commercial tower is used in an optimization process instead of using an empirical equation for the tower cost. Foundations were scaled as a function of hub height and rotor swept area, which is directly proportional to the tower overturning moment. The top and base diameters of the tower and their thickness are taken as design variables. The variations of diameters and thickness across the length of the tower are assumed as a linear distribution.

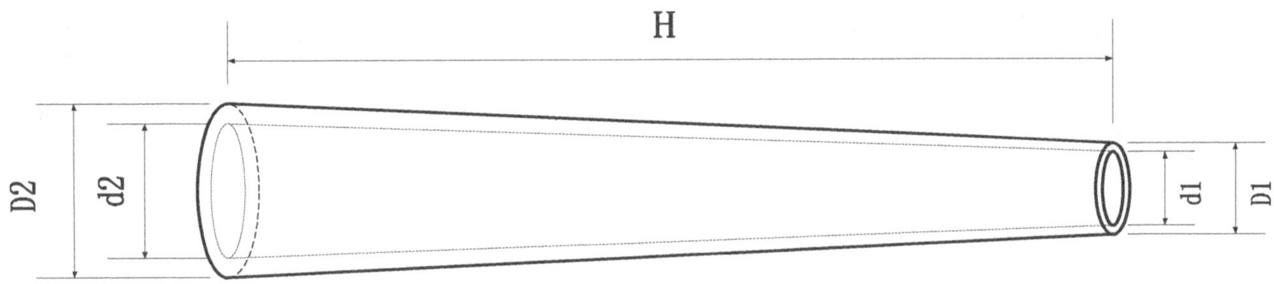

Figure 8. Schematic of tower structure.

4.2. Design Constraints

A real wind turbine must be designed to meet plenty of constraints. For the aerodynamic design, the blade root bending moment is always [14,17] set as a constraint for the blade design individually, but it is unnecessary for whole wind turbine design. The full scale design considers the whole structural stability instead of local aerodynamic performance. In this work, representative cases are used to meet the structural constraints. The wind turbine needs to meet the criteria of the structural strength in the process of operation. The ultimate strength analysis of the blade and the tower is performed in an extreme load condition. The 50-year extreme wind condition is defined as $V_{e50} = 1.4V_{ref}$ ($V_{ref} = 37.5$ m/s for Class III wind resources).

4.2.1. Blade Design Constraints

In order to verify the structural stability of the blade, the structural design needs to withstand the extreme wind condition. Some representative sections are selected to calculate the strain and verify the strength. These sections are at the blade root and the airfoils' locations along the blade span on both the upper and lower surfaces. A maximum strain condition is used as [17]:

$$\varepsilon_{50} \leq \frac{1}{\gamma_f \gamma_m}\varepsilon_{ult} \tag{4}$$

where the ε_{50} is the strain at 50-year extreme wind condition, the ε_{ult} is ultimate strain, the partial safety for loads (γ_f) is set as 1.35 and the partial safety factor for materials (γ_m) is set as 1.1 according to IEC 61400-1 requirement. The representative numbers of ultimate strain is set as 0.03. The maximum tip deflection of the blade under extreme loading is added to ensure adequate stiffness. In this work, considering the blade length is set as a variable, the maximum tip deflections with various blade lengths are set as:

$$\delta_{opt} \leq 1.05 \cdot \delta_{orig} \cdot \frac{R_{opt}}{R_{orig}} \tag{5}$$

where δ_{opt} is the maximum tip deflection of the optimal blade, the δ_{orig} is the maximum tip deflection of the original blade, R_{opt} is the length of the optimal blade, and R_{orig} is the length of the original blade.

4.2.2. Tower Design Constraints

The tower is the main support component of the wind turbine to withstand the aerodynamic loads and the total mass of the other components. The structural properties of the tower have a significant impact on the total cost of the wind turbine. An efficient, safe and economic design of the tower that meets all the design criteria is needed to minimize the total wind turbine cost. The loads acting on the tower mainly come from the aerodynamic loads on the rotor, and the wind loads on tower itself are also considered. The thrust force on the rotor is transferred to the tower top and causes tower deformation. In order to ensure structural stability and avoid large deformation, the maximum deformation at the top of the tower should not exceed the allowable deformation:

$$d_{\max} \leq d_{allow} \tag{6}$$

The d_{allow} is the allowable deformation and it can be determined according to the reference [18]. In order to ensure the tower has an adequate structural performance during optimization process, the allowable deformation is set as [8]:

$$d_{allow} \leq \frac{H_{opt}}{H_{orig}} d_{orig} \leq 1.25 \frac{H_{opt}}{100} \tag{7}$$

where H_{opt} is the height of the optimal tower, H_{orig} is the height of the original tower, and d_{orig} is the top deformation of the original tower.

In order to verify the structural stability of the tower, the structural design needs to withstand thrust from the wind turbine rotor. For a tapered tower, the location of the critical stress is at about the middle of the tower (the bending moment is largest at the base, but so is the moment of inertia) [17]. In this work, we assumed that the critical location was at the middle of the tower. Then, the stress at that location can be calculated as:

$$\sigma_{cr} = \frac{(m_{RNA} + m_{tower\_mid\_up})g}{A_{mid}} + \frac{THr_{mid}}{2I_{mid}} \tag{8}$$

where σ_{cr} is the critical stress, m_{RNA} is the mass of rotor and nacelle assembly, $m_{tower\_mid\_up}$ is the tower mass above the middle of the tower, A_{mid} is the area of the middle of the tower, T is the rotor thrust, H is the tower height, and I_{mid} is the area moment of inertia at the middle of the tower.

The stress σ_{cr} generated by the loads cannot exceed the allowable stress:

$$\sigma_{cr} < \sigma_{allow} \tag{9}$$

The σ_{allow} is the allowable stress and the value of it is given by:

$$\sigma_{allow} = \sigma_{ys} / \gamma_{tm} \tag{10}$$

where σ_{ys} is the yield strength and γ_{tm} is the material safety factor of the steel. The yield strength of the steel is 345 MPa for the wall thickness between 16 mm and 40 mm. The material safety factor is set as 1.1 according to IEC 61400-1.

4.3. The Optimization Algorithm

The purpose of this paper is to investigate the cost of energy for a low wind speed turbine considering rotor diameter and hub height as variables. The optimization progress aims to find the ideal blade shape parameters such as chord, twist, thickness distributions and the blade length, and to find the structural performance like webs locations, layer-up distributions and the tower parameters (height, taper and wall thickness distributions). The design process interfaces the particle swarm optimization (PSO) algorithm, BEM code, PreComp code and the COE model. There are three modules integrated in the PSO algorithm by using MATLAB: (1) the aerodynamic performance analysis based on BEM theory; (2) the structural performance analysis by PreComp code; and (3) the design constraints by beam theory and strength theory. The basic parameters of the PSO algorithm are defined as: the population size ($p = 80$), maximum number of generations ($g = 200$), learning factors ($c_1 = c_2 = 0.5$), variable dimensions ($n = n_{aerodynamic} + n_{structural} = 36$), and the inertia factor (0.85).

Figure 9 shows the flowchart of the optimization process. The aerodynamic and structural variables for blades and the tower are randomly generated as the initial population. The aerodynamic variables are used to control the blade shape, the structural variables are used to control the webs locations and number of layers along the blade, and the variables for the tower are used to control the tower height and structural parameters. BEM theory is applied to calculate the annual energy production and the aerodynamic loads which will cause the blade deflection and tower deformation. Moreover, the PreComp code is adopted to evaluate the mass and stiffness of the blade sections. The requisite parameters are airfoil profiles, chord length and twist angle, materials characteristics, internal configurations of blade sections and the lay-up distributions. When a wind turbine is acquired during the optimization process, the blade and tower constraints are validated. By analyzing the results, a new wind turbine design with minimum COE will be outputted.

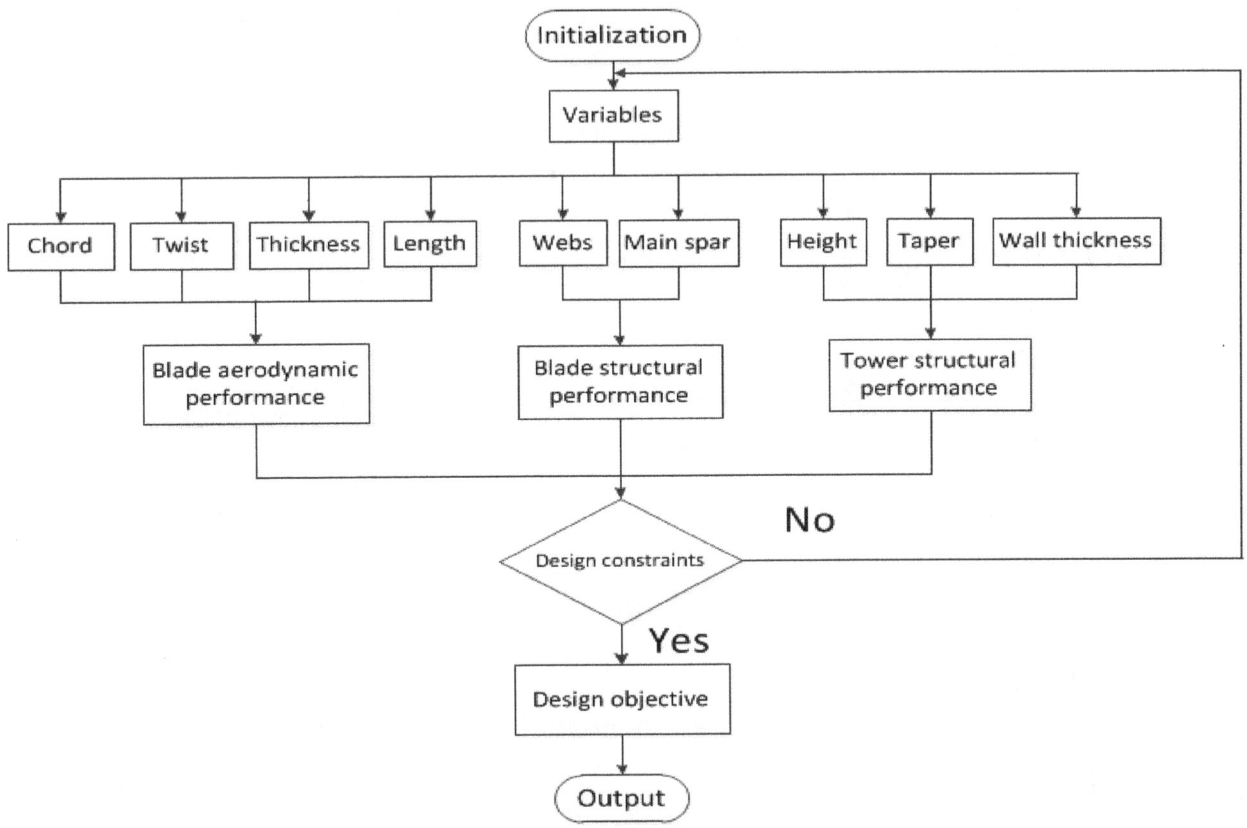

Figure 9. Flowchart of the optimization process.

5. Results and Discussion

In this section, the results are presented and analyzed. Most of the components are a function of the parameters such as rotor diameter and rated power, as shown in Figure 10. The relationship shows that the rated power also has a strong impact on COE. Thus, the optimization procedures considering rated power or not (set as 2.1 MW) are shown.

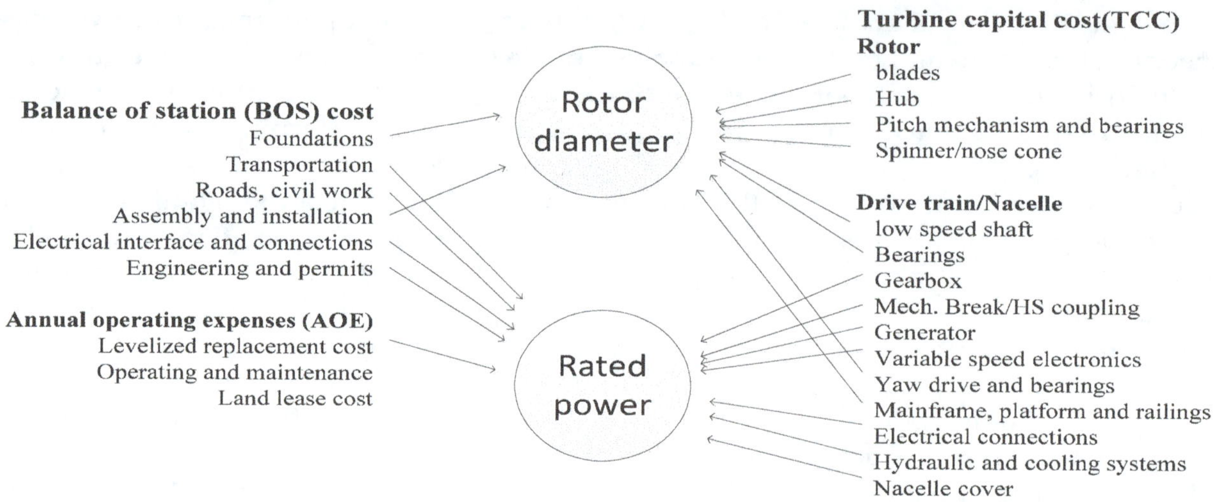

Figure 10. Relationship of rotor diameter and rated power to the components of wind turbine.

After the design procedure is finished, the iterative course of the optimization process by considering rated power is shown in Figure 11. It can be seen that the optimal results converge when the number of iteration steps reach 80.

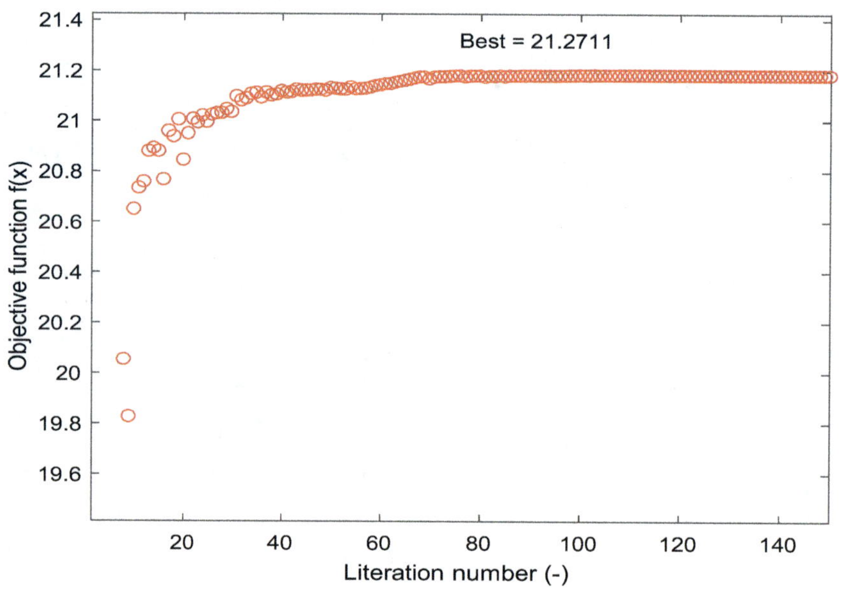

Figure 11. The iterative course for the optimization process considering rated power.

5.1. Blade Optimization Results

The chord and twist distributions of the new blade acquired by the optimization process are shown in Figure 12. The optimization process considering rated power is referred to as "CRP" for brevity. The chord distributions of the new blades decrease slightly, and the optimal chord distribution considering rated power is less than that without. The reason is that the optimal blade length acquired from the design considering rated power is less than that of the original blade and the optimal one without considering the rated power. Due to differences in aerodynamic loads, the structural requirements of the shorter blade are less than those of the longer one. The reduction of the chord distribution can result in a lighter blade mass and a lower thrust at the blade root. The twist distributions of the new blade are almost the same as the original one, the main reason is that the airfoil series adopted in this work are fixed, and the aerodynamic characteristics of the airfoils are the same. Another reason is that the twist distribution does not have a significant impact on COE compared to the chord distribution. The thickness distributions of the new blade acquired by the optimization process are shown in Figure 13. The main reason why the thickness distributions do not change much could be that the lengths of the optimal blades do not change significantly. In addition, when the thickness reduces, the stiffness would reduce too. In order to meet the design constraints, the thickness distributions do not change much.

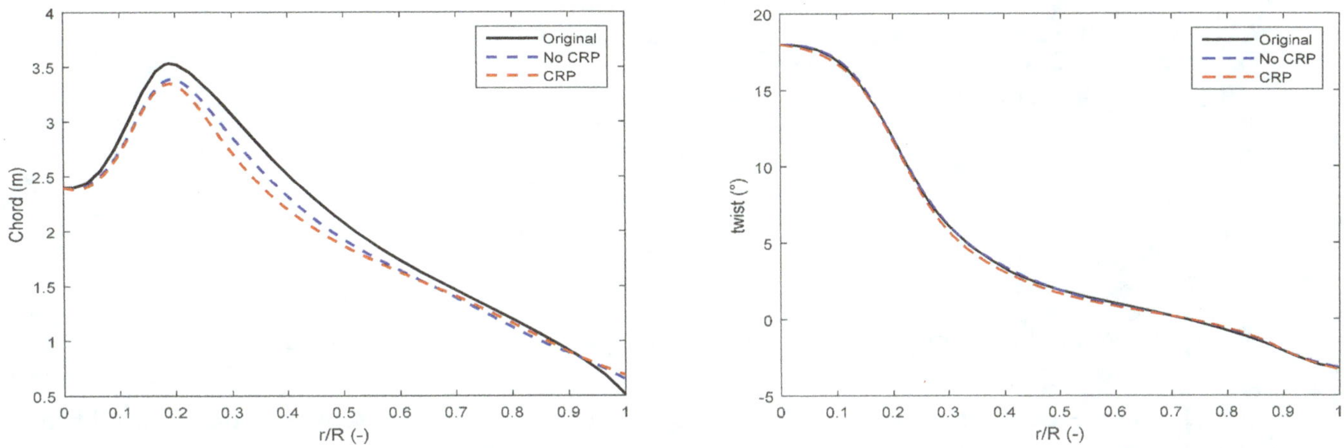

Figure 12. The comparison of chord and twist distributions between the optimal blades and the original one.

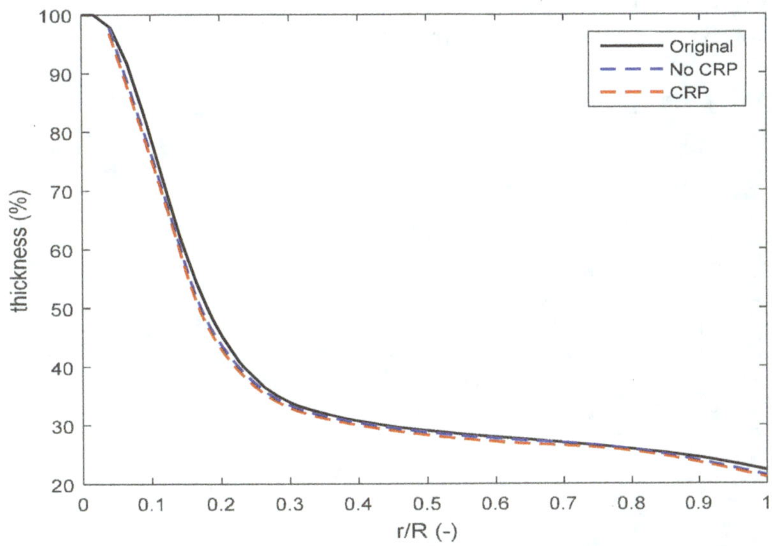

Figure 13. The comparison of thickness distributions between the optimal blades and the original one.

The Weibull distributions with annual average wind speed according to corresponding hub height (95, 117.2, 119.8 m) are shown in Figure 14. The annual average wind speed of 6 m/s of Weibull scaling factor (A = 6.8) and form factor (k = 1.91) determined from meteorological data of inland China is set as the annual average wind speed at hub height of 95 m. The annual average wind speed at optimal hub heights considering rated power (119.8 m) or not (117.2 m) are 6.3288 m/s and 6.2969 m/s, respectively. The Weibull form factor is assumed to be fixed as 1.91. The Weibull distributions show that the optimal hub heights have more probability with higher wind speed.

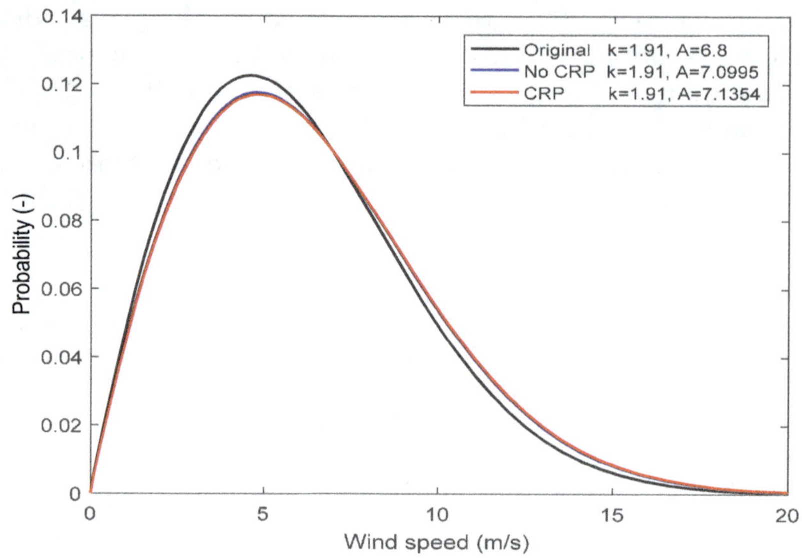

Figure 14. Weibull probability distribution of wind speed (6 m/s, 6.2969 m/s and 6.3288 m/s).

Figure 15 shows the comparison of the annual energy production between the optimal blades and the original one. Due to the differences of the Weibull distribution of the average wind speed at the hub height, the shape curve of the energy production between the original wind turbine and the optimal ones is not proportional. The energy production of the optimal wind turbine is slightly less than the original blade with a range of 3 m/s to about 8 m/s wind speed and is much higher from a speed of about 8 m/s to cut out speed (20 m/s). The results indicate that the AEP of the wind turbine (6.0921 GW) without considering rated power is higher than the original one (5.783 GW), but the one considering rated power (5.8364 GW) is similar to the original one.

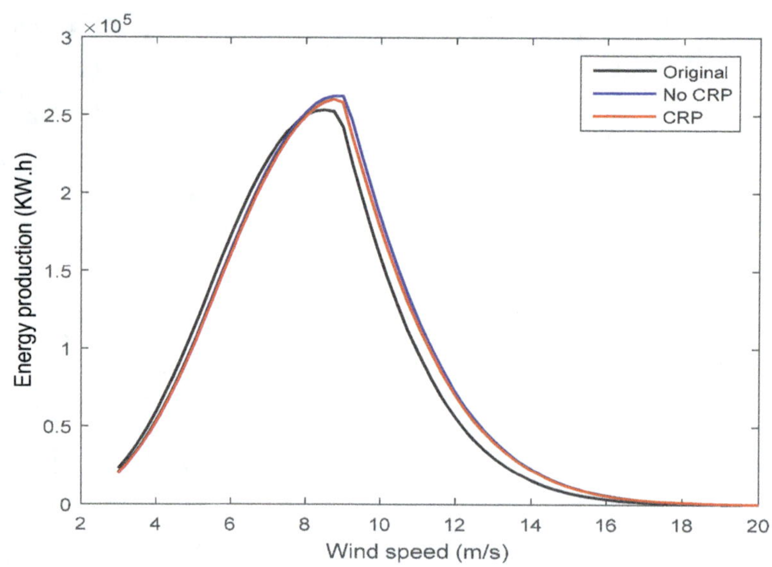

Figure 15. The comparison of the annual energy production (AEP) between the optimal blades and the original one.

Figure 16 shows that the comparison of the power curves between the optimal blades and the original one. The results indicate that the new blade without considering the design value of rated power as the variable can achieve rated power of 2.1 MW at the rated wind speed of about 9.2 m/s. The design value of rated power of the optimal wind turbine by considering rated power as the variable is 1.9 MW, and can achieve rated power at the rated wind speed of about 9.1 m/s. Both designs need higher wind speeds to achieve the rated power compared to the rated wind speed of 9 m/s of the original wind turbine. The main reason is that the lengths of the new blades are shorter than those of the original. This will result in less capacity in energy capture.

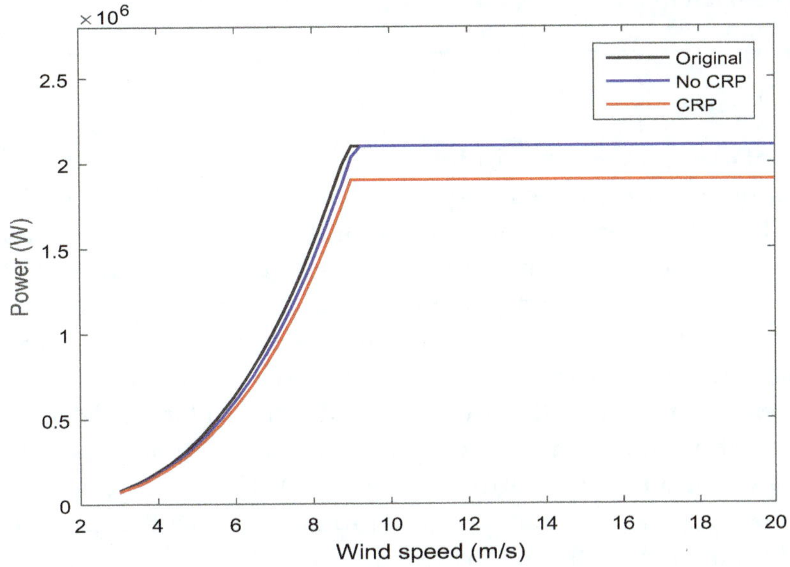

Figure 16. The comparison of the power curves between the optimal blades and the original one.

Figure 17 shows the distributions for spar caps and trailing edge reinforcement. The thickness of the spar caps and trailing edge reinforcement has increased compared to that of the baseline blade. The main reason for this is that the optimal blade needs to meet the structural constraints due to the large decrease of chord length distributions.

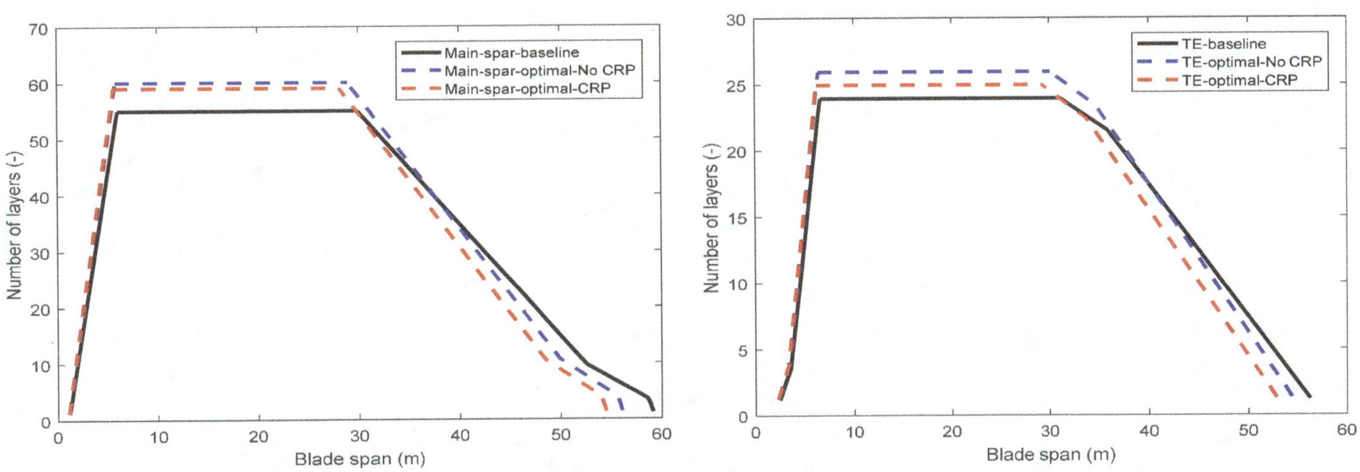

Figure 17. Structural thickness distributions of the baseline and the optimized blades.

The detailed results of the design objectives of minimizing the cost of energy are shown in Table 5. The longest blade is the original one and the shortest blade is the optimization considering rated power. The optimal wind turbines have COE of 0.0478 \$/kWh and 0.047 \$/kWh (CRP), compared to that of 0.057 \$/kWh of the initial one.

Table 5. Optimization design results of the blade.

Parameter	Initial Value	Optimal Value	Optimal Value (Rated)
Length (m) /Mass (kg)	59.8/13,240	57.84/12,540	56.2/11,620
Tip deflection (m)	7.576	6.92	6.31
AEP (GW)	5.7830	6.0921	5.8324
Cost of energy ($/kWh)	0.057	0.0478	0.047
Tip speed ratio	10.6	9.8	9.6

5.2. Tower Optimization Results

The distributions of outer diameters and wall thicknesses along the tower height are shown in Figure 18. Both of the optimal towers have larger outer diameter and thickness distributions to meet the structural constraints. Table 6 shows the comparison of the optimization results between the optimal towers and the original one. Both of the optimal towers are taller than the original one. With the increase of tower height, the annual average wind speed also increases. Hence, the rotor can capture more wind power to improve the AEP. However, with the increase of wind speed, the aerodynamic loads acting on the rotor and hence the thrust will also increase. The tower's structural design needs to consider that an increase in thrust on the tower top will cause a larger deformation. The tower mass increases when the tower height increases. Compared to the original tower, the two optimal towers have an increase in mass of about 43.65% and 42.49% (CRP). The main reason is that the tower height with 118m and 120m has higher annual average wind speed, and the greater rotor thrust requires a better structural performance with thicker walls. The height of the optimal tower considering rated power is higher but its mass is less than the one without considering rated power. The main reason for this is that the wind turbine with longer blades has larger thrust acting on the tower top.

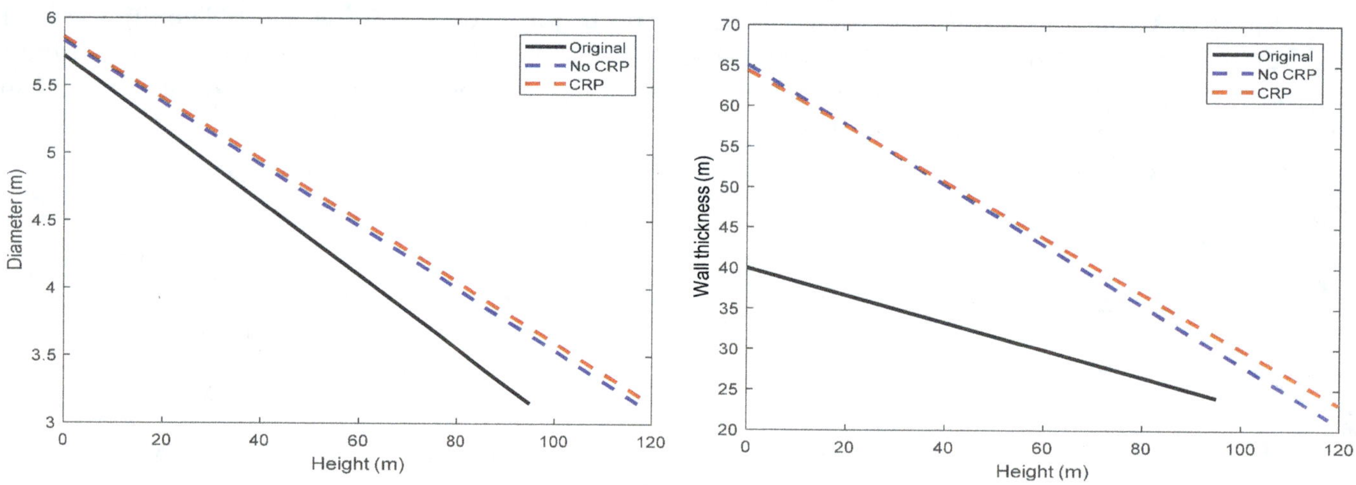

Figure 18. Tower outer diameter and wall thickness distributions of the baseline and the optimized towers.

Table 6. Optimization design results of the tower.

Parameter	Initial Value	Optimal Value	Optimal Value (Rated)
Height (m) /Mass (kg)	95/186,780	117.2/268,320	119.8/266,140
Top deformation (m)	0.65	0.828	0.816

Table 7 shows the costs of rotor, tower, and foundation in the cases of the original design, the optimal design (without CRP) and the optimal design (with CRP). The rotor costs of the optimal wind turbines are reduced compared with that of the original one. The main reason is that the optimal blades have a shorter length and lighter blade mass. The tower costs of the optimal wind turbines are increased compared with the original one due to the higher height and thicker wall distributions. Because of the shorter blade length and significant decrease in chord length distributions, the rotor thrust of the optimal wind turbine is less than that of the original one. Thus, the foundation costs do not change much, although the tower of the optimal wind turbine is higher than that of the original one.

Table 7. Costs of rotor, tower, and foundation.

Parameter	Rotor Cost ($1000)	Tower Cost ($1000)	Foundation Cost ($1000)
Original wind turbine	866.56	280.17	83.79
Optimal design (without considering rated power (CRP))	787.81	402.48	88.84
Optimal design (with CRP)	723.92	399.21	87.62

6. Conclusions

In this work, a wind turbine optimization for low wind speed areas is performed to investigate the relationship between rotor diameter and hub height. With the characteristic of high aspect ratio of the blade used in regions of wind Class III, the deflection of the longer blade is larger than the blade used in regions of wind Class I. According to this characteristic, the design needs to pay more attention to structural performance. In order to minimize the COE, the yield by increasing rotor diameter is less efficient than that by adding hub height, despite the fact that there is a large initial capital cost for the tower.

The AEP is evaluated by the determination of the electric energy generation as a function of average annual wind speed. The capacity of power generation is related to the aerodynamic performance of the rotor which is calculated by BEM theory and the annual average wind speed based on the hub height. The structural performance of the blade is calculated by the CLT method and the blade stability is estimated by the imposed constraints. The hub height has an impact on the annual average wind speed as well as AEP. A wind turbine with higher hub height has better wind resources but needs more efficient structural design. Although the larger rotor diameter and higher tower can generally produce more energy, the capital cost for that is not worthwhile. In low wind speed areas, it is not suitable to increase the rotor diameter alone and a solution needs to combine the whole system. For the purpose of minimizing the COE, increasing the rotor diameter for more annual energy production is not sufficient to outweigh increasing the cost of the wind turbine. The rated power is set as a variable to investigate its impact on COE. The results show that the design value of the rated power has a strong impact on COE.

Author Contributions: J.C. conceived and designed the project; H.Y. performed the computations and analyzed the results; H.Y., J.C. and X.P. wrote the paper.

References

1. International Electrotechnical Commission. *IEC 61400-1: Wind Turbines Part 1: Design Requirements*; International Electrotechnical Commission: Geneva, Switzerland, 2005.
2. Cristina, V.H.; Thomas, T.; Anahí, V.P. *JRC Wind Energy Status Report—2016 Edition*; Market, Technology and Regulatory Aspects of Wind Energy; Publications Office of the European Union: Luxembourg, 2017; pp. 10–11.
3. Wichser, C.; Klink, K. Low wind speed turbines and wind power potential in Minnesota, USA. *Renew. Energy* **2008**, *33*, 1749–1758. [CrossRef]
4. Wang, X.; Shen, W.-Z.; Zhu, W.-J.; Sørensen, J.-N.; Jin, C. Shape optimization of wind turbine blades. *Wind Energy* **2009**, *12*, 781–803.

5. Lee, J.; Lee, K.; Kim, B. Aerodynamic optimal blade design and performance analysis of 3 MW wind turbine blade with AEP enhancement for low-wind-speed-sites. *J. Renew. Sustain. Energy* **2016**, *8*, 1–12. [CrossRef]
6. Barnes, R.H.; Morozov, E.V.; Shankar, K. Improved methodology for design of low wind speed specific wind turbine blades. *Compos. Struct.* **2015**, *119*, 677–684. [CrossRef]
7. Barnes, R.H.; Morozov, E.V. Structural optimisation of composite wind turbine blade structures with variations of internal geometry configuration. *Compos. Struct.* **2016**, *152*, 158–167. [CrossRef]
8. Nicholson, J.C. Design of Wind Turbine Tower and Foundation Systems: Optimization Approach. Master's Thesis, University of Iowa, Iowa City, IA, USA, 2011.
9. Wang, L.; Kolios, A.; Luengo, M.M. Structural optimisation of wind turbine towers based on finite element analysis and genetic algorithm. *Wind Energy Sci. Discuss.* **2016**. [CrossRef]
10. Wright, A.K.; Wood, D.H. The starting and low wind speed behaviour of a small horizontal axis wind turbine. *J. Wind Eng. Ind. Aerod.* **2004**, *92*, 1265–1279. [CrossRef]
11. Pourrajabian, A.; Ebrahimi, R.; Mirzaei, M. Applying micro scales of horizontal axis wind turbines for operation in low wind speed regions. *Energy Convers. Manag.* **2014**, *87*, 119–127. [CrossRef]
12. Singh, R.K.; Ahmed, M.R. Blade design and performance testing of a small wind turbine rotor for low wind speed applications. *Renew. Energy* **2013**, *50*, 812–819. [CrossRef]
13. Salih, N.A.; Mohammed, A.H.; Talha, A.; Kamel, A.K. Experimental and theoretical investigation of micro wind turbine for low wind speed regions. *Renew. Energy* **2018**, *116*, 215–223.
14. Maki, K.; Sbragio, R.; Vlahopoulos, N. System design of a wind turbine using a multi-level optimization approach. *Renew. Energy* **2012**, *43*, 101–110. [CrossRef]
15. Fingerish, L.; Hand, M.; Laxson, A. *Wind Turbine Design Cost and Scaling Model*; Technical Report; National Renewable Energy Laboratory: Golden, CO, USA, 2006.
16. Bortolotti, P.; Bottasso, C.L.; Croce, A. Combined preliminary–detailed design of wind turbines. *Wind Energy Sci.* **2016**, *1*, 71–88. [CrossRef]
17. Ning, A.; Damiani, R.; Moriarty, P.J. Objectives and constraints for wind turbine optimization. *J. Sol. Energy Eng.* **2014**, *136*, 041010. [CrossRef]
18. Bir, G.S. *User's Guide to PreComp: Pre-Processor for Computing Composite Blade Properties*; National Renewable Energy Laboratory: Golden, CO, USA, 2005.
19. Hansen, M.O.L. *Aerodynamics of Wind Turbines*, 2nd ed.; Earthscan: Abingdon, UK, 2008.
20. Timmer, W.A.; van Rooij, P.R.J.O.M. Summary of the Delft University wind turbine dedicated airfoils. *J. Sol. Energy Eng.* **2003**, *125*, 48–96. [CrossRef]
21. Drela, M. *XFoil: An Analysis and Design System for Low Reynolds Number Airfoils*; Technical Report; MIT Dept. of Aeronautics and Astronautics: Cambridge, MA, USA, 1989.
22. Reddy, J.N. *Mechanics of Laminated Composite Plates and Shells: Theory and Analysis*; CRC: New York, NY, USA, 2003.
23. Timoshenko, S.; Woinowsky-Krieger, S.; Woinowsky, S. *Theory of Plates and Shells*; McGraw-Hill: New York, NY, USA, 1959.

9

Development of an Advanced Fluid-Structure-Acoustics Framework for Predicting and Controlling the Noise Emission from a Wind Turbine under Wind Shear and Yaw

Mingyue Zhou [1], Matias Sessarego [1], Hua Yang [2] and Wen Zhong Shen [1,*]

1 Fluid Mechanics Section, Department of Wind Energy, Technical University of Denmark, Nils Koppels Allé, Building 403, 2800 Lyngby, Denmark; 1993.zhou@163.com (M.Z.); msessare@gmail.com (M.S.)

2 School of Hydraulic Energy and Power Engineering, Yangzhou University, Yangzhou 225127, China; yanghua@yzu.edu.cn

* Correspondence: wzsh@dtu.dk

Abstract: Noise generated from wind turbines is a big challenge for the wind energy industry to develop further onshore wind energy. The traditional way of reducing noise is to design low noise wind turbine airfoils and blades. A wind turbine operating under wind shear and in yaw produces periodic changes of blade loading, which intensifies the amplitude modulation (AM) of the generated noise, and thus can give more annoyance to the people living nearby. In this paper, the noise emission from a wind turbine under wind shear and yaw is modelled with an advanced fluid-structure-acoustics framework, and then controlled with a pitch control strategy. The numerical tool used in this study is the coupled Navier–Stokes/Actuator Line model EllipSys3D/AL, structure model FLEX5, and noise prediction model (Brooks, Pope and Marcolini: BPM) framework. All simulations and tests were made on the NM80 wind turbine equipped with three blades made by LM Wind Power. The coupled code was first validated against field load measurements under wind shear and yaw, and a fairly good agreement was obtained. The coupled code was then used to study the noise source control of the turbine under wind shear and yaw. Results show that in the case of a moderate wind shear with a shear exponent of 0.3, the pitch control strategy can reduce the mean noise emission about 0.4 dB and reduce slightly the modulation depth that mainly occurs in the low-frequency region.

Keywords: aeroacoustics; wind turbine; noise modelling; noise control

---

1. Introduction

As a clean and renewable energy, wind power has the advantages of low cost, low environmental pollution and wide availability. These unique advantages of wind power make it an important part of the sustainable energy mix in many countries. However, wind energy also has some drawbacks that hinder its global use. The noise caused by wind turbines has become one of the primary sources polluting the urban environment today. Wind turbine noise caused by the movement of the blades through the air is often seen as an essential aspect causing great annoyance as compared with other noise sources [1]. In Danish regulation [2], a modern wind turbine should be set up at least four times its tip height away from residential areas. Even segregated by such a long distance, a turbine still produces a sound pressure level (SPL) of more than 40 dB (A), which almost equals some common home appliances [3]. A wind turbine operating under wind shear and in yaw produces periodic changes of blade loading, which intensifies the amplitude modulation (AM) of the generated noise, and thus can give more annoyance to the people living nearby. If the aerodynamic noise of a wind turbine can be reduced without limiting its rotational speed, wind energy can be utilized with a high efficiency.

Therefore, in this paper, aerodynamic characteristics and noise features of a wind turbine in wind shear and yaw is studied.

Noise from a wind turbine or wind farm often has a feature of amplitude modulation with a varying sound pressure amplitude during its operation. There are two types of AM noise: the one is created by changing the blade noise directivity to receiver [4] and the other is caused by non-uniform inflow and operation, for example the operation in wind shear and yaw. There are three methods (Japanese F-S method [5], UK method [6] and min-max method [7]) to characterize the modulation depth. Recently, Barlas et al. [8] studied the amplitude modulation caused by a wind turbine wake using an advanced Computational Fluid Dynamics (CFD) (EllipSys3D/AL) and a sound propagation model based on solving parabolic wave equation and found that the modulation depth can reach 4–5 dB even with the same strength of noise source.

The prediction formulae of trailing edge noise were summarized first using data from an experiment by Schlinker and Amiet in 1981 [9]. In 1985, Grosveld [10] used the helicopter noise formulae synthesized by Schlinker to calculate the emission noise of several two-blade, low-power downwind horizontal axis wind turbines, and the results were in good agreement with the measured data. In 1986, De Wolf continued the research based on a similar model [11]. Back to 1981, Viterna [12] utilized an approach for the low-frequency noise of a wind turbine. A rather complex airfoil self-noise prediction model (known as the Brooks, Pope and Marcolini: BPM model) using wind tunnel acoustic measurements of National Advisory Committee for Aeronautics (NACA) 0012 airfoils was published by Brooks et al. [13] in 1989. In 1996, Fuglsang and Madsen [14] used the BPM model and the inflow noise model by Lowson [15] modified from Amiet's model [16] to calculate the noise emission of a Vestas V27 wind turbine. In 2003, the National Renewable Energy Laboratory (NREL) [17] developed its NREL AirFoil Noise (NAFNoise) based on the BPM model. The BPM was improved by including the real blade geometry for predicting wind turbine noise generation by Zhu et al. from Technical University of Denmark (DTU) in 2005 [18]. In 2009, Bowdler [19] studied wind shear effects on noise at different locations and at different time instants of the day and of the year using field data collected at wind farm sites. Oerlemans [20] investigated the influence of wind shear on the amplitude modulation of wind turbine noise in 2015. Wind shear and atmospheric turbulence effects on wind turbine noise using the Monin–Obukhov similarity theory was studied in 2016 by Yuan and Cotté [21]. To predict the trailing-edge noise, a more advanced trailing-edge noise model (TNO) was developed by Parchen [22] using a relation between the sound pressure level at far field and the surface pressure spectrum at trailing edge and related the far field sound spectrum as a function of turbulent boundary layer quantities. A refined TNO model using CFD was made by Kamruzzaman et al. [23] to consider the non-isotropic issue in the trailing-edge boundary layer. Tian [24] extended Amiet's model for simulating noise from a wind turbine.

This paper deals with the development of an advanced fluid-structure-noise technique for simulating noise generated from a wind turbine and controlling the noise generation of the wind turbine in wind shear and yaw by adjusting the pitch angle of its blades. A coupled flow and acoustics code that simultaneously predicts the noise and aerodynamic outputs of a wind turbine is applied in this study.

The paper is organized as follows. Section 2 presents the related theory and methodology. Validation and control results are presented in Section 3. Finally, conclusions are drawn in Section 4.

2. Theory and Methodology

The flow past a wind turbine is modelled with the actuator line (AL) method introduced by Sørensen and Shen [25], which was implemented in DTU's in-house finite volume code EllipSys3D. The actuator line method was coupled successfully with a high-order spectral method by Kleusberg et al. [26,27]. EllipSys3D was developed by the co-operation of the Department of Mechanical Engineering at DTU [28] and the Department of Wind Energy at Risø National Laboratory [29] that is now the Department of Wind Energy at DTU. The AL method [25,30] represents the aerodynamic loads of wind turbine blades by using a body force distributed along rotating lines. The body force imposed in the Navier–Stokes equations is computed by the aero-elastic code FLEX5 [31] also developed at DTU. The coupling of FLEX5 and EllipSys3D/AL methods is performed at every time-step. The flow data at the blades from EllipSys3D/AL are interpolated from the flow mesh and then fed into FLEX5. Using these information, the angle of attack and relative velocity including blade deformation/motions are calculated in FLEX5 and the aerodynamic loads on the blades are obtained further by using tabulated airfoil lift, drag and moment vs. angle of attack (AoA) data including dynamic stall. The new blade positions and loads from FLEX5, which includes the effect of blade bending and motion of the rotor (up/down/side-to-side/rotations) are then fed into EllipSys3D/AL as a body force on the moving actuator lines. With these new blade positions, the new loads are redistributed on the CFD mesh and a new solution is obtained in EllipSys3D/AL. In a standard aero-elastic code, these effects and motions are calculated by using the blade momentum theory (BEM) as the aerodynamic model. Here, the high-fidelity Navier–Stokes AL method was used to solve the fluid-structure interaction problem instead. To perform with large eddy simulation, the mixed scale model of Ta Phuoc [32,33] was used. To calculate the noise emission from the blades, the BPM code was used. The variables of angle of attack and relative velocity were calculated in FLEX5 and fed to BPM. A flow chart of the combined EllipSys3D/AL, FLEX5 and BPM framework is shown in Figure 1.

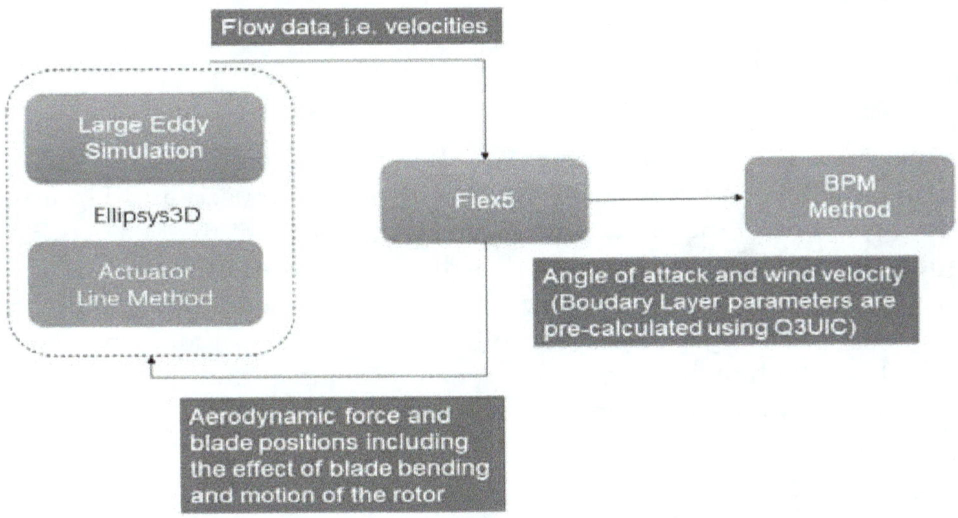

Figure 1. Flow chart of the combined EllipSys3D/AL, FLEX5 and BPM program.

In EllipSys3D, the solution to the incompressible Navier–Stokes equations is advanced in time using an iterative time-stepping method. The velocity-pressure coupling equations at each time-step are solved iteratively by sub-iterations with the usage of under-relaxation. First, the code solves the momentum equations as a predictor so that the solution can be advanced in time. The pressure correction equation, i.e., the rewritten continuity equation in satisfying the local mass flux conservation in the discretized form, is solved as a corrector making the final flow field satisfy the continuity constraint. This is a two-step procedure corresponding to a single sub-iteration, and the process is repeated until the solution becomes convergent within sub-iterations. The variables are then updated after the convergent solution is updated, and then followed by the next time step.

The aero-elastic code FLEX5 was designed to simulate a wind turbine's dynamic behavior regarding different wind conditions. FLEX5 operates in the time domain, and the output is the time-series of simulated loads and deflections. The detailed information of nine modules in FLEX5 can be found in [34].

The acoustic prediction BPM model used in this work takes the detailed geometrical and flow information into account, e.g., tip shape, blade geometry (chord and twist distributions) as well as instantaneous wind speed and direction. The velocity at each blade segment is computed by accounting for the induction effect and the vibration velocity of the blades. Furthermore, the geometry contours of airfoil sections of blade segments are also included in the prediction as important inputs, such as the boundary thickness and displacement thickness. The boundary layer parameters for a NACA 0012 was benchmarked with the viscous-inviscid interaction code XFOIL [35] and the strong viscous-inviscid interactive coupling code Q^3UIC [36], and compared with the measured values [13]. The comparisons (not shown) indicated that XFOIL under-predicts the boundary layer thicknesses about 30% and Q^3UIC gave a good agreement with the measurement. In this study, we compute the boundary quantities of airfoils using Q^3UIC. For more advanced calculations, the structure deformation in the blade torsion direction can also be considered, which will lead to changes in angle of attack. More details about the noise prediction calculation can be found in [18,37].

In wind turbine control systems, the proportional–integral–derivative (PID) controller is widely applied to control the generator speed and blade pitch angle. Obviously, the ideal controller should be developed by using the PID controller theory. In the present study, given the practical complexity, two control strategies, of (1) full step pitch input control and (2) interpolated pitch input control, are developed, which are P-type controls.

For the full step pitch input control, the blade pitch β is controlled at each time-step with an objective of AoA at one or more selected cross-sections, α_i, equal to its mean value in a cycle at one or more selected cross-sections, $\alpha_{i,mean}$, as follows

$$\beta = \frac{1}{N} \sum_{i=1,\dots,N} \alpha_i - \alpha_{i,mean}$$

where N is the number of selected cross-sections. After approximately 100 s, the solution becomes stable. Alternatively, the interpolation pitch input control can be used and Figure 2 illustrates the concepts. The idea is to control the blade pitch in the way that (i) at the peaks or troughs of one or more cross-sections, the blade pitch is controlled as the one in the full step pitch input control; (ii) at the azimuth positions in between the blade pitch is controlled by a pitch angle interpolated from the pitch angles at the peaks and troughs in (i) according to its azimuth position. Concretely, the peaks and troughs are extracted first with their corresponding azimuth angles and the mean AoAs are obtained from the maximums and minimums (red circles in Figure 2):

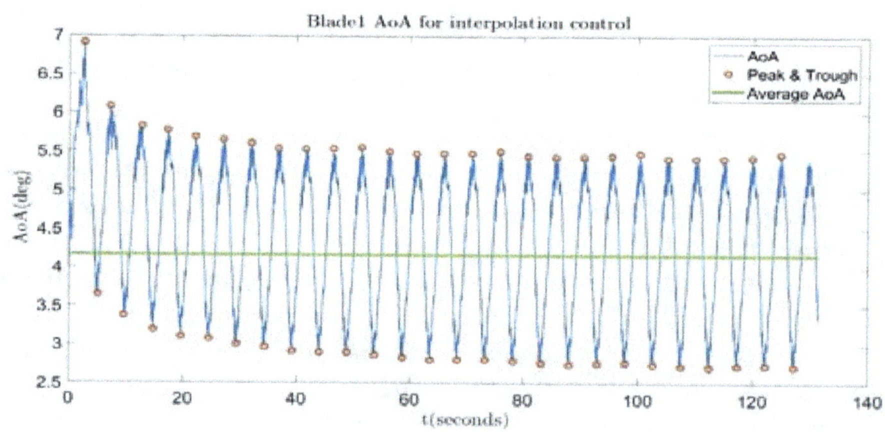

Figure 2. Pitch input setup based on angle of attack (AoA).

3. Numerical Results

In this section, the accuracy of the used numerical code is first assessed by performing a validation against field measurements performed in the DanAero project [38]. Results from the control study under the same conditions are presented afterwards.

3.1. *Validation of the Simulation Framework*

Before going further to discuss the performance of a wind turbine in different operation setups, the accuracy of the simulation code should be verified first. The 2.3 MW NM80 wind turbine was used here.

The design geometry of the LM 38.8 blade and aerodynamic airfoil data for the generic 2.3 MW variable speed and pitch controlled NM80 wind turbine were provided through a confidential agreement of using the DANAERO database. The airfoil data were used directly in AL/FLEX5, and the structural data originally prepared for HAWC2 were used to produce an equivalent FLEX5 input file. We assume the wind turbine operates in an ideal situation, so the tilt angle and tower shadow effects are neglected. Some key parameters of the wind turbine are listed in Table 1.

Table 1. Key wind turbine parameters.

Parameters	Value	Parameters	Value
Rotor diameter	80 m	Rated wind speed	16 m/s
Hub height	60 m	Max rotor speed	16.2 rpm
Rated power	2.3 MW		
Cut-in wind speed	4.5 m/s	Cut-out wind speed	25 m/s

Specific parameters for the two verification cases are listed below in Table 2.

Table 2. Description of Case I and Case II.

Parameters	Case I	Case II
Pitch angle	−4.75°	−4.75°
Rotor speed	16.2 rpm	16.2 rpm
Hub height wind speed	9.792 m/s	8.429 m/s
Shear exponent	0.249	0.262
Air density	1.22 kg/m^3	1.22 kg/m^3
Yaw angle	−6.02°	−38.34°
Ambient pressure	1005 Pa	1020 Pa

Figure 3 describes the configuration of the wind turbine and some parameters defined in Table 2.

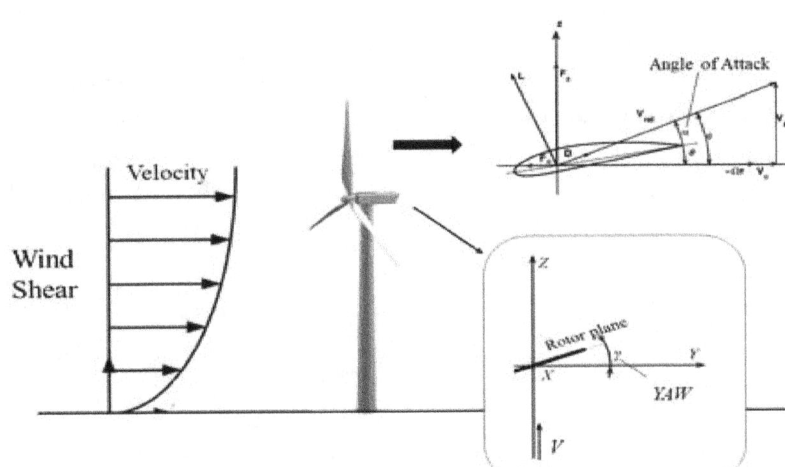

Figure 3. Configuration of a wind turbine under wind shear and yaw.

In the computations, a relatively fine mesh of about 19 million cells was used with $192 \times 192 \times 512$ cells in the transversal, vertical and streamwise directions in a domain of $[-16R, 16R] \times [0R, 19R] \times [-16R, 34R]$ (R is the rotor radius), respectively. A time-step based on rotor radius and inflow wind speed of 0.002 was used. A resolution of 30 cells per rotor radius was used in the rotor plane and 18 cells per rotor radius were uniformly distributed from 1D (rotor diameter) in front of the turbine to 11D behind the turbine in order to have a good resolution in the wake region. The airfoil data used in the computations were 2-dimensional airfoil data provided from the International Energy Agency (IEA) Task 29 consortium. To take into account the dynamic stall effects, the Øye dynamic stall model [34] was used as this is a part of FLEX5.

The boundary condition used in the computations is that at the inlet (min z), spanwise boundary (x direction) and up-boundary (max y), a sheared wind velocity profile is used, at the ground (y = 0) no-slip condition is used, and at the outlet (max z) the convective boundary condition is used. In this study, no synthetic inflow turbulence is used.

Figures 4 and 5 illustrate the comparisons between the measured data and simulation results. The simulation results were extracted from a cycle after the flow was stabilized. The forces normal and tangential to the local airfoil chord at 33%, 48%, 76% and 92% rotor radius noted as Fn33, Ft33, Fn48, Ft48, Fn76, Ft76, Fn92, Ft92, respectively, are shown below:

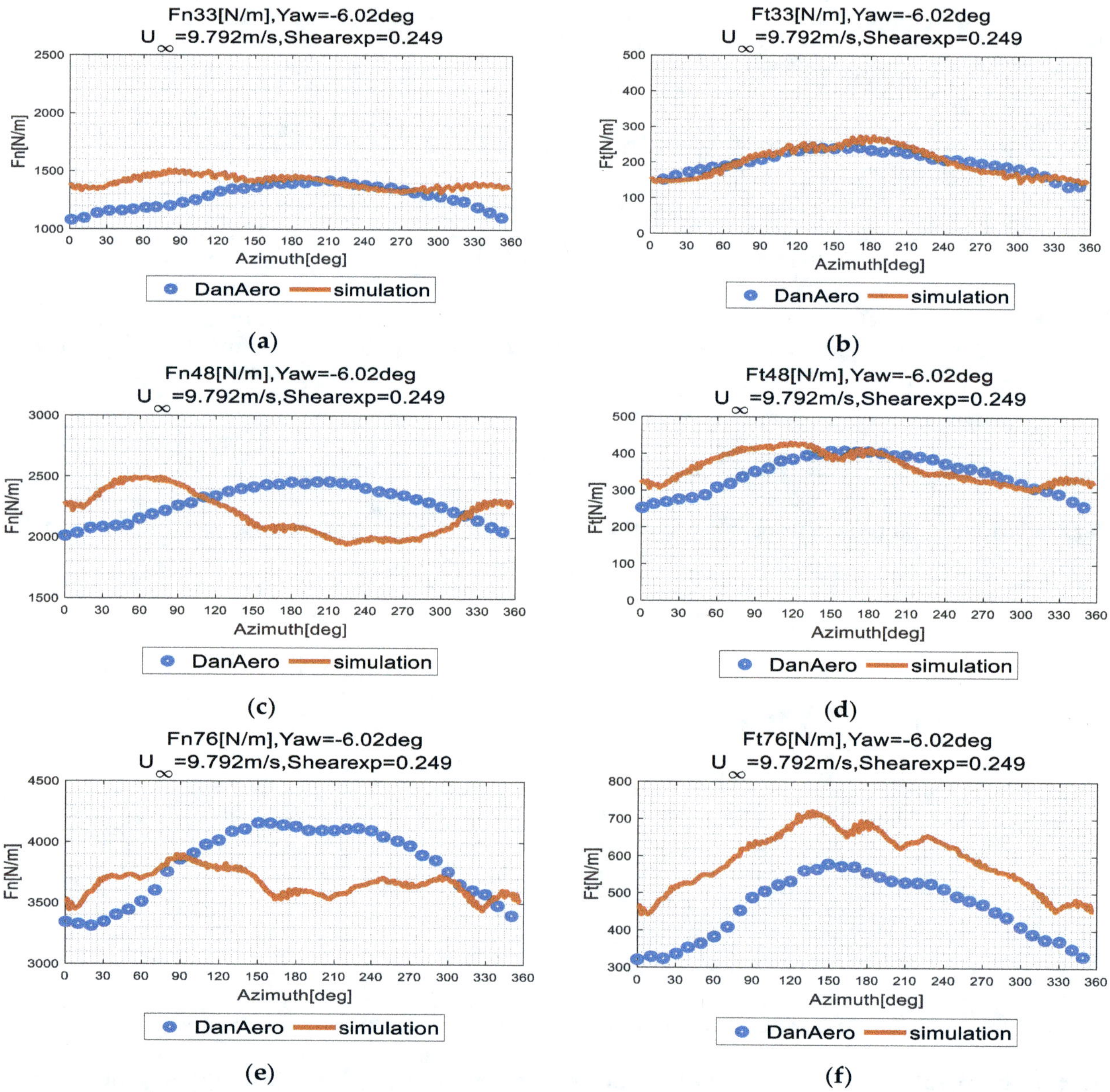

Figure 4. *Cont.*

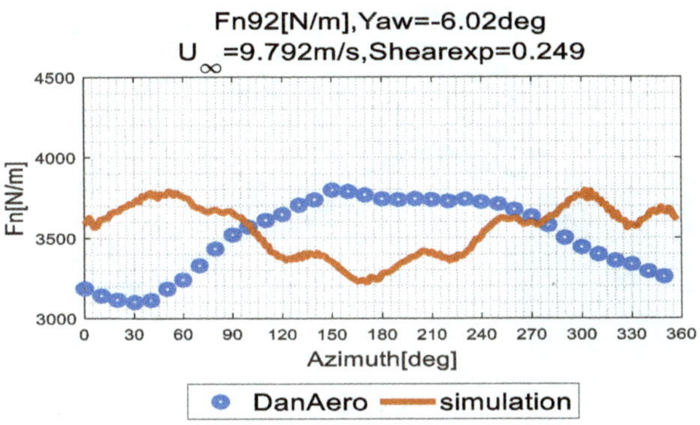

(g)

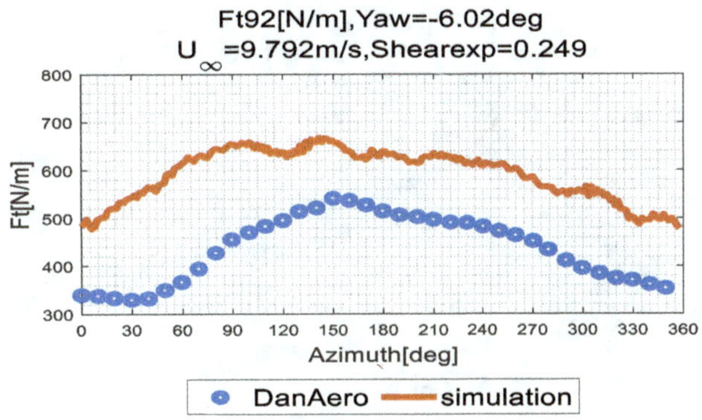

(h)

Figure 4. Forces normal and tangential to the local chord at 33%, 48%, 76% and 92% rotor radius for Case I (**a**) Fn33; (**b**) Ft33; (**c**) Fn48; (**d**) Ft48; (**e**) Fn76; (**f**) Ft76; (**g**) Fn92; (**h**) Ft92.

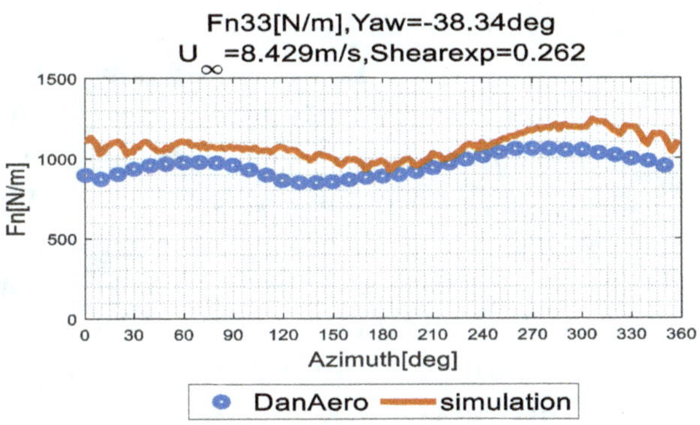

(a)

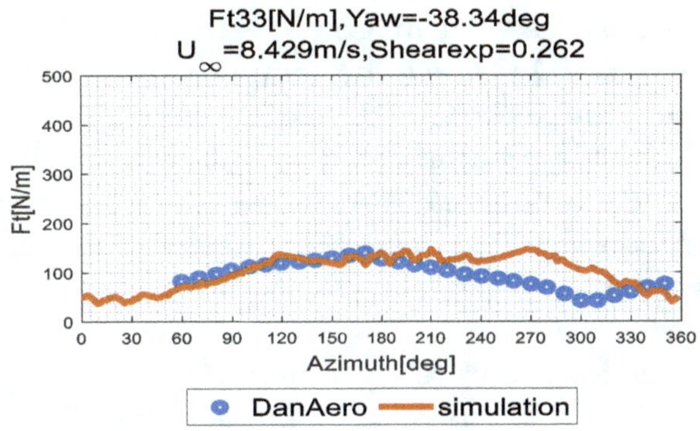

(b)

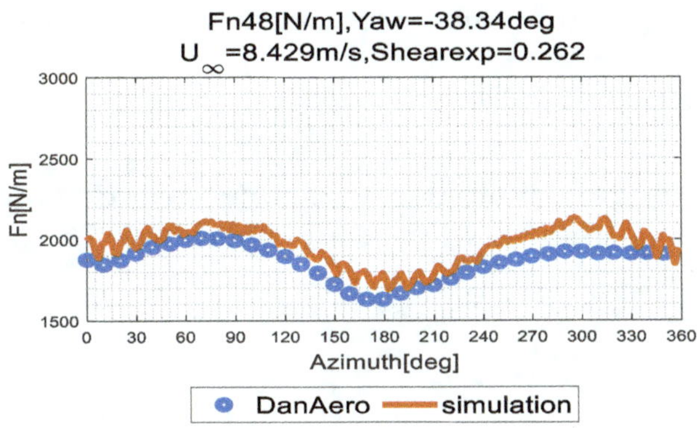

(c)

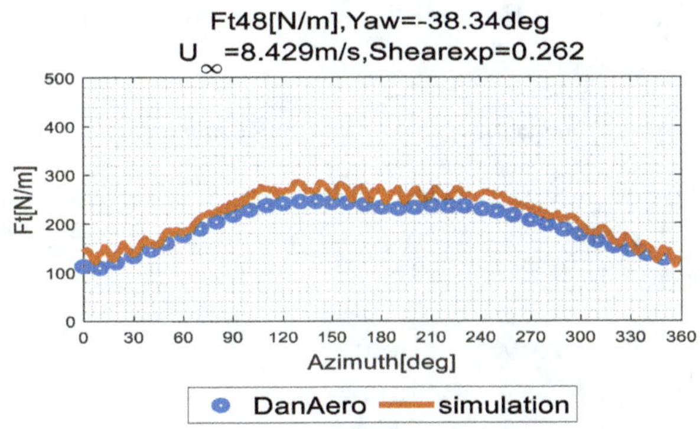

(d)

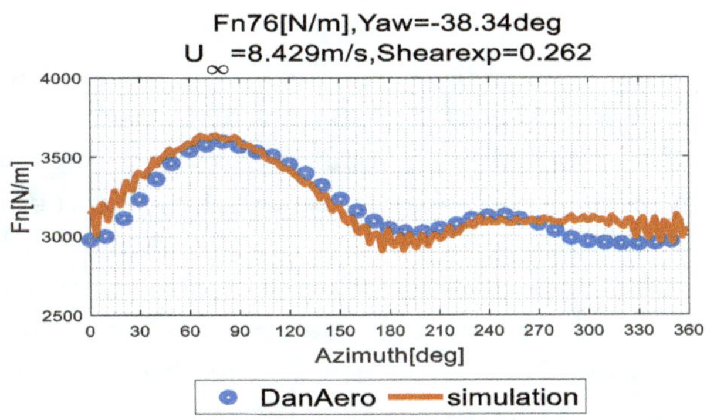

(e)

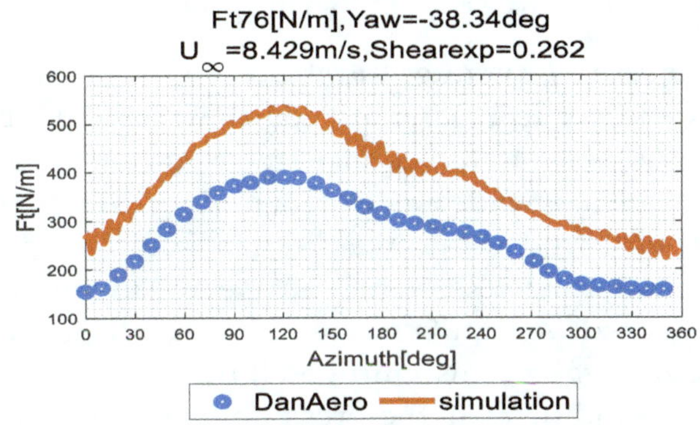

(f)

Figure 5. *Cont.*

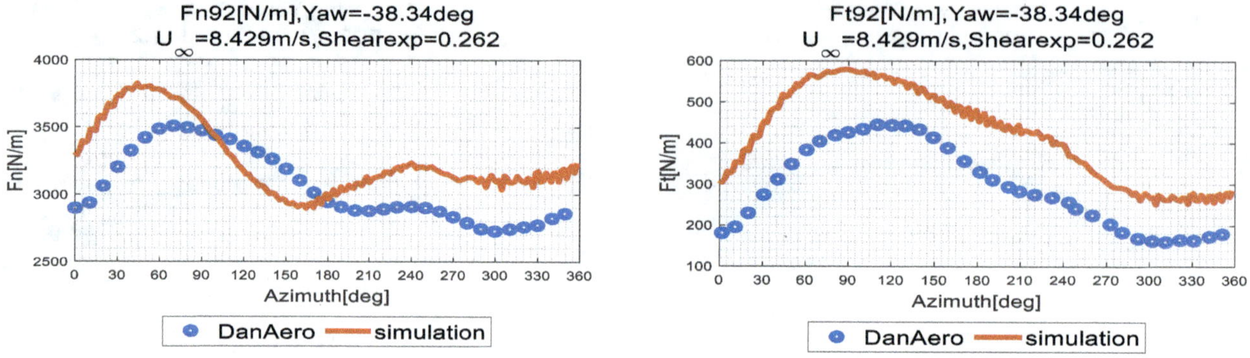

(g) (h)

Figure 5. Forces normal and tangential to the local chord at 33%, 48%, 76% and 92% rotor radius for Case II (**a**) Fn33; (**b**) Ft33; (**c**) Fn48; (**d**) Ft48; (**e**) Fn76; (**f**) Ft76; (**g**) Fn92; (**h**) Ft92.

As observed, the magnitudes between the measurements and simulations are similar for most of the quantities. In Case I, the computed normal force does not follow with the measured one and this is probably due to the Øye dynamic stall model used in the computations, which does not perform well at small angles of attack. On the other hand, the tangential force agrees well with the measurements. In Case II, the normal force is seen to be very well captured, while the tangential force is slightly over-predicted. The trends in function of azimuth angle (with 0° when the blade is down-pointing) followed the numerical results well. Since the tangential force is sensitive to the local angle of attack and airfoil data, calibrating airfoil data and its 3-dimensional corrections to rotational effects is a challenge for accurately predicting the performance of a wind turbine under wind shear and yaw. From the comparisons, it is also seen that the agreement is fairly good for both small (6°) and large (38°) yaw angles under a moderate wind shear. Therefore, it can be concluded from the validation that the simulation technique used in the present paper can predict the wind turbine performance relatively well.

3.2. Power Performance and Load Simulation

Since a wind turbine is rarely working with a large yaw angle, we define scenarios for the control study according to Case I in the validation. The specifics of the four different scenarios are described below:

Case 1: No yaw, no shear;
Case 2: 10° yaw, no shear;
Case 3: No yaw, 0.3 wind shear power law exponent;
Case 4: 10° yaw, 0.3 wind shear power law exponent.

Figure 6 illustrates the streamwise velocity and pressure field for Case 1 and Case 4. Horizontal axis x and vertical axis y are non-dimensional with the rotor radius of R = 40 m, which means the real values of them should equal to the axis values times R. The rotor center is located at x = 1.5 and y = 1.5. In Figure 6a, a velocity slowdown (induction effect) created by the rotor when exacting wind energy can be observed in the areas both in front of and behind the rotor, where the impact of induction effect is larger in the wake. When no shear is included, no wind speed stratification above the ground can be observed. A significant wind speed stratification above the ground can be observed in Figure 6b, while the wake is also mitigated by the lower longitudinal wind speed due to the shear. Similarly, Figure 6c,d illustrate the pressure field for Case 1 and Case 4, respectively. In general, the pressure in front of the rotor is larger than the pressure behind the rotor. By introducing the wind shear and yaw, the pressure both in front of and behind the rotor is changed, especially for the pressure in front of the rotor, which becomes more homogeneous.

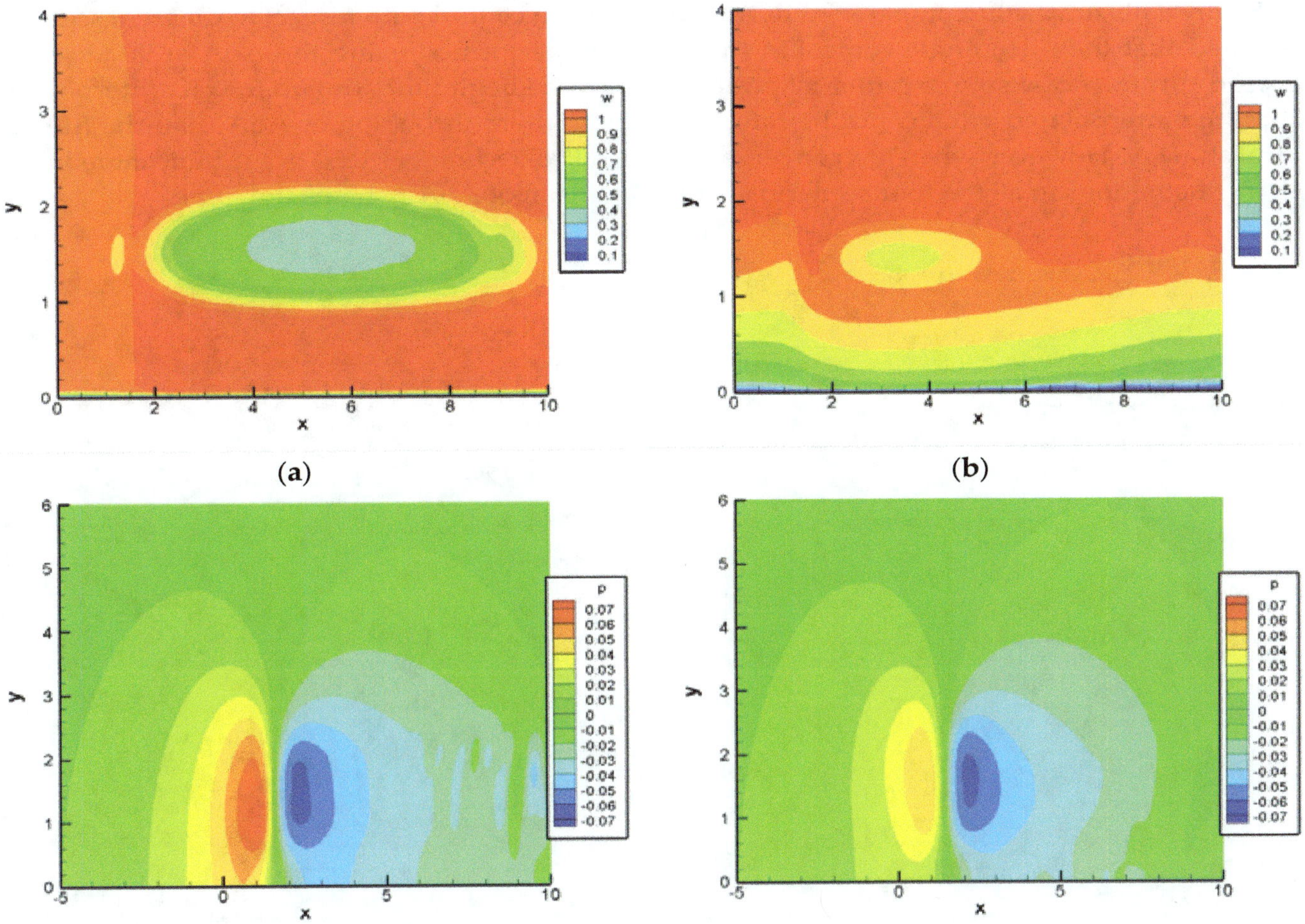

(a) (b) (c) (d)

Figure 6. Streamwise velocity field and pressure field for Case 1 and Case 4: (**a**) streamwise velocity for Case 1; (**b**) streamwise velocity for Case 2; (**c**) pressure for Case 1; (**d**) pressure for Case 2.

The angle of attack (AoA) at 90% length of the blade for the four scenarios with no control is investigated. The comparison of the AoA can be found in Figure 7. In general, it can be observed that the fluctuations of AoA become much larger by introducing yaw or shear. In this study, the 0.3 wind shear power law exponent leads to a more significant amplitude change than the 10-degree yaw angle.

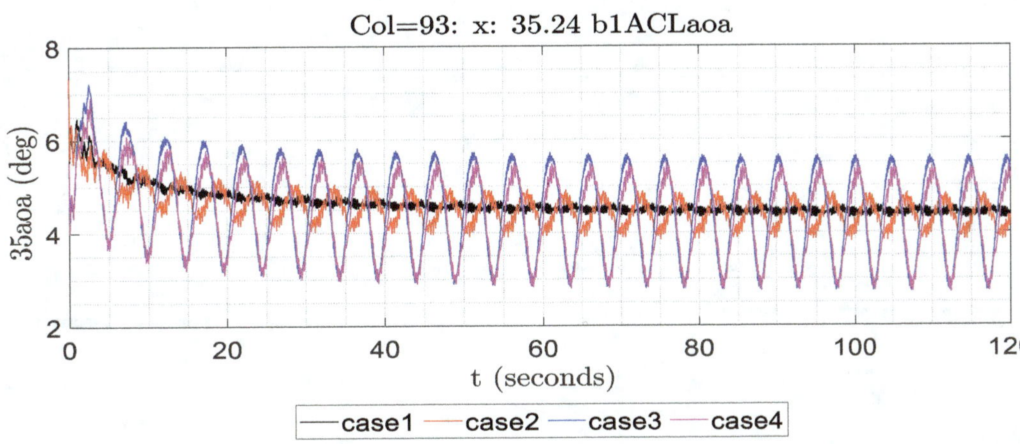

Figure 7. Angle of attack in a cross-section of 35.24 m (90%R) without control strategy.

Figure 8 illustrates the comparisons of AoA under pitch control with respect to different scenarios. It is noted that the pitch angle control in this study is performed using the AoA at 90%R where the wind turbine noise source center is located [37]. In the beginning, due to inexact interpolation values, the interpolated pitch control strategy results in more throbbing of AoA than the full step control strategy. However, when it is converged, both control techniques lead to a significant reduction of AoA fluctuations in all the four scenarios, while the mean of AoA remains the same.

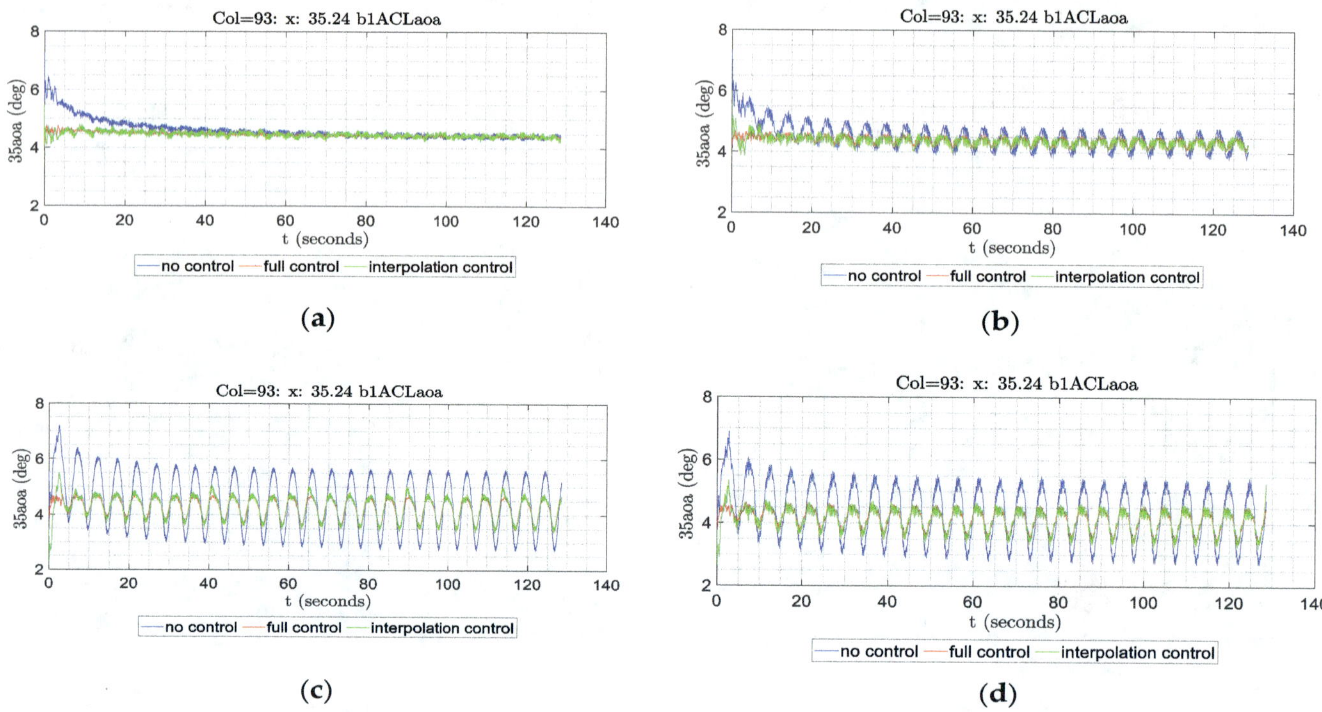

Figure 8. Angle of attack in section 35.24 m (90%R) with control strategy: (**a**) Case 1; (**b**) Case 2; (**c**) Case 3; (**d**) Case 4.

Figure 9 shows the pitch input values based on the results of AoA mentioned above. Pitch angles for two control strategies are calculated by using the angle of attack at 90%R of the no control case. It can be observed that the pitch is constant with 0.15° pitch offset when no control is performed. In terms of pitch in all the four cases, no significant difference can be found between control strategies I and II.

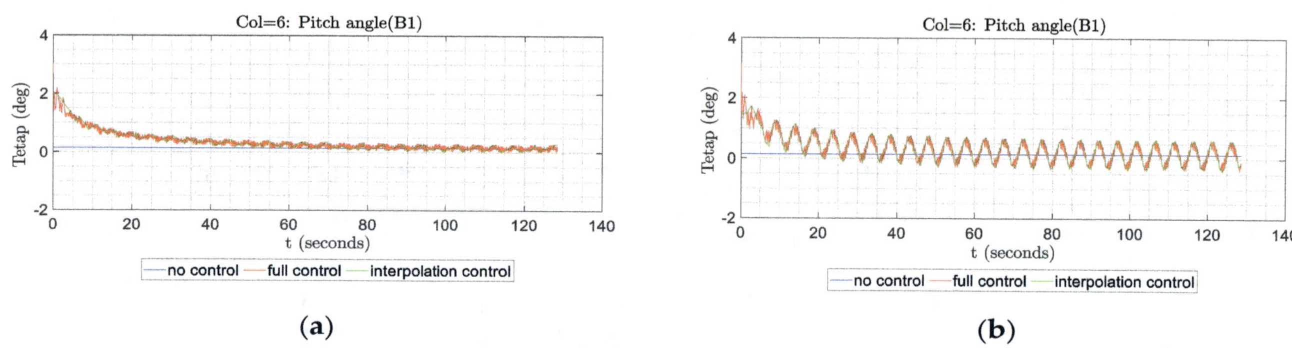

Figure 9. *Cont.*

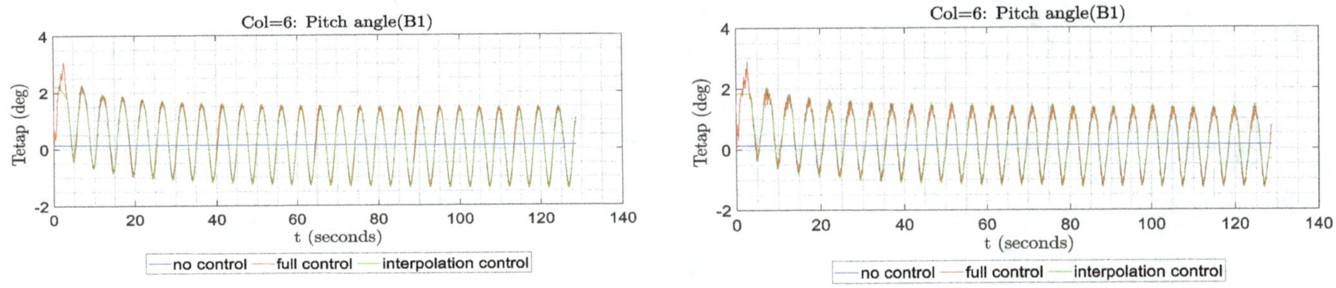

(c) (d)

Figure 9. Pitch in a cross-section of 35.24 m (90%R) with and without control strategy: (**a**) Case 1; (**b**) Case 2; (**c**) Case 3; (**d**) Case 4.

In Figure 10, power and thrust coefficients, noted as C_p and C_T, respectively, are plotted for the four scenarios. It can be seen that C_p is similar in all the four scenarios whereas introducing the shear and yaw slightly reduces C_p especially when C_p becoming more stationary, i.e., after 110 s. The reduction of the mean C_p is mainly due to the yaw misalignment, where the axial wind speed is smaller. It also should be noted that the scenarios with a yaw angle, C_p has more periodical fluctuations due to the yaw angle. A similar trend like C_p can be found in the thrust coefficient C_T, shown in Figure 10b, where the reduction of C_T is realized by the shear and yaw. The most significant difference happened in the scenario with both yaw and wind shear. Moreover, it can be observed that the difference between the yaw versus the yaw plus shear scenarios is relatively small. The reduction of C_T in Case 2, Case 3 and Case 4 is due to the reduction of the axial wind speed caused by the yaw.

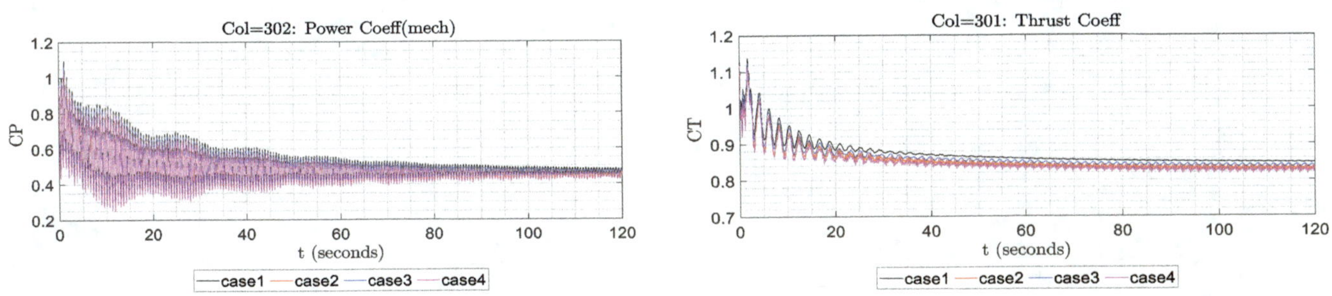

(a) (b)

Figure 10. C_p and C_T without control strategy: (**a**) power coefficient; (**b**) thrust coefficient.

By implementing the pitch adjustment, more fluctuations in Figure 11 are observed in the C_p and C_T signals while the mean C_p and C_T almost remain the same. By introducing the control methods, the fluctuation magnitudes of both control strategies increase, especially in control technique I. Some jumps of C_p and C_T values occur in the transient period as well. Moreover, the transient periods when performing both control techniques are longer, as compared to no control. For control technique I, the unexpected large fluctuations in C_T and C_p might be explained by the large pitch changes at each time step. For control technique II, similar results are observed when the fluctuation magnitudes are much smaller. The possible explanation is that there are fewer input pitch values and the corresponding interpolation enables better convergence of the elastic code.

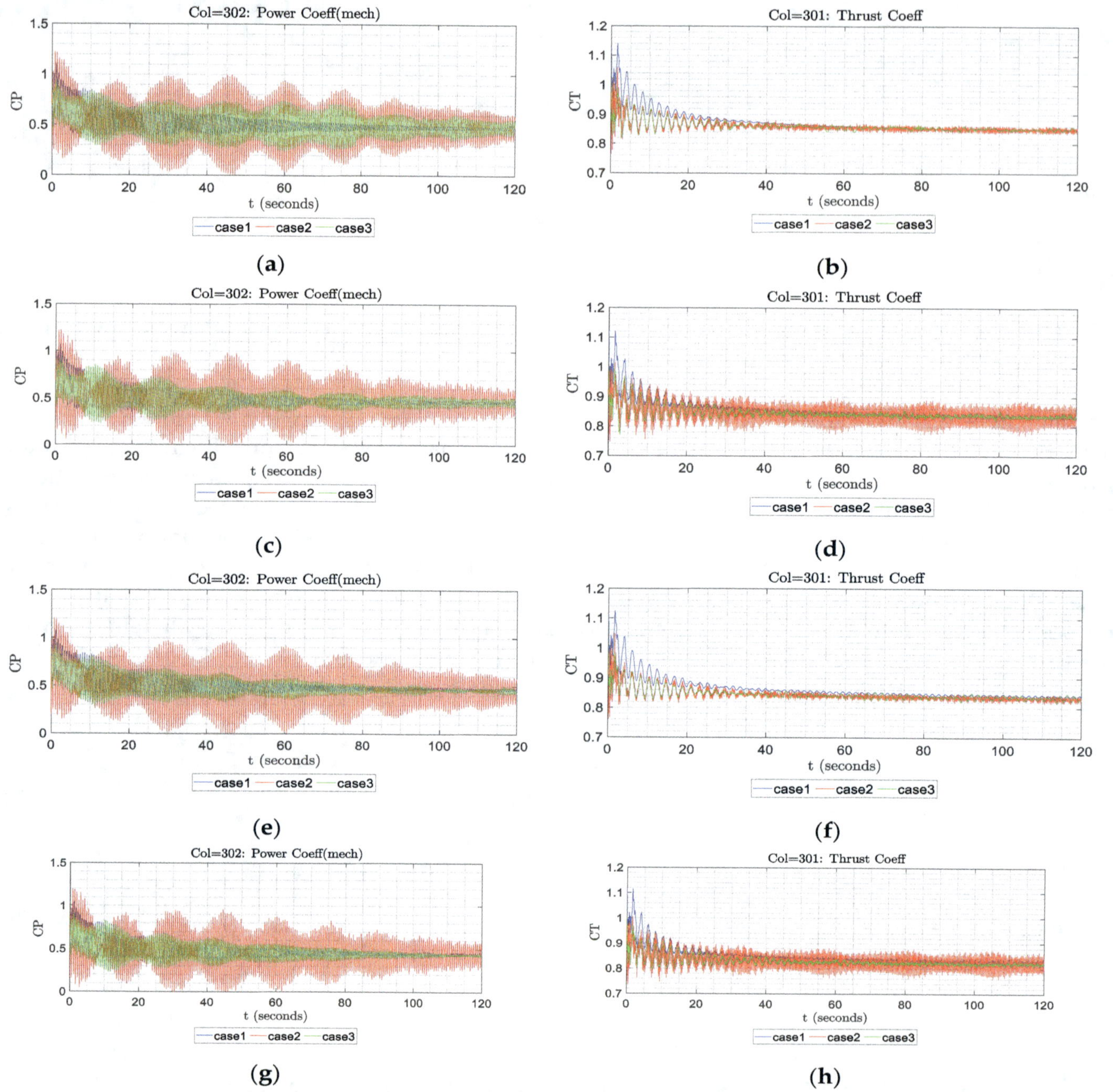

Figure 11. Power and thrust coefficients with a pitch adjustment: (**a**) C_p for Case 1; (**b**) C_T for Case 1; (**c**) C_p for Case 2; (**d**) C_T for Case 2; (**e**) C_p for Case 3; (**f**) C_T for Case 3; (**g**) C_p for Case 4; (**h**) C_T for Case 4.

3.3. Acoustical Analysis

Effects of Wind Shear and Yaw on Sound Emission

Figure 12 illustrates the polar plot of the sound pressure level in the function of azimuth angle of blade 1 for the four cases, which is called the amplitude modulation (AM) noise. AM noise is often a big issue because it gives more perception to the people living nearby. However, it should be noted that there are two different phenomena that make AM noise: one is caused by the blade noise directivity change during its operation [4] and another is caused by the non-uniform flow and operation, such as wind shear and yaw. In the present study, the latter is focused. The observer is located at 1.5 m height and a distance of 500 m downstream from the turbine (500 m away from the turbine is often a place where people are living although the sound propagation effects due to ground and atmospheric conditions are not considered here). The choice of the 500 m distance downstream location is to minimize the AM noise created by the blade noise directivity change. The speed of sound used in the calculations is 340 m/s. The turbulence level and turbulence length scale are set as 3% and 100 m, respectively. This choice was made in accordance with the 0.3 wind shear exponent defined in the scenarios. A more realistic choice can be made when the detailed flow information is measured. The 0° azimuth angle is defined when blade 1 is vertically down and coincides with the tower. All these four scenarios are performed with no control. As can be observed in Cases 1 and 3, the sound pressure level is slightly larger when the shear is included, while it significantly decreases with a 10° yaw, i.e., for both Cases 2 and 4. Normally, the SPL from one blade is highest when it is pointing upwards. The reason can be seen from the BPM sound prediction equations that the SPL is related to the fifth power of the oncoming velocity. For the case of three blades, at zero azimuth angle (blade 1 is pointing downwards), the other two blades are pointing at 120° and 240° when the noise is almost highest during its operation for Cases 3 and 4.

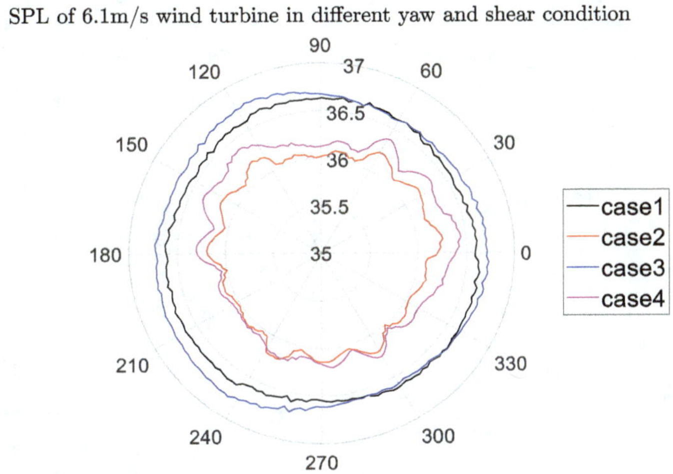

Figure 12. Sound pressure level in function of blade azimuth angle without control at 1.5 m height and 500 m distance downstream from the turbine at a wind speed of 6 m/s.

Figure 13 illustrates the comparison of SPL regarding different control scenarios. It can be observed that compared with the non-controlled case, the controlled SPL results, i.e., with control techniques I and II, are similar for Case 1 and Case 2. For Case 1, the peaks for both control strategies are in step, whereas the magnitude of the no control results is slightly smaller. For Case 2, the fluctuations are also generally in step, and both the time and magnitude of the peaks are consistent. Therefore, the implementation of the control techniques may not lead to noise reductions for the no-shear scenarios. The main reason for the negligible difference in Cases 1 and 2 is that the modifications of the pitch and the corresponding AoA are relatively small where the fluctuations of noise level are mainly due to blade deflections. A larger difference is expected for a larger pitch. It should be noted that Case 2 corresponds to a slightly lower SPL as compared to Case 1, due to the yaw-caused lower axial wind speed. For Cases 3 and 4, a larger decrease of SPL, i.e., about 0.4 dB, can be observed by introducing either control strategy I or II, whereas the trend remains the same. No significant difference can be found between the results of the two control techniques. It can be concluded that by introducing the control techniques, the reduction of SPL can be realized for the shear scenarios. In comparison to the angle of attack plots in Figure 8 with those plots of SPL in Figure 13, a non-linear effect between these two quantities is seen due to the blade structural responses. The AoA variation of the interpolation control in Figure 8d is reduced about 1°. According to the rule of a 1° AoA reduction resulting in a 1 dB noise reduction [39], the noise change in the present study is within this range. For comparison, it can be observed that the differences between the controlled and uncontrolled results of A-weighted SPL regarding all the cases are negligible. The reason is that the

A-weighted SPL accounts for more of the high-frequency noise to which humans are more sensitive and de-emphasizes the low-frequency contribution. In our case, the reduction of the noise mainly occurs in the low-frequency part and thus the A-weighting SPL is similar in all the cases. It should be noted that the low-frequency noise is important in far-field as it can propagate in a long distance with very small air absorption. To check the AM noise, the modulation depth for Case 4 is calculated according to the method proposed in [7] and shown in Table 3. From the table, it is seen that the modulation depth is slightly reduced.

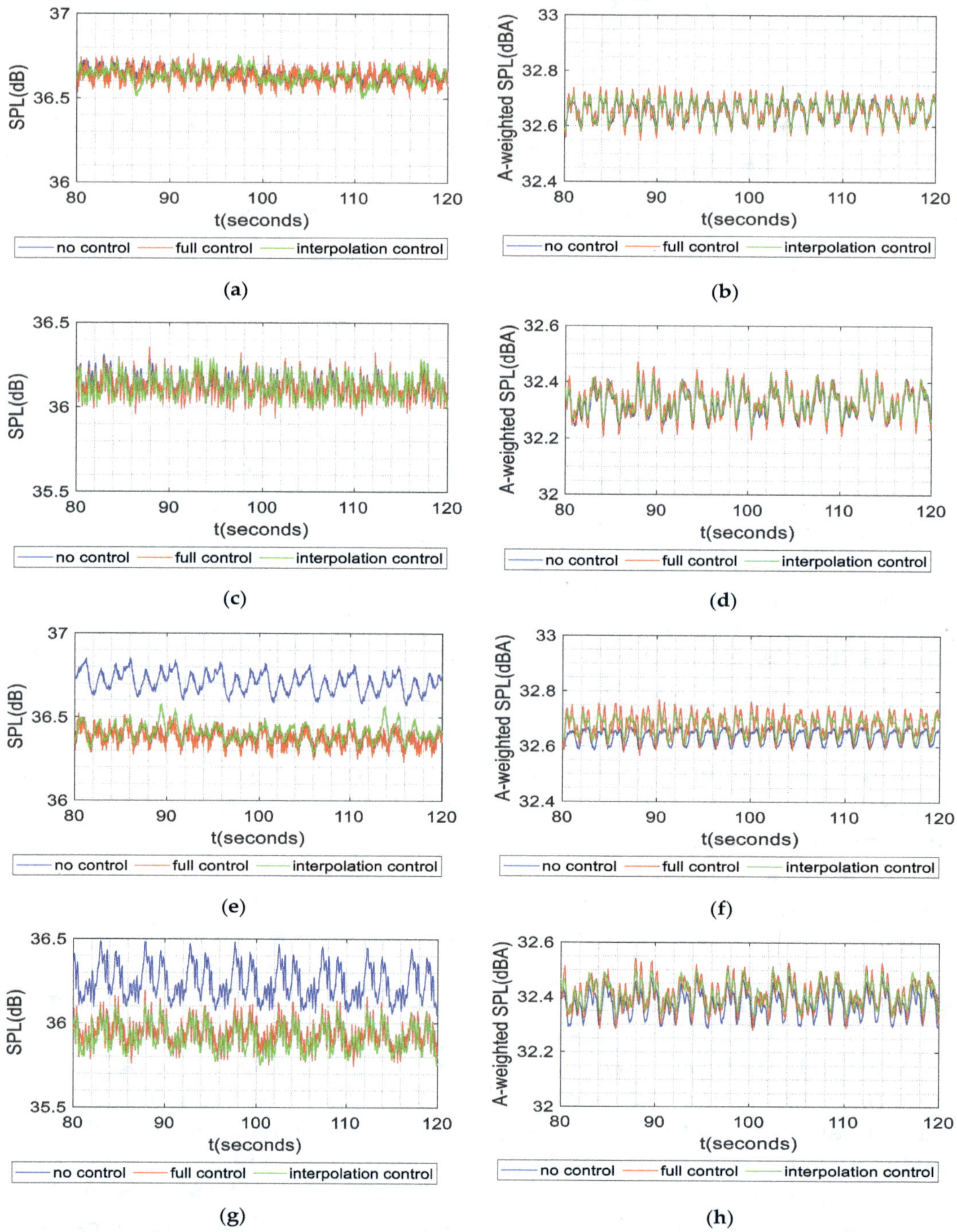

Figure 13. Sound pressure level and A-weighted sound pressure level of the turbine at 6 m/s wind speed: (**a**) sound pressure level (SPL) for Case 1; (**b**) A-weighted SPL for Case 1; (**c**) SPL for Case 2; (**d**) A-weighted SPL for Case 2; (**e**) SPL for Case 3; (**f**) A-weighted SPL for Case 3; (**g**) SPL for Case 4; (**h**) A-weighted SPL for Case 4.

Table 3. Modulation depth for Case 4.

condition	6 m/s no control	6 m/s full control	6 m/s inter control
modulation depth (dB)	0.3331	0.3127	0.2736
condition	10 m/s no control	10 m/s full control	10 m/s inter control
modulation depth (dB)	0.4308	0.3834	0.3692

Moreover, this phenomenon can be verified in Figure 14, where the contribution of SPL in different frequencies at the time instant (120 s) is shown. From the figure, it is seen that at the low frequency (which is less than 315 Hz), SPL is reduced obviously by the pitch control strategy.

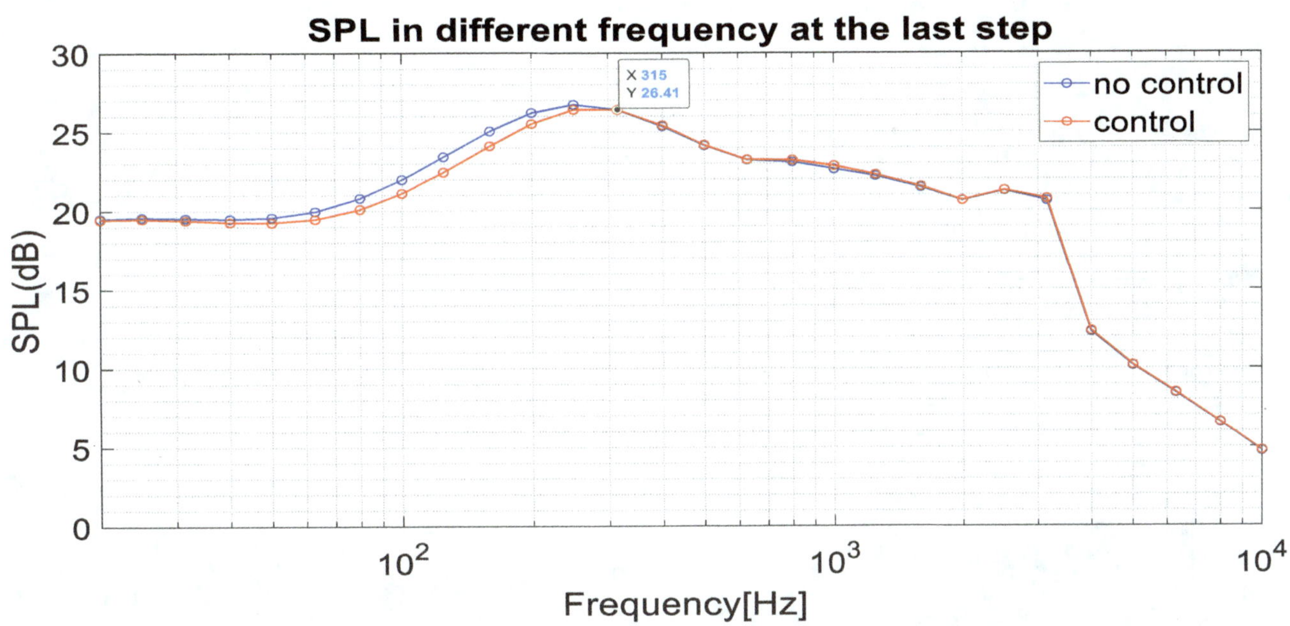

Figure 14. SPL spectra of case 4 of the turbine at 6 m/s wind speed at a time instant of 120 s after control.

Figure 15 illustrates the SPL and A-weighted SPL in Cases 1, 2 and 4, but with a lager inflow wind speed of 10 m/s. With the lager wind speed, the SPL increases about 8.5 dB in each case. The interesting thing deserved to be noticed is that the pitch control methods at 10 m/s display a similar effect in Case 4 as at 6 m/s but the variations in amplitude are more important. Since the A-weighted SPL was designed according to human hearing and emphasizes the contribution from the high frequency sound centered at 1000 Hz, it can give an impression that the control methods give a small effect on A-weighted SPL. The modulation depth is listed in Table 3 and a small reduction is seen. It is worth noting that the yaw and shear considered in this study are relatively small and these effects can be much bigger in large shear and yaw cases. Moreover, the noise propagation effects due to changing atmospheres and wakes [8] are not included here. In the future, the present noise source-modelling framework will be further coupled with the propagation model [8] in order to control wind turbine noise at far field.

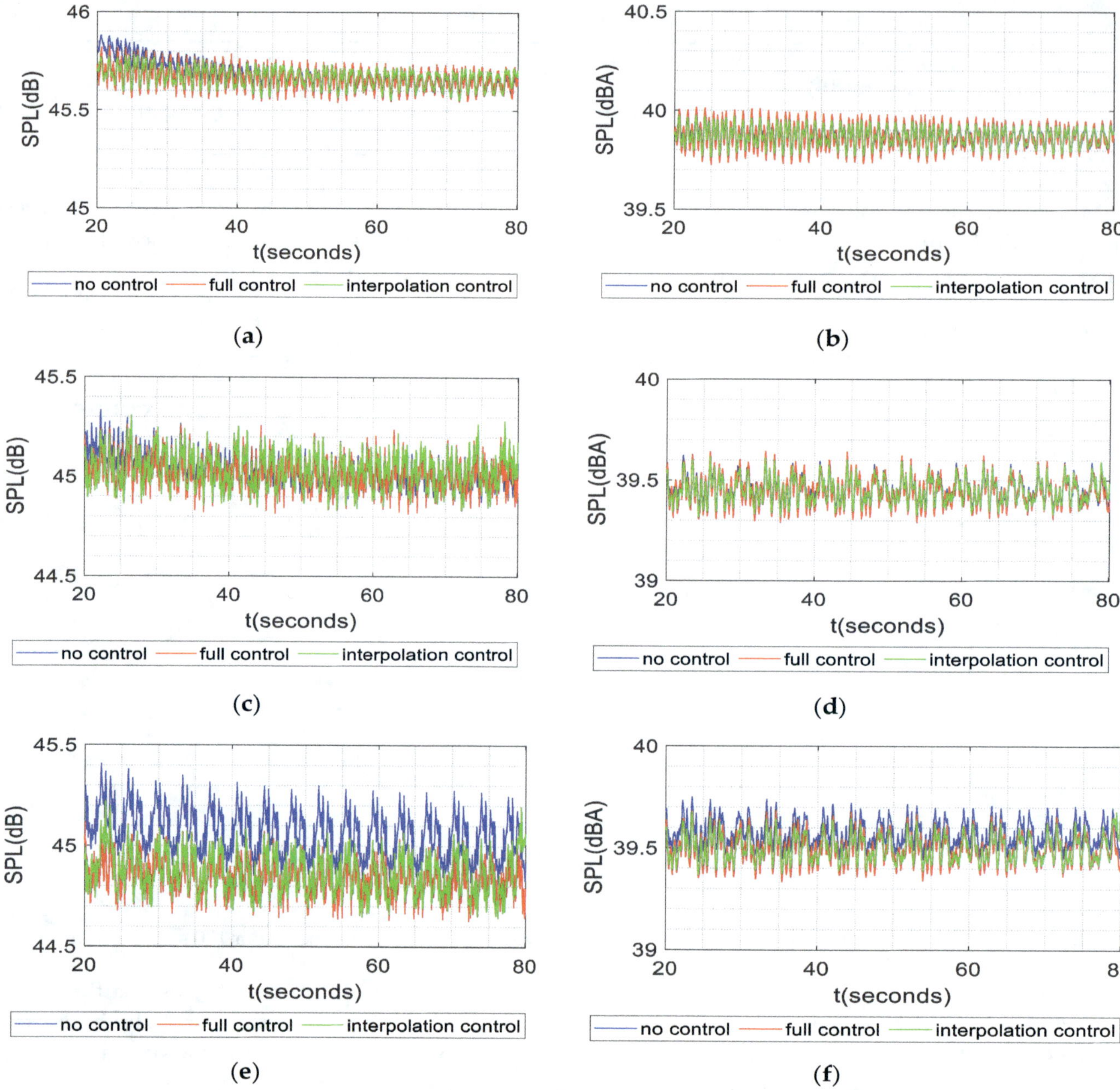

Figure 15. Sound pressure level and A-weighted Sound pressure level from the wind turbine at 10 m/s wind speed: (**a**) SPL for Case 1; (**b**) A-weighted SPL for Case 1; (**c**) SPL for Case 2; (**d**) A-weighted SPL for Case 2; (**e**) SPL for Case 4; (**f**) A-weighted SPL for Case 4.

4. Conclusions

In this paper, an advanced flow-structure-acoustics framework has been developed and validated against load field measurements for the case of the 2.3 MW NM80 wind turbine under a moderate wind shear and a small and a large yaw angle. The code has been further used for studying and controlling noise generation from a wind turbine under a moderate wind shear and a small yaw, which is often the case when a wind turbine operates. Through analysis of the results, several conclusions are summarized here:

1. Simulations based on the coupled Ellipsys3D/AL/Flex5 framework agreed well with field load measurements on the 2.3 MW NM80 turbine.

2. It was observed that the wind shear had a significant impact on the flow field, which led to a fainter wake. The longitudinal wind speed was reduced by introducing yaw, which accordingly decreased the sound pressure level.
3. The reduction of sound pressure level under a moderate wind shear situation with a shear exponent of 0.3 was undertaken by a pitch adjustment, while the power and thrust coefficients generally became more undulant. The pitch adjustment techniques were only useful for the case with a wind shear.
4. Given the miscellaneous reasons causing the wind turbine noise and the complex relations in SPL prediction formulations, it was obviously found that SPL and AoA were not linearly related. This non-linear relationship was also confirmed by comparing the variation of AoA and SPL.
5. Through the comparison of SPL and A-weighted SPL, the pitch control implemented in this paper mainly affected the wind turbine noise at low frequencies, while unapparent effects occurred with high-frequency noise.
6. The mean SPL was reduced with 0.4 dB and the modulation depth was reduced slightly.

Author Contributions: Conceptualization, W.Z.S.; Formal analysis, M.Z.; Investigation, M.Z.; Methodology, W.Z.S.; Project administration, W.Z.S.; Resources, W.Z.S.; Software, M.S. and W.Z.S.; Supervision, M.S., H.Y. and W.Z.S.; Visualization, M.Z.; Writing—original draft, M.Z.; Writing—review & editing, M.S., H.Y. and W.Z.S. All authors have read and agreed to the published version of the manuscript.

Abbreviations

BPM	a semi-empirical noise prediction model developed by Brooks, Pope and Marcolini
AL	actuator line
BEM	blade momentum theory
AoA	angle of attack
SPL	sound pressure level
PID	proportional–integral–derivative controller

References

1. Van den Berg, F.; Pedersen, E.; Bouma, J.; Bakker, R. *Project WINDFARM Perception: Visual Acoustic Impact of Wind Turbine Farms on Residents—Final Report. Report FP6–2005-Science-and-Society-20*; University Medical Centre Groningen: Groningen, Netherlands, 2008.
2. *Danish Regulation for Wind Turbine Noise*; Ministry of Environment and Food of Denmark: Copenhagen, Denmark, 2019. Available online: https://mst.dk/service/nyheder/nyhedsarkiv/2019/feb/ny-bekendtgoerelse-om-stoej-fra-vindmoeller/ (accessed on 13 February 2020).
3. How Loud Is a Wind Turbine. Available online: https://www.windpowerengineering.com/loud-wind-turbine-really/ (accessed on 27 October 2020).
4. Oerlemans, S. Detection of Aeroacoustic Sound Sources on Aircraft and Wind Turbines. Ph.D. Thesis, University of Twente, Enschede, the Netherlands, 2009.
5. Fukushima, A.; Yamamoto, K.; Uchida, H.; Sueoka, S.; Kobayashi, T.; Tachibana, H. Study on the amplitude modulation of wind turbine noise: Part 1—Physical investigation. In Proceedings of the INTER-NOISE, Innsbruck, Austria, 15–18 September 2013.
6. Coles, D.; Levet, T.; Cand, M. Application of the UK IOA Method for Rating Amplitude Modulation. In Proceedings of the 7th International Conference on Wind Turbine Noise, Rotterdam, the Netherlands, 2–5 May 2017.
7. Fernandez, F.A.; Burdisso, R.A.; Arenas, J.P. Indoor simulation of amplitude modulated wind turbine noise. *Wind Energy* **2017**, *20*, 507–519. [CrossRef]
8. Barlas, E.; Zhu, W.J.; Shen, W.Z.; Kelly, M.; Andersen, S.J. Effects of wind turbine wake on atmospheric sound propagation. *Appl. Acoust.* **2017**, *122*, 51–61. [CrossRef]

9. Schlinker, R.; Amiet, R. Helicopter rotor trailing edge noise. In Proceedings of the 7th Aeroacoustics Conference, Palo Alto, CA, USA, 5–7 October 1981; NASA CR-3470. pp. 1–145. [CrossRef]
10. Grosveld, F.W. Prediction of Broadband Noise from Horizontal Axis Wind Turbines. *J. Propuls. Power* **1985**, *1*, 292–299. [CrossRef]
11. De Wolf, W.B. EEN Predictiemethode voor het Aerodynamische Geluid van Windturbines met Horizontale AS. In *NLR TR 870181L*; Netherlands Aerospace Centre: Amsterdam, the Netherlands, 1986.
12. Viterna, L.A. The NASA-LeRC Wind turbine sound prediction code. In *NASA CP-2185*; NASA Lewis Research Center: Cleveland, OH, USA, 1981.
13. Brooks, T.F.; Pope, D.S.; Marcolini, M.A. Airfoil self-noise and prediction. In *NASA RP-1218*; NASA Langley Research Center: Hampton, VA, USA, 1989.
14. Fuglsang, P.; Madsen, H.A. Implementation and verification of an aeroacoustic noise prediction model for wind turbines. In *Risø-R-867(EN)*; Risø National Laboratory: Roskilde, Denmark, 1996.
15. Lowson, M.V. Assessment and prediction of wind turbine noise. In *Flow Solutions Report 92/19—ETSU W/13/00284/REP*; Harwell Laboratory, Energy Technology Support Unit: Harwell, UK, 1992.
16. Amiet, R.K. Acoustic radiation from an airfoil in a turbulent stream. *J. Sound Vibr.* **1975**, *41*, 407–420. [CrossRef]
17. Moriarty, P.; Migliore, P. Semi-empirical aeroacoustic noise prediction code for wind turbines. In *NREL/TP-500-34478*; National Renewable Energy Lab.: Golden, CO, USA, 2003; pp. 1–39.
18. Zhu, W.J.; Heilskov, N.; Shen, W.Z.; Sørensen, J.N. Modeling of aerodynamically generated noise from wind turbines. *J. Solar Energy Eng.* **2005**, *127*, 517–528. [CrossRef]
19. Bowdler, D. Wind shear and its effect on noise assessment. In Proceedings of the Third International Conference on Wind Turbine Noise, Aalborg, Denmark, 17–19 June 2009. Available online: https://www.dickbowdler.co.uk/content/publications/Wind-Shear-and-its-Effect-on-Noise-Assessment.pdf (accessed on 26 October 2020).
20. Oerlemans, S. Effect of wind shear on amplitude modulation of wind turbine noise. *Int. J. Aeroacoust.* **2015**, *14*, 715–728. [CrossRef]
21. Tian, Y.; Cotté, B. Wind turbine noise modeling based on Amiet's theory: Effects of wind shear and atmospheric turbulence. *Acta Acust. United Acust.* **2016**, *102*, 626–639. [CrossRef]
22. Parchen, R. Progress report DRAW, a prediction scheme for trailing-edge noise based on detailed boundary-layer characteristics. In *TNO-Report HAG-RPT-980023, Report*; TNO Institute of Applied Physics: The Hague, the Netherlands, 1998.
23. Kamruzzaman, M.; Lutz, T.; Würz, W.; Shen, W.Z.; Zhu, W.J.; Hansen, M.; Bertagnolio, F.; Madsen, H.A. Validations and improvements of aeroacoustics models using detailed experimental data. *Wind Energy* **2012**, *15*, 45–61. [CrossRef]
24. Tian, Y. Modeling of Wind Turbine Noise Sources and Propagation in the Atmosphere. Ph.D. Thesis, ENSTA Paris Tech, Palaiseau, France, 2016.
25. Sørensen, J.N.; Shen, W.Z. Numerical modeling of wind turbine wakes. *J. Fluids Eng.* **2002**, *124*, 393–399. [CrossRef]
26. Kleusberg, E.; Mikkelsen, R.F.; Schlatter, P.; Ivanell1, S.; Henningson, D.S. High-order numerical simulations of wind turbine wakes. *J. Phys. Conf. Ser.* **2017**, *854*, 012025. [CrossRef]
27. Kleusberg, E. Wind-Turbine Wakes—Effects of Yaw, Shear and Turbine Interaction. Ph.D. Thesis, Royal Institute of Technology, Stockholm, Sweden, 2019.
28. Michelsen, J.A. Basis3D-a platform for development of multiblock PDE solvers. In *Technical Report AFM 92-05*; Technical University of Denmark: Copenhagen, Denmark, 1992.
29. Sørensen, N.N. General purpose flow solver applied to flow over hills. In *Risø-R-827(EN)*; Risø National Lab.: Roskilde, Denmark, 1995.
30. Shen, W.Z.; Zhu, W.J.; Sørensen, J.N. Actuator line/Navier–Stokes computations for the MEXICO rotor: Comparison with detailed measurements. *Wind Energy* **2012**, *15*, 811–825. [CrossRef]
31. Øye, S. FLEX Simulation of wind turbine dynamics. In *State of the Art of Aeroelastic Codes for Wind Turbine Calculations*; Pedersen, M.B., Ed.; The Technical University of Denmark: Lyngby, Denmark, 1996.
32. Ta Phuoc, L. Modèles de Sous Maille Appliqués aux Ecoulements Instationnaires Décollés. In *Aérodynamique Instationnaire Turbulente-Aspects Numériques et Expérimentaux*; DGA/DRET: Paris, France, 1994.

33. Shen, W.Z.; Zhu, W.J.; Sørensen, J.N. Aero-acoustic computations for turbulent airfoil flows. *AIAA J.* **2009**, *47*, 1518–1527. [CrossRef]
34. Øye, S. *FLEX5 Manual*; Department of Wind Energy, Technical University of Denmark: Lyngby, Denmark, 1999.
35. Drela, M. XFOIL, an analysis and design system for low Reynolds number airfoils. In Proceedings of the Conference on Low Reynolds Number Aerodynamics, University of Notre Dame, Notre Dame, IN, USA, 5–7 June 1989; pp. 1–12.
36. Garcia, N.R.; Sørensen, J.N.; Shen, W.Z. A strong viscous-inviscid interaction model for rotating airfoils. *Wind Energy* **2014**, *17*, 1957–1982. [CrossRef]
37. Leloudas, G.; Zhu, W.J.; Sørensen, J.N.; Shen, W.Z.; Hjort, S. Prediction and reduction of noise from a 2.3 MW wind turbine. *J. Phys. Conf. Ser.* **2007**, *75*, 012083. [CrossRef]
38. Madsen, H.A.; Bak, C.; Troldborg, N. The DanAero experiments—A detailed investigation of MW wind turbine aerodynamics and aeroacoutics. *Wind Energy* **2020**. submitted.
39. Wagner, S.; Bareib, R.; Guidati, G. *Wind Turbine Noise*, 1st ed.; Springer: Berlin/Heidelberg, Germany, 1996.

10

Comparison of the Blade Element Momentum Theory with Computational Fluid Dynamics for Wind Turbine Simulations in Turbulent Inflow

Sebastian Ehrich [1,*], Carl Michael Schwarz [1], Hamid Rahimi [1,2], Bernhard Stoevesandt [2] and Joachim Peinke [1,2]

1 ForWind, University of Oldenburg, Institute of Physics, Küpkersweg 70, 26129 Oldenburg, Germany; carl.michael.schwarz@uni-oldenburg.de (C.M.S.); hamid.rahimi@uni-oldenburg.de (H.R.); joachim.peinke@uni-oldenburg.de (J.P.)
2 Fraunhofer IWES, Küpkersweg 70, 26129 Oldenburg, Germany; bernhard.stoevesandt@iwes.fraunhofer.de
* Correspondence: sebastian.ehrich@uni-oldenburg.de

Abstract: In this work three different numerical methods are used to simulate a multi-megawatt class class wind turbine under turbulent inflow conditions. These methods are a blade resolved Computational Fluid Dynamics (CFD) simulation, an actuator line based CFD simulation and a Blade Element Momentum (BEM) approach with wind fields extracted from an empty CFD domain. For all three methods sectional and integral forces are investigated in terms of mean, standard deviation, power spectral density and fatigue loads. It is shown that the average axial and tangential forces are very similar in the mid span, but differ a lot near the root and tip, which is connected with smaller values for thrust and torque. The standard deviations in the sectional forces due to the turbulent wind fields are much higher almost everywhere for BEM than for the other two methods which leads to higher standard deviations in integral forces. The difference in the power spectral densities of sectional forces of all three methods depends highly on the radial position. However, the integral densities are in good agreement in the low frequency range for all methods. It is shown that the differences in the standard deviation between BEM and the CFD methods mainly stem from this part of the spectrum. Strong deviations are observed from 1.5 Hz onward. The fatigue loads of torque for the CFD based methods differ by only 0.4%, but BEM leads to a difference of up to 16%. For the thrust the BEM simulation results deviate by even 29% and the actuator line by 7% from the blade resolved case. An indication for a linear relation between standard deviation and fatigue loads for sectional as well as integral quantities is found.

Keywords: wind turbine; aerodynamics; turbulent inflow; Computational Fluid Dynamics; blade element momentum theory; actuator line method; Fatigue Loads

---

1. Introduction

One of the most common tools for wind turbine design and load calculation is the Blade Element Momentum method (BEM). Although BEM based tools are known to be very efficient, they are based on many simplifications. Among others, some worth mentioning are stationary environment, 2D airfoil characteristics and inviscid fluid behavior [1]. While preserving its efficiency, over the years a lot of correction models have been developed to improve the accuracy, e.g., by using tip loss or dynamic stall models. A comprehensive overview of those models can be found in the work of Schepers [1]. However, especially for turbulent inflow scenarios, the reliability of such estimations is still unknown. This is mainly due to the relatively coarse discretization of wind fields and rotor blades, the incorrect dynamic aerodynamic response to complex inflow, the unresolved flow field but also the assumption of inviscid and stationary flow. Nevertheless, wind turbines are operating under atmospheric wind conditions and the dynamics of the forces that are acting on the blades are still unknown. Most of the effects were measured in the wind tunnel with uniform or very simplified changing inflow conditions. However, turbulent inflow scenarios are relevant to the design and certification process, e.g., to the design load cases 1.1–1.3 of the IEC standard [2]. The research conducted in this field is limited with recent studies done by Madsen et al. [3] and in the AVATAR project [4]. In both cases BEM is compared to higher fidelity codes. In the work by Madsen et al. a 2.5MW turbine was simulated with BEM and Blade Resolved (BR) Computational Fluid Dynamics (CFD) with a turbulent inflow based on the model by Mann [5]. The wind fields for BEM were extracted at the rotor position in the empty CFD domain. Additionally, measurement data of this turbine and BEM results for the same data were compared. In both cases spectra of sectional forces were investigated. In the AVATAR project similar quantities were in the focus of the study, but one Mann box was generated and used for the CFD as well as the BEM simulation. In CFD the natural decaying process of the wind fields leads to a decreased turbulence intensity and different turbulent structures at the turbine compared to BEM. Therefore the simulation results of BEM and CFD are not comparable. Also global quantities like thrust and torque as well as their corresponding fatigue loads were not investigated. Although those works are giving a valuable insight to BEM modeling uncertainties when combined with turbulent inflow, there is still a lack of comparable studies and a best practice guide for the comparison of CFD and BEM simulations is still missing.

The present work aims to compare BEM based simulations for turbulent inflow conditions with two higher fidelity CFD methods, namely the BR simulation and the Actuator Line (AL) model [6]. New in this work is the combination of the investigation of frequency resolved integral forces, the analysis of fatigue loads and a bridge between BR and BEM calculations, namely the AL method. The goal of this study is to shed light into the difficulties and expectations of the different methods proposed and information for future comparisons of CFD and BEM based simulations shall be provided.

This paper is organized as following. The basic setup of the turbine, the turbulent wind fields, the CFD and BEM simulations, and the fatigue load estimation are addressed in Section 2. In Section 3 key aerodynamic quantities of the turbine such as power, thrust and sectional forces and the corresponding equivalent fatigue loads as well as spectral properties are investigated. Finally a conclusion including future work is presented in Section 4.

2. Methodology

In this section the basic methodology used in this work concerning the wind turbine, the wind fields and the numerical setup as well as a short introduction to the fatigue load estimation procedure is presented.

2.1. The Wind Turbine Model

The numerical studies are performed on the NREL-5MW reference wind turbine with a rotor diameter of 126 m [7]. This three bladed horizontal axis wind turbine with a blade length of 61.5 m was designed by the National Renewable Energy Laboratory. It is close to the current average rotor size of newly installed wind turbines which accounts to 100–120 m. More details on the reference turbine can be found in Table 1.

Table 1. Overview of NREL-5MW turbine specification.

Parameter	Value
Number of rotor blades	3
Rotor diameter	126 m
Rated aerodynamic power	5.3 MW
Rated wind speed	11.4 m/s

Table 1. *Cont.*

Parameter	Value
Rated rotational speed	12.1 rpm
Blade cone angle	2.5°
Shaft tilt angle	5.0°
Blade length	61.5 m
Blade mass	17,740 kg

2.2. The Wind Fields

It is assumed that the specific inflow turbulence model does not change the general trend of the outcome of the study presented in this paper as long as some main properties of the wind fields are fulfilled like spatial and temporal correlations, Gaussianity and stationarity. However, the model used here should also be as simple as possible with a very low memory consumption which is for spectral methods usually very high if the time and space domains are large. Therefore velocity fluctuations are represented by a Gaussian process realized at the inflow patch of the CFD domain. In particular, coupled stochastic Ornstein-Uhlenbeck processes of the form

$$\frac{\mathrm{d}u_i{}^{(k)}(t)}{\mathrm{d}t} = -\gamma\left(u_i{}^{(k)}(t) - \bar{u}^{(k)}\right) + \sqrt{D}\sum_j^N H_{ij}\Gamma_j{}^{(k)}(t) \tag{1}$$

with an exponentially decaying spatial and temporal correlation function for all wind components are used without taking shear into account. In this stochastic differential equation each subscript $i, j = 1 \ldots N$ denotes a specific cell on the inlet patch where the turbulence is generated and each superscript $k = x, y, z$ corresponds to the longitudinal, lateral or vertical velocity component. $\gamma = 0.97\ \mathrm{s}^{-1}$ is a damping factor, $\bar{u}^{(y,z)} = 0\ \mathrm{ms}^{-1}$ and $\bar{u}^{(x)}$ the mean velocities in x, y and z direction, $D = 0.62\ \mathrm{m}^2\mathrm{s}^{-3}$ a diffusion constant, H the lower Cholesky-decomposed correlation matrix and Γ a Gaussian, delta-correlated random noise vector with variance 2.

Fields with a mean wind speed of $\bar{u}^{(x)} = 11.4\ \mathrm{ms}^{-1}$ corresponding to the rated operating condition of the wind turbine and a Turbulence Intensity (TI) of 20% are fed into the CFD domain. This relatively high TI had to be chosen, because of the natural decaying process leading to much smaller turbulence intensities. namely 11%, in the vicinity of the turbine in the absence of shear. To make the comparison between different codes in a similar manner, the pitch and rotor speed are fixed and the effect of structural deformations, tilt and cone as well as the tower and nacelle are neglected.

2.3. Numerical Methods

In the following the BR, AL and BEM method are described in more detail:

- BR: The blade resolved simulations are performed with the CFD software OpenFOAM V4.1 [8], which is an open-source package with a collection of modifiable libraries written in C++. The simulation domain is discretized using the bladeblockMesher [9] and windTurbine mesher tools [10] with a combination of structured and unstructured grids consisting of a total of 36M cells. To exclude influences of the mesh on the solution, a mesh study has been done in our previous work [11]. The grid of the complete simulation domain and the rotating part are shown in Figure 1 and a sectional view on the mesh near the blade in Figure 2. The cell length at the inlet, which is three rotor diameters upstream of the rotor, is 1m. This mesh is uniformly extruded up to the near blade mesh where a refinement takes place and the smallest cell length of approximately 0.2 m is reached. In order to limit the cell aspect ratio in radial direction, adaptive wall functions are being applied. The solution for the flow field is achieved by solving the incompressible Delayed Detached Eddy Simulations equations. It is assumed that the flow is fully turbulent, i.e., the laminar-turbulent transition in the boundary layer is not considered, because the inflow itself is turbulent and therefore the laminar boundary layer is not expected to have a strong influence on the blade. The closure problem is treated by use of the turbulence model proposed by Spalart et al. [12] with the proposed standard parameters. In order to solve the pressure-velocity coupling, the PIMPLE algorithm, is used, which is a combination of the loop structures of SIMPLE [13] and PISO [14]. Integration in time was performed using the implicit second order backward scheme together with the second order linear upwind scheme for discretizing the convection terms. For these simulations 360 cores were used for approximately 20 days resulting in 600 s simulation time with a time step of 0.01 s. This corresponds to 121 revolutions of the rotor.

- AL: The rotor is simplified by an AL model [6], where all rotor blades are reduced to rotating lines with a radial distribution of body forces at 60 positions along the lines. A Gaussian filtering kernel with the dimensionless smoothing parameter ε equal to 4 was used to obtain well fitting sectional forces and power output to the BR case. Based on the angle of attack at the given positions on the line, body forces are gained from lookup tables containing polars of two dimensional airfoils. A Prandtl tip [15] and root loss correction, as well as a 3D correction proposed by Du and Selig [16] are applied to the airfoil data. Except for a coarser resolution in the vicinity of the rotor with an average cell length of 1m the domain is exactly the same as for the BR case leading to a domain size of 22M cells. The flow field is simulated by Large Eddy Simulations (LES) with a Smagorinsky subgrid-scale model [17]. These simulations have been conducted on 360 cores for 25 h real time reaching 600 s simulation time with a time step of 0.01 s.
- BEM: This method uses wind fields recorded in the CFD domain on an equidistant 31×31 grid with a grid cell size of 4.5 m in both directions. In this case the fields are obtained at the position of the rotor from the same mesh as for the AL method but in an empty CFD domain. This seems to be a reasonable choice, because under this condition the TI at the rotor will be comparable for the CFD and BEM simulations. The BEM based tool FAST v8 [18] in combination with AeroDyn v15 [19] with the Beddoes Leishman type unsteady aerodynamics model [19,20] and an equilibrium wake model is used in this work. The BEM model relies on the very same airfoil data and correction models as the AL, but with only 17 blade sections. Because of the simplicity of the BEM approach the simulation with the same time stepping and total time as for the higher fidelity approaches only took two minutes on one core.

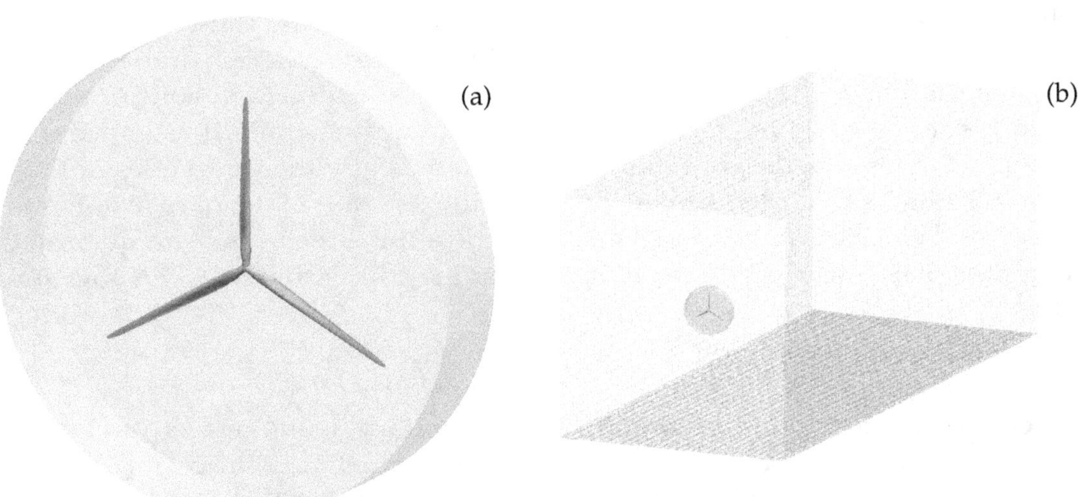

Figure 1. Grids of (**a**) rotating part and (**b**) full domain of Blade Resolved (BR) simulation.

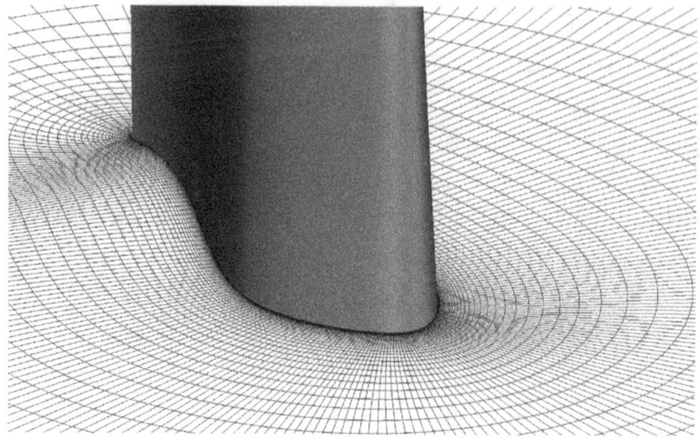

Figure 2. View onto the grid at the pressure side of the rotor blade.

2.4. Fatigue Load Estimation

One of the most important quantities relevant for wind turbine designs are the fatigue loads on different parts of the turbine estimated by damage equivalent loads. They are also known as Equivalent Fatigue Loads (EFL), which will play a role in the load analysis of the three methods investigated here. EFL can be defined as

$$EFL = \left(\frac{\sum n_i r_i^m}{T}\right)^{\frac{1}{m}}, \quad (2)$$

where r_i are the specific load ranges, n_i the corresponding amount of load ranges, T the reference number of ranges (here the dimensionless time in seconds) and $m = 8$ the Wöhler exponent. More information on the estimation of the EFL and the underlying rainflow algorithm can be found in [21,22].

3. Results

In this section the results of this study are presented. The focus lies on the wake structure, average and standard deviation of sectional and integral forces, as well as fatigue loads and Power Spectral Densities (PSD).

3.1. Wake Dynamics

Snapshots of the velocity magnitude for the BR and AL cases are shown in Figure 3. Despite the differences in the methods both simulations show similar fields in terms of the velocity magnitude upstream of the turbine. This leads to the conclusion that the induction zone of the turbine is also similar for both simulations. For a more detailed view of the similarities and differences of the two compared CFD methods the wake structure is analyzed by the time series of the main stream component at 0.5, 1 and 2 diameters at the centerline downstream of the nacelle in Figure 4. It can be seen that the wind velocity is in good agreement in most time intervals. As a quantitative measure of similarity of the wake structure, the coefficients for the correlation between both simulations are calculated by

$$\rho(x) = \frac{\langle (u_{AL}(x,t) - \bar{u}_{AL}(x))\,(u_{BR}(x,t) - \bar{u}_{BR}(x))\rangle_t}{\sigma_{AL}(x)\sigma_{BR}(x)} \quad (3)$$

where we used the time averaging operation $\langle . \rangle_t$, the standard deviations σ_{AL} and σ_{BR} and the local mean values $\bar{u}_{AL}$ and $\bar{u}_{BR}$ of the time series of AL and BR simulations. The correlation is 0.54, 0.41 and 0.51 for 0.5, 1 and 2 diameters downstream, respectively. This shows that the wind field dynamics are rather similar and not uncorrelated at different distance from the rotor for both simulation methods. However, the BR simulation is resolving the tip and root vortex in a more physical way which results in smaller structures in the near wake.

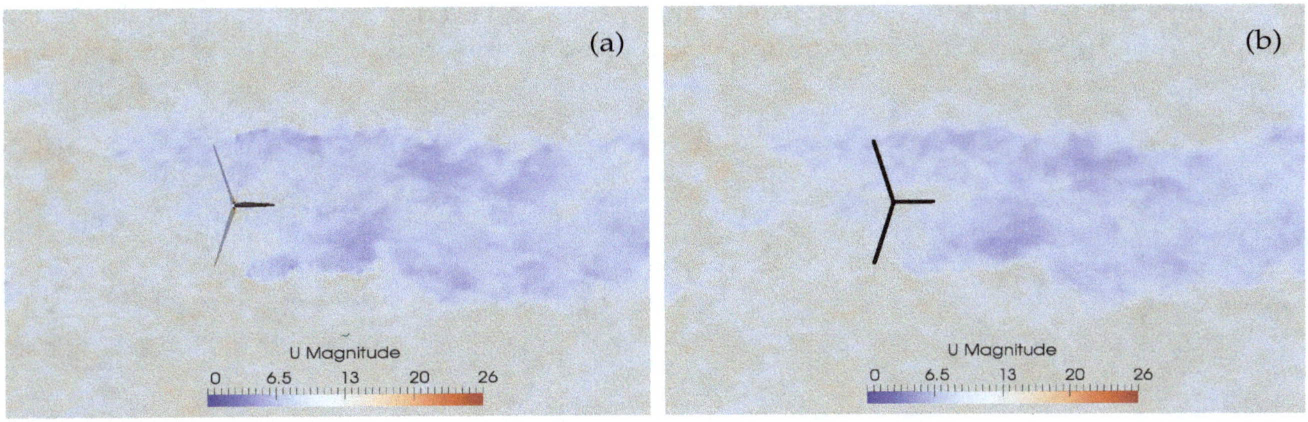

Figure 3. Snapshots of the velocity magnitude for (**a**) BR and (**b**) Actuator Line (AL) simulations.

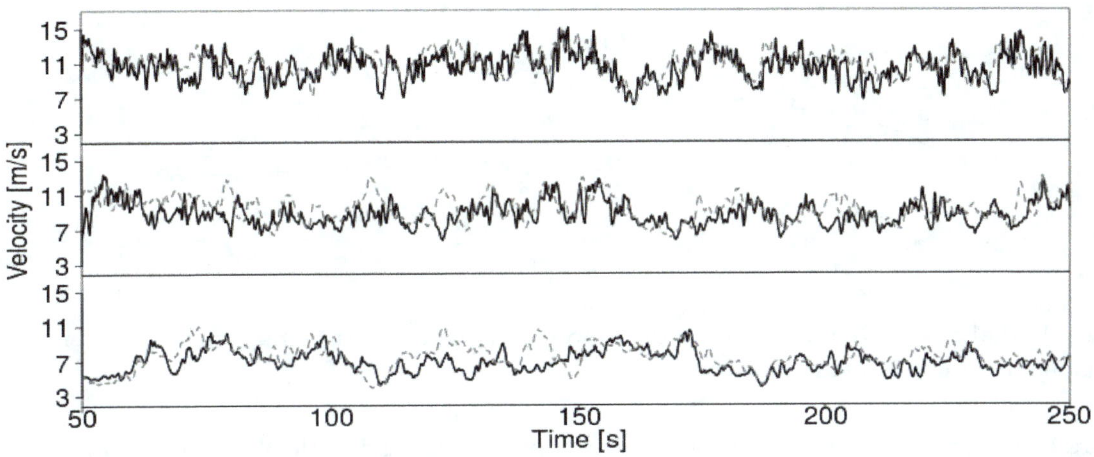

Figure 4. Wind velocity time series of BR (black, solid) and AL (grey, dashed) simulation for 0.5, 1 and 2 diameters (from top to bottom) at the centerline downstream of the rotor nacelle.

3.2. Sectional Forces

All three simulation methods (BR, AL and BEM) are compared based on averaged sectional quantities, namely axial and tangential forces, the corresponding standard deviations and the EFLs. The sectional axial and tangential forces on the blade do not reveal big differences for the average between all cases in the mid-span as shown in Figure 5. However at the root (up to 30% of span) and the very tip (from 80% of span onwards) some differences can be observed. BEM and AL rely on similar root and tip correction models and the very same airfoil data which explains their good agreement. In comparison to the more realistic BR case, where 3D effects are modeled physically, the forces at the root and tip section are far off supporting the well known issues with 3D effects for BEM and AL. The more developed vortices in the BR simulation induce much higher loads in the root and tip region than for the other methods.

The standard deviations for the tangential and axial force distributions, shown in more detail in Figure 6, are very different for all three methods. With the BR case as the reference the standard deviation of sectional forces tend to be lower at the root of the blade for the BEM and AL methods whereas they are larger in the mid-span. Close to the tip at approximately 95% of the blade the AL method gives much smaller values while BEM is in the range of the BR simulation. The corresponding errors are estimated by following three steps. In the first step the time series for the specific forces at each radial position are separated into five distinct sub time series. In the second step the standard deviation for each sub time series is calculated. Finally the standard deviation of the set of the standard deviations from the second step is computed which provides the error for the whole time series.

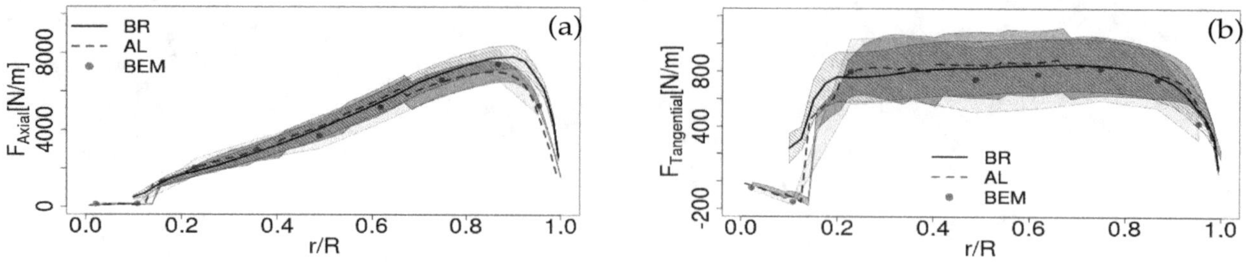

Figure 5. Radial distributions of (**a**) axial and (**b**) tangential forces per unit span. The shaded areas illustrate one standard deviation of the fluctuations.

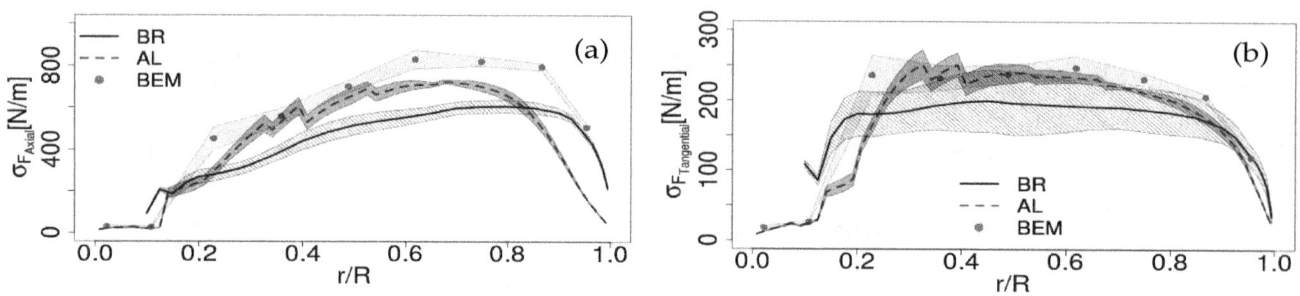

Figure 6. Radial distribution of standard deviation of (**a**) axial and (**b**) tangential forces per unit span. The shaded areas correspond to errors.

A more detailed insight in the standard deviation is gained by looking at the PSDs of axial and tangential forces at 22%, 50%, 75% and 95% radial position, shown in Figures 7 and 8.

All blade sections have in common that the spectra of the highly resolved BR case drop fastest between 2 and 10 Hz to a constant noise level. This is due to the expected bigger structures generated by the blade geometry but also due to the DDES model which is treated as RANS in the boundary layer acting as a filter.

For the AL method the energy drop happens at much higher frequencies because of the missing geometry and the LES model, which generally filters at smaller scales. The noisy behavior at high frequencies is part of every simulation of this kind and is due to numerical instabilities resulting from the convolution of the forces given at the airfoil sections with a Gaussian kernel [23].

The BEM spectra show filtering effects at higher frequencies corresponding to a multiplication with a *sinc*-like-function but with different frequencies for different radial positions. The origin of this effect could not be found and has to be investigated further. Due to the missing geometry and the lack of unsteady wake effects BEM is also expected to make worse predictions than the higher fidelity methods in the high frequency range. Therefore this high frequency part has to be taken with care like for the other two methods with their aforementioned problems of smoothing for the AL and the RANS treatment for the BR case and none of them can be considered trustworthy in this range.

In the low frequency range, for axial as well as tangential forces it can be observed that in comparison to the other two methods BEM has a much higher energy content for the inboard section. This is responsible for the high standard deviation at this section.

At radial position of 50% and 75% of the blades all the spectra are very similar in the low frequency range but it is still obvious that BEM has a higher energy content up to the 1P frequency, especially for the axial forces. Furthermore all three methods are able to capture the 1P frequency which is based on the passing of the blades through long time correlated structures.

At the outboard section at 95% the energy content of the BEM results in low frequencies is in the same range as for the BR case, but the AL method shows significant smaller values and the 1P frequency is not captured anymore, especially for the axial forces, which is due to a smeared out tip vortex.

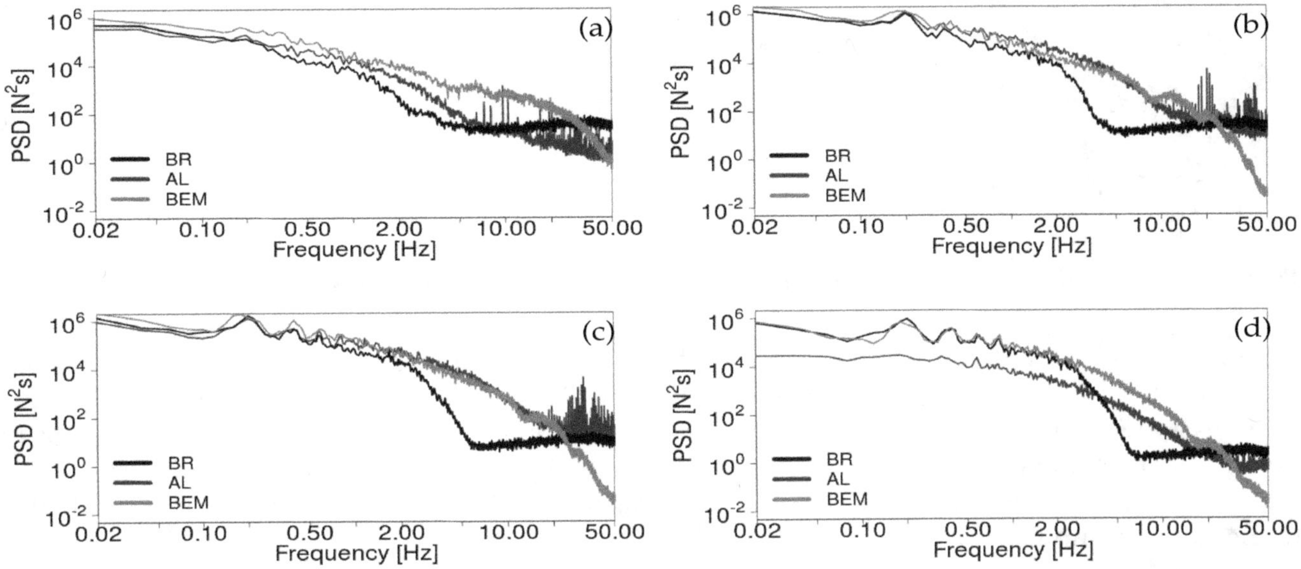

Figure 7. Power Spectral Density (PSD) of axial forces at (**a**) 22.5%; (**b**) 50%; (**c**) 75% and (**d**) 95% radial position.

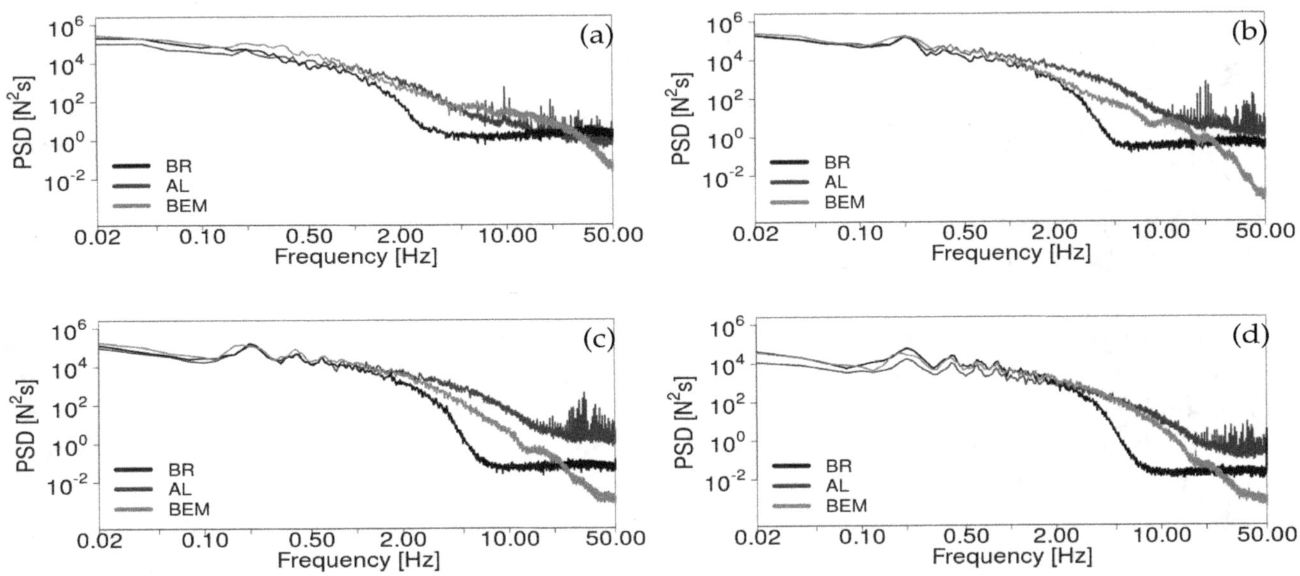

Figure 8. PSD of tangential forces at (**a**) 22.5%; (**b**) 50%; (**c**) 75% and (**d**) 95% radial position.

To illustrate the different flow structure sizes in the different simulation methods, the two point correlation of the sectional forces along the blade are measured. The argument for this approach is that the correlation of sectional forces is similar to the correlation of the wind velocity close to the blade surface, which is strongly connected to the integral scales of wind. In Figure 9 the correlation at each radial position with the forces at a reference position, here 22.5% radial location, is shown. This specific position is in principle arbitrary, but if the focus lies on the measurement of big structures, this point should not be too close to the tip or root where smaller flow structures dominate. However, even if the tendency for the correlation functions of the three different models is the same, the larger correlation values for the BR case along the whole blade indicate that the structure sizes are also bigger than for the AL and BEM cases.

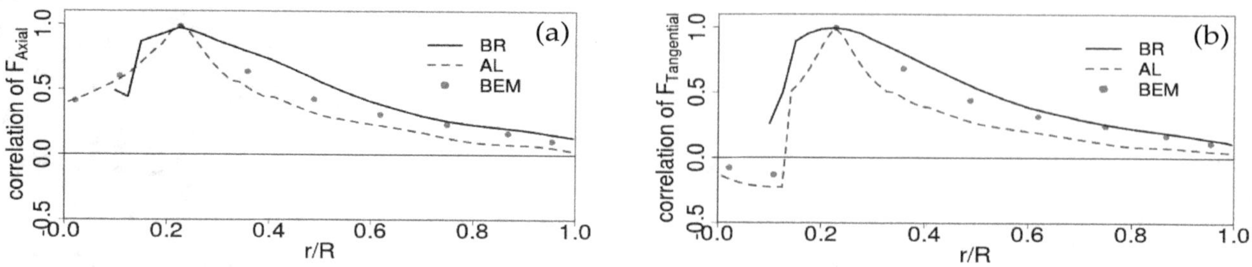

Figure 9. Correlation of (**a**) axial and (**b**) tangential forces at different radial positions with 22.5% position.

To give a more detailed description of the effect of forces on the rotor blades, fatigue loads at each section have been investigated. As observed in Figure 10, the EFL distribution over the whole blade is similar to the distribution of standard deviation in Figure 6. The BEM method gives higher EFL for every section for axial forces and almost everywhere higher values for tangential forces. All other points mentioned for sectional standard deviations also hold for sectional EFL. The errors are calculated with the same method as for Figure 6.

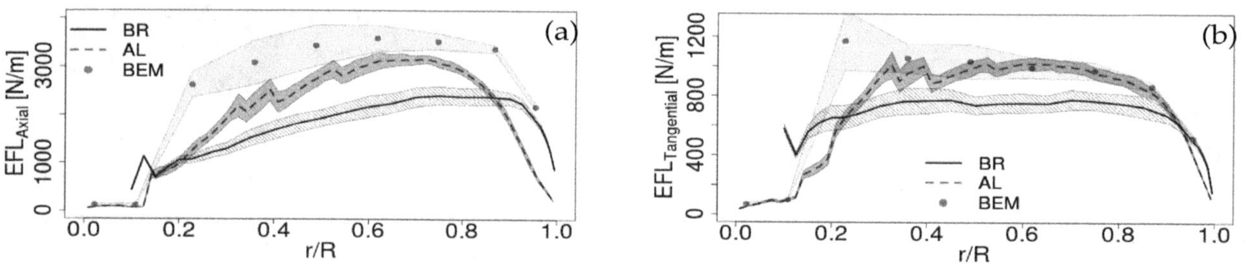

Figure 10. Radial distribution of (**a**) axial and (**b**) tangential equivalent fatigue loads per unit span. The shaded areas correspond to the errors.

The sectional EFL are linearly dependent on the sectional standard deviation for all simulation methods as seen in Figure 11. This raises the point that the dynamics quantified by two different properties of the load time series, namely standard deviation and EFL, are similar for all three methods. With the Wöhler Exponent $m = 8$ and the wind fields used here, this linear relation follows approximately $EFL = 4\sigma$ for axial and tangential forces.

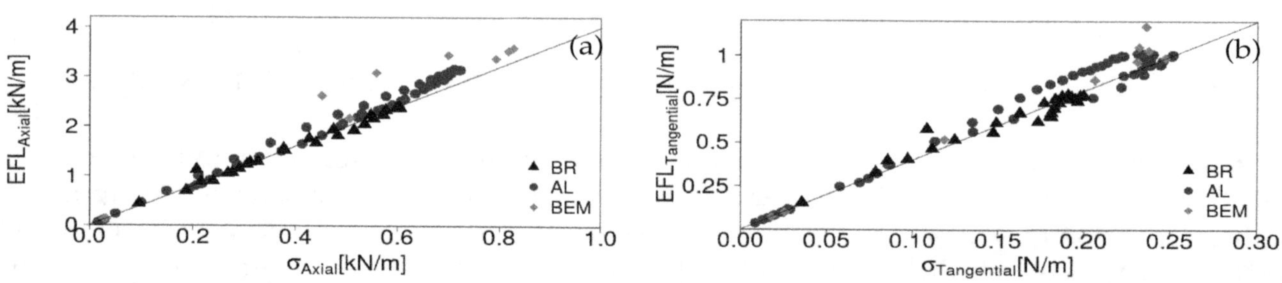

Figure 11. Equivalent Fatigue Loads (EFL) dependence on standard deviation for (**a**) axial and (**b**) tangential forces for all three methods. The diagonal line is governed by $EFL = 4\sigma$.

3.3. Integral Forces

In the following, integral forces resulting from sectional forces from Section 3.2 are discussed. Time series of the aerodynamic thrust and torque are shown in Figure 12. As can be observed, the thrust as well as the torque evolution is very similar for BR and AL. With the definition of the correlation coefficient

$$\rho = \frac{\langle (F_1(t) - \bar{F}_1)(F_2(t) - \bar{F}_2) \rangle_t}{\sigma_1 \sigma_2} \tag{4}$$

where F_1 and F_2 are the considered forces and σ_1 and σ_2 the corresponding standard deviations, the correlation is calculated as 0.90 and 0.93 for thrust and torque respectively, even if the distribution of sectional forces and the wake dynamics are more different (see Figures 4–6). A comparison between BR and BEM reveals a much smaller correlation with values of 0.08 and 0.09 for thrust and torque. This supports the argument that the wind dynamics are handled in a very different way for BEM and CFD simulations for the current setup.

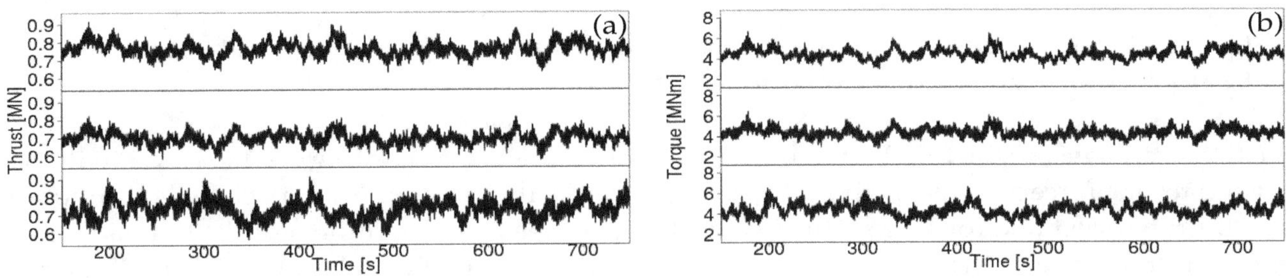

Figure 12. Time series for **(a)** thrust and **(b)** torque for Blade Resolved (BR), Actuator Line (AL) and Blade Element Momentum (BEM) methods (from top to bottom).

In Table 2 the means, standard deviations and EFLs of torque and thrust for all three methods are illustrated with their corresponding errors. Those errors are obtained by separating the corresponding time series' into five distinct equally sized ensembles and calculating the ensemble standard deviation. For comparison reasons the ratios of the calculated quantities for AL and BR as well as BEM and BR are shown, because the BR case is assumed to be the highest fidelity approach. For aerodynamic thrust and torque AL and BEM show lower values than the BR case with a minimum ratio of 91.7%. which can also be estimated from the sectional forces in Figure 5.

Table 2. Thrust and torque comparison by mean, standard deviation σ and EFL between BR, AL and BEM.

		BR	AL	BEM	AL/BR [%]	BEM/BR [%]
Thrust [kN]	mean	761 ± 7.3	698 ± 4.0	729 ± 9.6	91.7	95.8
	σ	37.9 ± 4.3	33.9 ± 2.6	49.7 ± 6.5	89.4	131.1
	EFL	147.6 ± 9.7	136.7 ± 4.2	190.1 ± 8.0	92.6	128.8
Torque [MNm]	mean	4.43 ± 0.10	4.32 ± 0.08	4.35 ± 0.13	97.5	98.2
	σ	0.495 ± 0.044	0.452 ± 0.038	0.599 ± 0.061	91.3	121.0
	EFL	1.90 ± 0.15	1.90 ± 0.13	2.29 ± 0.12	99.6	116.0

Analog to the relation between mean sectional forces and mean integral forces, the standard deviations of tangential and axial forces in Figure 6 correspond to a standard deviation in the thrust and torque. The standard deviation for the AL case is for both quantities approximately 10% smaller than for the BR simulation, because of the missing variance in the root and tip region. Especially for the torque the tip region plays a significant role. In contrast to that, for the BEM simulation an approximately 20 to 30% increase in the standard deviation is obtained which is also very clear from the much higher sectional standard deviation except at the root. A reason for the high standard deviation could be the equilibrium wake model used here, where the wake dynamics are not captured in a proper way. The relation between EFL and the standard deviation of integral forces is shown in Figure 13 where we additionally added the diagonal line governed by $EFL = 4\sigma$ like for sectional forces in Figure 11. Sectional as well as integral forces seem to follow the same linear trend.

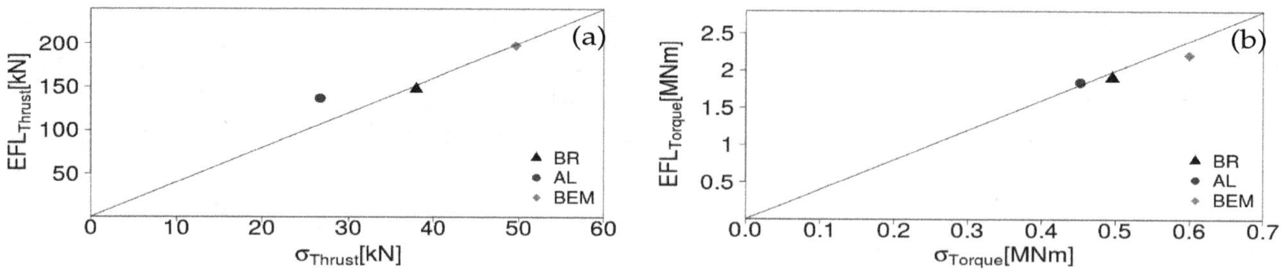

Figure 13. EFL dependence on standard deviation for (**a**) thrust and (**b**) torque for all three methods. The diagonal line is governed by $EFL = 4\sigma$.

Further analysis of the EFL by the underlying range histograms in Figure 14 shows that for the thrust and torque force all three methods have in general a similar trend. But especially in the moderate ranges for the thrust and for large ranges in both quantities BEM has more counts than AL and BR. Due to their large weight the most important part for the EFLs are the high ranges. Those high ranges must not be neglected, even if they have very few counts, because very important quantities like the difference of global maxima and minima and the standard deviation contribute to them. Also the similarity of the inflow conditions justifies the comparison of the loads. The contribution of all ranges to the EFLs in Equation (2), namely $n \cdot r^m$, is shown in Figure 15, where the big contribution of the highest ranges to fatigue loads is very clear.

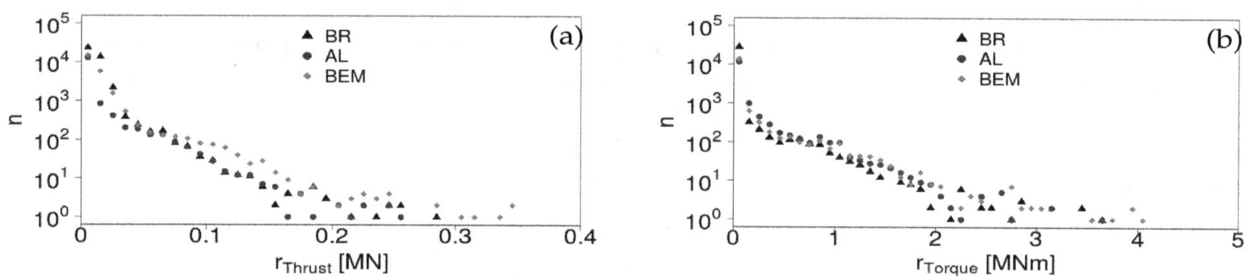

Figure 14. Range histograms for (**a**) thrust and (**b**) torque.

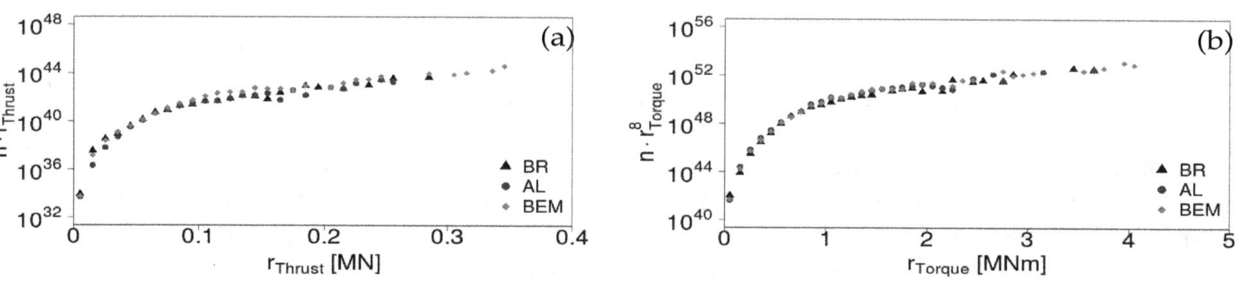

Figure 15. Contribution of ranges to EFL for (**a**) thrust and (**b**) torque.

For an analysis of the dynamics of torque and thrust, PSDs are investigated and shown in Figure 16. All three methods show a very good agreement in the shape of spectra for low frequencies in region I up to 1.5 Hz including the 3P frequency and the first harmonic. For higher frequencies up to approximateley 4 Hz, here region II, BEM and AL are very similar but the BR method drops very quickly and saturates in a white noise like behaviour for higher frequencies in region III. The reason for this is the big structures generating blade geometry and the RANS model working in the boundary layer of the blade while the AL does not have a geometry and it uses LES everywhere and therefore does not smooth out the wind field in the vicinity of the blade as has been discussed already for the sectional forces. It again has to be amplified that the BR case has the highest resolution in the blade region and therefore has the most reliable results. The difference between BEM and AL in the high frequency range in region III could be according to the missing damping, filtering and dynamic wake effects in the BEM case. The AL method uses a Gaussian filter to smooth out the forces along the blade and the wake is much more complex than in the static wake model of BEM. It has to be amplified, that especially for the torque, a transition from BEM to AL can not be made just by filtering the signals, because filtering corresponds to a drop in the PSD, which is not observed in the mid range.

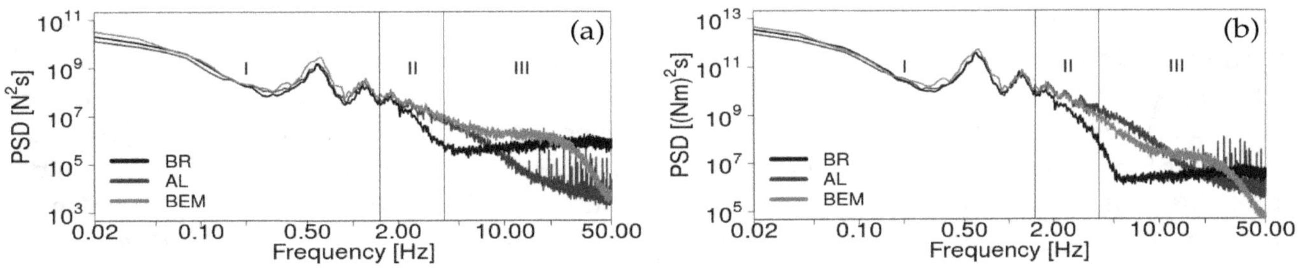

Figure 16. PSD of (**a**) thrust and (**b**) torque.

For a more detailed analysis, the cumulative frequency dependent variance normalized by the total variance, i.e.,

$$\hat{\sigma}^2(f) = \frac{\int_0^f PSD(f')\mathrm{d}f'}{\sigma^2} \tag{5}$$

has been investigated in Figure 17. This is equivalent to the energy content of the spectrum at frequencies smaller than f, normalized by the total energy contained in the spectrum. The energy content for the different regions can be determined as $\hat{\sigma}^2_{\text{region}} = \hat{\sigma}^2(f_{\max}) - \hat{\sigma}^2(f_{\min})$ with $f_{\max}$ the upper frequency limit and $f_{\min}$ the lower frequency limit of the region. For all three method the conclusion can be drawn that by far the most energy is contained in region I, in particular for the thrust: BR 95.4%, AL 93.3% and BEM 94.9%. Region II also contains a small amount of energy with 2.3% for BR, 5.8% for AL and 3.2% for BEM. The torque is distributed similarly with 97.7% for BR, 91.7% for AL and 91.1% for BEM in region I and 2.2%, 7.1% and 3.5% in region II. The variance contained in region III is neglegible for both quantities. Based on the high percentage of variance contained in region I it can be concluded that the big difference in the total variance between BEM and the other two methods mainly stems from this frequency range. Therefore the extraction of wind fields at the position of the rotor plane has to be seen as a first step for this comparison and it has to be investigated if it is more appropriate to extract the fields downstream of the rotor position or to artificially damp the wind input in BEM on large scales.

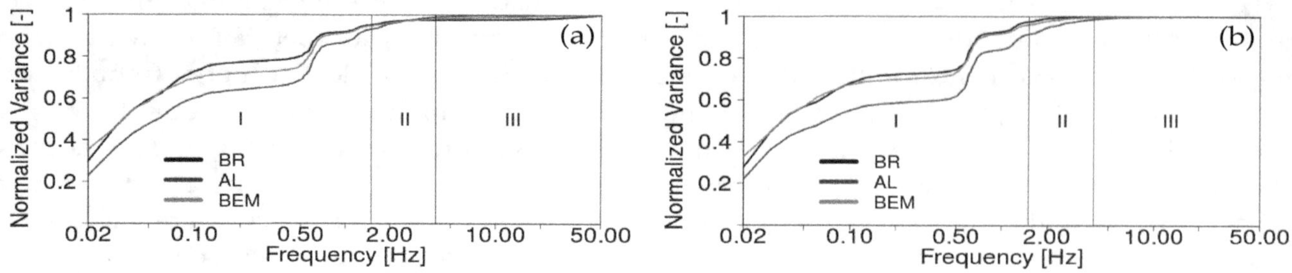

Figure 17. Normalized cumulative frequency dependent variance of (**a**) thrust and (**b**) torque.

4. Discussion

From this work it can be concluded that the AL, BR and the BEM method with extracted wind fields at the rotor plane from an empty domain are very similar with respect to averaged tangential and axial forces, except close to the root and the tip. However, a comparison of BEM with the CFD methods with respect to the standard deviations of sectional and integral forces shows very different results. For the integral quantities thrust and torque, BEM tends to overestimate the standard deviation compared to the other two methods. It was found out that the standard deviation differences between all methods mainly come from the different amplitudes in the first part of the spectrum and not from the filtering effects on larger frequencies. This leads to the conclusion that, if similar results for the standard deviation for BEM and CFD should be achieved, either the wind fields have to be extracted downwind of the rotor position from the empty domain simulation or a correction of wind fields for the BEM case has to be considered. A good starting point could be to substitute the equilibrium wake model by a dynamic wake model which could damp the oscillations and with it the standard deviation. This and extracting the wind fields in an empty domain downwind from the position where the turbine is normally located will be part of the future work. Another contribution to the differences, which also should not be neglected, comes from the fact that the turbulent wind field decays across the CFD domain due to the missing energy injection. In an empty domain, the turbulence intensity is higher upstream and lower downstream of the position, where the turbine would be located. This decay is not considered within the BEM simulation and could be a reason for the differences.

An indication for a linear dependence between standard deviation and EFL could be found for sectional as well as integral forces which seem to follow the same trend. This shows that the dynamics characterized by the EFL and by the standard deviation are similar for all three methods. However, this dependence is assumed to be existent only if the wind fields are comparable with respect to temporal and spatial correlations to assure similar turbulent structures which play a big role for EFL estimations. Out of scope of this paper, an advantage of this linear relationship could be that the computational expensive rainflow count algorithm is not needed if one is interested in equivalent fatigue loads. A simple calculation of the standard deviation could give this information as well. Additionally the equivalent fatigue loads could also be estimated if the same simulation is redone with a different standard deviation for the wind velocities.

Even if no experiments for comparison were performed, the BR simulation with its high resolution close to the blades and the full geometry consideration can be expected to be the most physical and therefore it should be seen as a benchmark case. However, a further part of the future work should be a fair comparison with experiments where the focus should be set on the statistical equivalence of the wind fields at least across the full rotor size in the CFD domain, in the BEM simulation and in the experiments.

Author Contributions: Conceptualization S.E., C.M.S. and H.R.; Software, S.E.; Methodology, S.E., C.M.S. and H.R.; Formal Analysis, S.E.; Writing-Original Draft Preparation, S.E. and H.R.; Writing-Review & Editing, J.P.; Supervision, J.P.; Project Administration, B.S.

Acknowledgments: The simulations were performed at the HPC Cluster EDDY, located at the University of Oldenburg (Germany) and funded by the Federal Ministry for Economic Affairs and Energy (Bundesministeriums für Wirtschaft und Energie) under grant number 0324005.

Abbreviations

The following abbreviations are used in this manuscript:

AL	Actuator Line
BEM	Blade Element Momentum
BR	Blade Resolved
CFD	Computational Fluid Dynamics
EFL	Equivalent Fatigue Loads
PSD	Power Spectral Density
TI	Turbulence Intensity

References

1. Schepers, J.G. Engineering Models in Wind Energy Aerodynamics. Ph.D. Thesis, Delft University of Technology, Delft, The Netherlands, 2012. [CrossRef]
2. International Electrotechnical Commission. *WIND TURBINES—PART 1: DESIGN REQUIREMENTS*; Standard IEC 61400-1; International Electrotechnical Commission: Geneva, Switzerland, 2005.
3. Madsen, H.A.; Sørensen, N.N.; Bak, C.; Troldborg, N.; Pirrung, G. Measured aerodynamic forces ona full scale 2MW turbine in comparison with EllipSys3D and HAWC2 simulations. *J. Phys. Conf. Ser.* **2018**, *1037*. [CrossRef]
4. Kim, Y.; Lutz, T.; Jost, E.; Gomez-Iradi, S.; Muñoz, A.; Méndez, B.; Lampropoulos, N.; Sørensen, N.; Madsen, H.; van der Laan, P.; et al. AVATAR Deliverable D2.5: Effects of Inflow Turbulence on Large Wind Turbines. AVATAR Deliverable D2.5, Technical Report. Available online: http://www.eera-avatar.eu/fileadmin/avatar/user/avatar_D2p5_revised_20161231.pdf (accessed on 4 December 2018).
5. Mann, J. Wind field simulation. *Probabilistic Eng. Mech.* **1998**, *13*, 269–282. [CrossRef]
6. Sørensen, J.N.; Shen, W.Z. Numerical Modeling of Wind Turbine Wakes. *J. Fluids Eng.* **2002**. *124*, 393–399. [CrossRef]
7. Jonkmann, J.; Butterfield, S.; Musial, W.; Scott, G. *Definition of a 5-MW Reference Wind Turbine for Offshore System Development*; Technical Report TP-500-38060; National Renewable Energy Laboratory: Golden, CO, USA, 2009.
8. OpenCFD. *OpenFOAM—The Open Source CFD Toolbox—User's Guide*; 1.4 ed.; OpenCFD Ltd.: Reading, UK, 2007.
9. Rahimi, H.; Daniele, E.; Stoevesandt, B.; Peinke, J. Development and application of a grid generation tool for aerodynamic simulations of wind turbines. *Wind Eng.* **2016**, *40*, 148–172. [CrossRef]
10. Rahimi, H. Validation and Improvement of Numerical Methods for Wind Turbine Aerodynamics. Ph.D. Thesis, University of Oldenburg: Oldenburg, Germany, 2018.
11. Rahimi, H.; Garcia, A.M.; Stoevesandt, B.; Peinke, J.; Schepers, J.G. An engineering model for wind turbines under yawed conditions derived from high fidelity models. *Wind Energy* **2018**, *21*, 618–633. [CrossRef]
12. Spalart, P.R.; Deck, S.; Shur, M.L.; Squires, K.D.; Strelets, M.K.; Travin, A. A New Version of Detached-eddy Simulation, Resistant to Ambiguous Grid Densities. *Theor. Comput. Fluid Dyn.* **2006**, *20*, 181. [CrossRef]
13. Patankar, S.; Spalding, D. A calculation procedure for heat, mass and momentum transfer in three-dimensional parabolic flows. *Int. J. Heat Mass Transfer* **1972**, *15*, 1787–1806. [CrossRef]
14. Issa, R. Solution of the implicitly discretised fluid flow equations by operator-splitting. *J. Comput. Phys.* **1986**, *62*, 40–65. [CrossRef]
15. Prandtl, L. *Ludwig Prandtl Gesammelte Abhandlungen: Zur Angewandten Mechanik, Hydro- und Aerodynamik*; Springer: Berlin, Germany, 2013.
16. Du, Z.; Selig, M. A 3-D stall-delay model for horizontal axis wind turbine performance prediction. In Proceedings of the 1998 ASME Wind Energy Symposium, Reno, NV, USA, 12–15 January 1998; p. 21.
17. Smagorinsky, J. General Circulation Experiments with the Primitve Equations. *Mon. Weather Rev.* **1963**, *91*, 99–164. [CrossRef]

18. Jonkman, B.; Jonkman, J. *FAST v8.16.00a-bjj*; Technical Report; NREL: Golden, CO, USA, 2016.
19. Jonkman, J.; Hayman, G.; Jonkman, B.; Damian, R. *AeroDyn v15 User's Guide and Theory Manual (Draft Version)*; Technical Report; NREL: Golden, CO, USA, 2016.
20. Leishman, J.G.; Beddoes, T.S. A Semi-Empirical Model for Dynamic Stall. *J. Am. Helicopter Soc.* **1989**, *34*, 3–17. [CrossRef]
21. Madsen, P.H. *Recommended Practices for Wind Turbine Testing and Evaluation. 3. Fatigue Loads*; Riso National Laboratory: Roskilde, Denmark, 1990.
22. Rychlik, I. A new definition of the rainflow cycle counting method. *Int. J. Fatigue* **1987**, *9*, 119–121. [CrossRef]
23. Martínez-Tossas, L.A.; Churchfield, M.J.; Leonardi, S. Large eddy simulations of the flow past wind turbines: Actuator line and disk modeling. *Wind Energy* **2015**, *18*, 1047–1060. [CrossRef]

11

Aerodynamics and Motion Performance of the H-Type Floating Vertical Axis Wind Turbine

Ying Guo, Liqin Liu, Xifeng Gao * and Wanhai Xu

State Key Laboratory of Hydraulic Engineering Simulation and Safety, Tianjin University, Tianjin 300072, China; yynocry@tju.edu.cn (Y.G.); liuliqin@tju.edu.cn (L.Q.L.); xuwanhai@tju.edu.cn (W.H.X.)

* Correspondence: gaoxifeng@tju.edu.cn

Abstract: Aerodynamics and motion performance of the floating vertical wind turbine (VAWT) were studied in this paper, where the wind turbine was H-type and the floating foundation was truss spar type. Based on the double-multiple-stream-tube theory, the formulae were deduced to calculate the aerodynamic loads acting on the wind turbine considering the motions of the floating foundation. The surge-heave-pitch nonlinear coupling equations of the H-type floating VAWT were established. Aerodynamics and motion performance of a 5 MW H-type floating VAWT was studied, and the effect of the floating foundation motions on the aerodynamic loads was analyzed. It is shown that the motions of the floating foundation on the aerodynamics cannot be ignored. The motion of the H-type floating VAWT was also compared with that of the Φ-type floating VAWT: they have the same floating foundation, rated output power, mooring system and total displacement. The results show that the H-type floating VAWT has better motion performance, and the mean values of surge, heave and pitch of the H-type floating VAWT are much smaller comparing with the Φ-type floating VAWT.

Keywords: H-type floating VAWT; truss Spar floating foundation; coupling of aerodynamics and hydrodynamics

---

1. Introduction

With the increase of unit capacity, size of the offshore wind turbine is getting bigger and bigger. The instability of floating horizontal axis wind turbine (HAWT) may become obvious as the huge drive system is placed on the top of the high tower [1]. The drive system of Vertical Axis Wind Turbine (VAWT) is installed on the bottom, which has little effect on the dynamic behavior of the tower. This makes floating VAWT more advantageous in the development of large-scale offshore wind power [2].

Darrieus wind turbine is the earliest lift type VAWT, and it has the higher power coefficient comparing with other VAWT [1]. According to the form of the blade, the Darrieus wind turbine includes curved blade and straight blade, withib which the Φ-type and the H-type are the most common. With the development of the offshore wind farm, the feasibility of the Darrieus wind turbine is assessed to be used in deep water and supported by the floating foundation.

The research on the floating VAWT has been carried out, including the aerodynamic load calculations of the floating VAWT, the motion comparison of the VAWT supported by different floating foundations, the influences between aerodynamic and hydrodynamic loads, and so on. Some methods and codes were developed to calculate the floating VAWT, including the FloVAWT code [3], the enhanced Hawc2 code [4], Simo-Riflex-DMS (AC) code [5,6], and a rigid-flexible coupling code [7]. The motion performances of the floating VAWT were investigated by these codes. Vita studied the feasibility of a 5 MW Φ-type wind turbine supported by a rotating spar type foundation by using the enhanced Hawc2 code [4]. Borg and Collu calculated the responses of vertical wind turbines supported by different floating foundations by using the FloVAWT code [8]. Cheng et al. implemented the AC method in the Simo-Reflex to conduct the fully coupled aero-hydro-servo-elastic simulation of the floating VAWT [6,9]. Cahay et al. put forward the concept of a three-blade H-type wind turbine installed on a semi-submersible floating foundation [10].

In the above research, the form of the floating foundations mainly includes Spar type, Semi-submersible type, and TLP type. Liu et al. studied the motion performance of a 5 MW VAWT supported by the truss Spar platform [11], which had shorter length than the existent spar type foundation [12–14] and can be used in the moderate water depth with good motion performance, and had lighter weight than the existing semi-submersible type foundation [15]. The above codes cannot be used to calculate the truss Spar type VAWT, as the nonlinear coupling between the heave and pitch of truss Spar foundation is important, similar to the deep see Spar platform [16,17], and this is not considered in the above methods.

Compared with Φ-type, H-type is easier to manufacture and has higher power output [18]. In this paper, we studied the motion performances of the H-type VAWT supported by a truss Spar type floating foundation. The aerodynamic loads were calculated considering the dynamic stall and motions of the floating foundation. The surge–heave–pitch coupling nonlinear motion equations of the floating VAWT were established, and the motions of the floating VAWT were assessed.

2. The Aerodynamic Loads of H-Type Floating VAWT

Based on the double multiple stream tube (DMST) theory, the aerodynamic loads acting on the blades were calculated considering dynamic stall, tip losses, aspect ratio, tower shadow effect and the influence of the surge–heave–pitch motions of the floater. The whole floating VAWT was dealt with as a rigid body.

The stream tubes were divided into upwind and downwind sectors, respectively. Coordinate system is shown in Figure 1, where the coordinate origin o is in the center of gravity of the floating VAWT, and ξ_1, ξ_3, and ξ_5 are the displacements of surge, heave and pitch of the floating VAWT, respectively. The positive directions of the surge and heave motions are along the positive directions of the x axis and z axis, respectively; the positive direction of the pitch motion is clockwise. The vector relations of the wind turbine are shown in Figure 2.

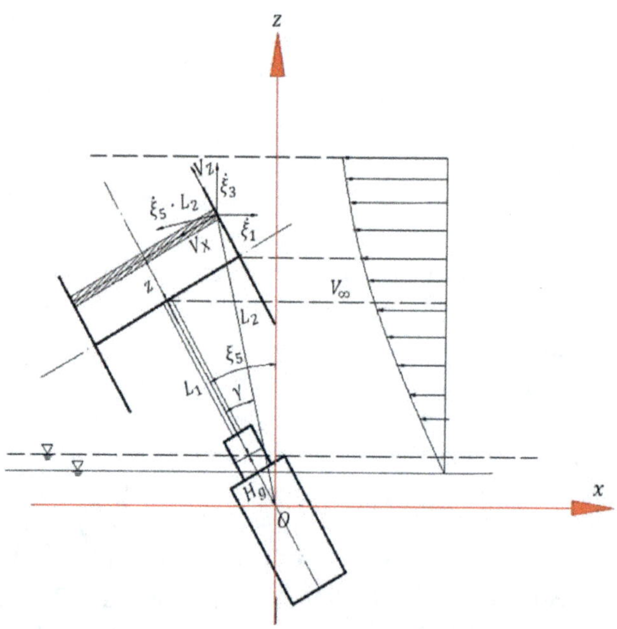

Figure 1. Coordinate system.

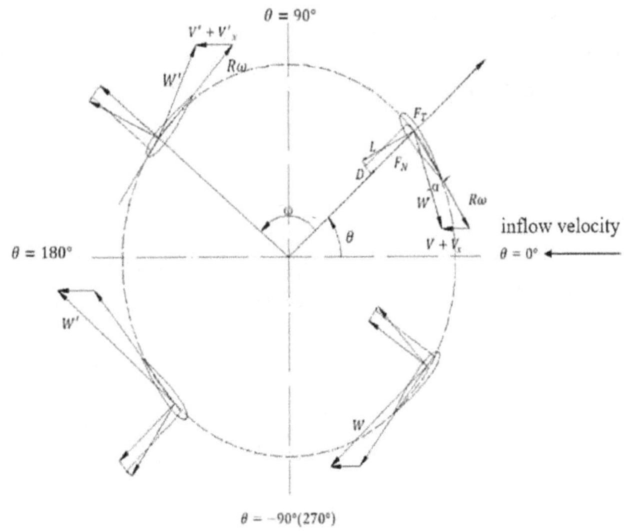

Figure 2. Equatorial velocity vector.

The steady wind was considered here, it was a function of the height above the water level, and for a local height z_i, the wind speed can be written as follow,

$$V_{\infty i}(z_i) = V_{\infty}(z_i/z_{ref})^a, \tag{1}$$

where V_{∞} is the average wind velocity at a reference height z_{ref}, the vertical distance from the center of the blades to the water level is taken as the reference height in this paper, and with power $a = 0.14$ [19].

Assuming the velocity parallel to the direction of the steam tube affects the induced velocities and the velocity perpendicular to the direction of the steam tube only affects the local relative velocities, the coordinate and the velocity of each stream tube were deduced considering the motions of the floating foundation in the following parts.

Induced velocities at the upwind and downwind sectors V and V', and the induced equilibrium velocity V_e are given by [20],

$$V = uV_{\infty i}\cos\xi_5, \tag{2}$$

$$V_{\mathrm{e}} = (2u-1)V_{\infty i}\cos\xi_5, \tag{3}$$

$$V' = u'V_e = u'(2u-1)V_{\infty i}\cos\xi_5. \tag{4}$$

where u is the interference factor, and u' is the second interference factor.

For the upwind sector, the distance from local blades to the gravity center of floating wind turbine system is expressed as,

$$L_2 = \sqrt{(R\cos\theta)^2 + (z + L_1 + H_g)^2}, \tag{5}$$

where R is the radius of rotor, θ is the azimuth of rotor, z is the distance from each stream tube to equator (below equator is negative direction), L_1 is the distance from equator to still water level, and H_g is the distance from still water level to the gravity center of floating wind system when the system is in a state of stillness. The angle between the local position of the blade and the tower is written as

$$\gamma = \arctan(\frac{R\cos\theta}{z + L_1 + H_g}), \tag{6}$$

Then, the local wind velocity of the blade is given by,

$$V_{\infty \mathrm{i}} = V_{\infty}\cdot(\frac{L_2\cos(\xi_5-\gamma) - (H_g-\xi_3)}{H + L_1})^{\mathrm{a}}, \tag{7}$$

where H is the half-height of rotor. The velocities of local blades can be derived from the surge, heave and pitch velocities of floater by using transformation matrix. Then, the resultant velocities, which are caused by the motion of the floating foundation, parallel to the direction of the steam tube for the local position can be written as,

$$V_x = -\dot{\xi}_1 \cos \xi_5 + \dot{\xi}_3 \sin \xi_5 - \dot{\xi}_5 \cdot L_2 \cdot \cos \gamma. \tag{8}$$

Using a similar deducing progress, we get the resultant velocities parallel to and vertical to the direction of the steam tube for the downwind sector, as follows,

$$V_x{}' = -\dot{\xi}_1 \cos \xi_5 + \dot{\xi}_3 \sin \xi_5 - \dot{\xi}_5 \cdot L_2 \cdot \cos \gamma. \tag{9}$$

For the rotor of upwind sector, namely, the range of azimuth θ is $[-\pi/2, \pi/2]$, and the local relative velocity of the rotor can be written as,

$$W = \sqrt{(V + V_x)^2 \cos \theta^2 + (R\omega - V \sin \theta)^2}, \tag{10}$$

where ω is the rotation velocity of the blade.

The local attack of angle is derived as,

$$\alpha = \arcsin[\frac{(V + V_x) \cos \theta}{W}]. \tag{11}$$

The azimuthal angle of upwind sector of the rotor is divided into several angular tubes, and the angle of each tube is $\Delta\theta$. Assuming the induced velocity for each angular tube is a constant, the interference factor of each angular tube is a constant, it can be presented as,

$$u(\theta) = \frac{KK_0}{KK_0 + \int_{\theta-\Delta\theta/2}^{\theta+\Delta\theta/2} f_{up}(\theta) d\theta}, \tag{12}$$

where K, K_0 and $f(\theta)$ can be given by following expressions,

$$K = 8\pi R/(nc), \tag{13}$$

$$K_0 = \sin(\theta + \Delta\theta/2) - \sin(\theta - \Delta\theta/2), \tag{14}$$

$$f_{up}(\theta) = (W/V)^2 [C_N \cos \theta + C_T \sin \theta], \tag{15}$$

where n is the number of blade, c is the chord length, and C_N and C_T are the normal force and tangential force coefficients, respectively, they can be given by,

$$C_N = C_L \cos \alpha + C_D \sin \alpha, \tag{16}$$

$$C_T = C_L \sin \alpha - C_D \cos \alpha, \tag{17}$$

where C_L and C_D are the lift coefficient and the drag coefficient, respectively. 2D static coefficient C_L^{2D} and C_D^{2D} can be obtained by interpolation of the experimental data [21] considering the local Reynolds number and the local attack angle α, and 3D static coefficient C_L^{3D} and C_D^{3D} can be derived by taking into account the influence of tip losses and finite aspect ratio [22]. Finally, the coefficients can be corrected by the Gormont–Berg method about the dynamic stall [23].

The local Reynolds number is presented as,

$$\mathrm{Re}_t = \frac{cW}{v_\infty}, \tag{18}$$

where $v_\infty = 1.47 \times 10^{-5} (m^2/s)$ is the coefficient of kinematic viscosity.

For the given the geometric lines of rotor, rotational speed and wind speed of each stream tube, the final induced factor $u(\theta)$ can be derived by iteration of Equations (2)–(18). Furthermore, the normal force and tangential force coefficients can be obtained.

The normal force F_N and tangential force F_T of the blade at different θ are further calculated by the following formulae,

$$F_N = 0.5\rho c H \int_{-1}^{1} C_N W^2 d\zeta, \tag{19}$$

$$F_T = 0.5\rho c H \int_{-1}^{1} C_T W^2 d\zeta, \tag{20}$$

where $\zeta = H_0/H$ is local height of the stream tube to equator plane of the rotor.

The thrust force F_X and side force F_Y of the blade are calculated as following,

$$F_X = F_N \cos\theta + F_T \sin\theta \tag{21}$$

$$F_Y = F_N \sin\theta - F_T \cos\theta \tag{22}$$

The power coefficient is presented as following,

$$C_{P1} = (R\omega/V_\infty) \times \frac{ncH}{2\pi S} \int_{-\frac{\pi}{2}}^{\frac{\pi}{2}} \int_{-1}^{1} C_T (W/V_\infty)^2 d\zeta d\theta. \tag{23}$$

Similarly, the normal force F'_N and the tangential force F'_T can be calculated for the downwind sector, and obtain the power coefficient C_{p2}. The power coefficient for one cycle can be written as,

$$C_P = C_{P1} + C_{P2}. \tag{24}$$

In this paper, the steady wind speed was used to calculate the aerodynamics loads. The lift and thrust forces acting on the heave and surge of the floating VAWT can be obtained by decomposing the normal and tangential forces. Calculating the moment of the thrust force about the center of gravity of the floating VAWT, the pitch moment of the aerodynamics acting of the pitch motions of the floating VAWT can be obtained.

Taking a fixed H-type wind turbine VAWT260 [24] as an example, the rotor power was calculated (the coupling of aerodynamics and hydrodynamics was not considered in this part). The VAWT260 H-rotor has two blades with chord length of 1.02 cm and airfoil of NACA0018. The lift and drag coefficients used in this paper are from static test, and iteration residual standard is set to 10^{-4}. The results of rotor power obtained by the numerical simulation were compared with those of model experiment [24], as shown in Figure 3.

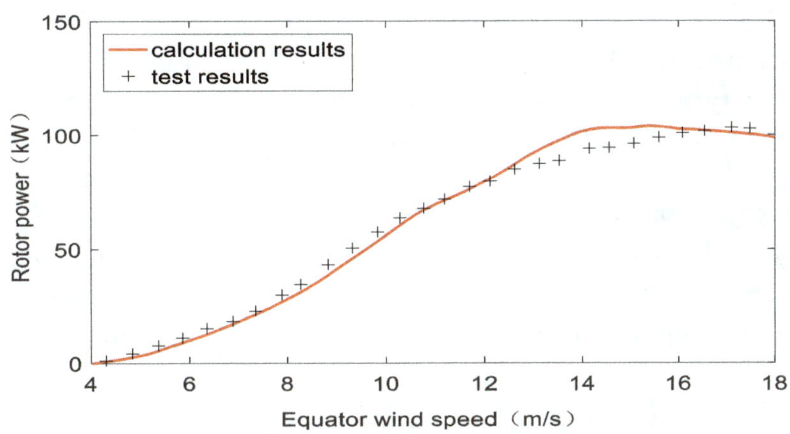

Figure 3. Validity checking of the aerodynamic loads.

It can be seen in Figure 3 that the rotor power obtained by the numerical simulation agrees well with the results of the experiment.

To verify the aerodynamic loads, comparisons of tangential and normal force coefficients with the calculation results of Paraschivoiu [25] are presented in Figure 4. A NACA0015 blade profile, a chord of 0.2 m and 6 m high H-type blade is used here. The rotational speed is 125 rpm, and the tip speed ratio is 5.2. The double multiple stream tube method with considering finite aspect ratio is used in [25].

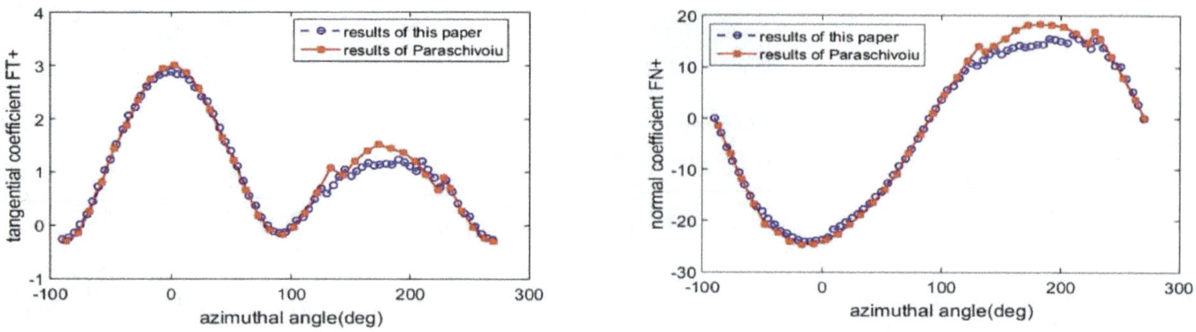

(**a**) Tangential force coefficient (**b**) Normal force coefficient

Figure 4. Validity checking of the aerodynamic loads.

Figure 4 shows that the aerodynamic load coefficients mainly agree well with the results of Paraschivoiu. The reason for the relatively obvious differences appearing downwind is that the tower effect is ignored in Paraschivoiu's calculation. The thrust force and side force used in the dynamical equation in Section 3 can be checked by Equations (21) and (22).

3. The Dynamical Equations of the Floating VAWT

3.1.The Nonlinear Coupled Motion Equations

Liu et al. [16] established the nonlinear coupled motion equations of the Spar platform, and verified the equations by comparing with test modal. In this paper, the method is used to analyze the dynamic motion of the floating VAWT. Regarding the whole floating VAWT as a rigid body, considering the inertia force (moment), damping force (moment), restoring force (moment), mooring force, wind and wave loads, the surge–heave–pitch coupled equations of the floating VAWT can be presented as follows,

$$(\boldsymbol{M} + \boldsymbol{A}_{\infty})\ddot{X}(t) + \int_0^t \boldsymbol{h}(t-\tau)\dot{X}(t)d\tau + K(X(t))X(t) = F(t, X, \dot{X}), \tag{25}$$

where M is the mass matrix of the floating VAWT, A_{∞} is the mass matrix at infinite frequency, and $h(t-\tau)$ is the retardation function accounting for the fluid memory effect X, $\dot{X}$ and $\ddot{X}$ represents the displacement, velocity and acceleration vectors of the floating VAWT at its center of gravity, respectively, and $X = [\xi_1\, \xi_3\, \xi_5]^T$. K is the restoring matrix, including the hydrostatic restoring, restoring of the mooring system, and the nonlinear hydrostatic restoring. $F(t, X, \dot{X})$ is the exciting force vector, which includes the Froud–Krylov force $F_{fk}(t)$, diffraction force $F_d(t)$, wind loads $F_w(X, \dot{X}, t)$ (including the aerodynamic loads $F_w{}^a(X, \dot{X}, t)$ and the wind loads $F_w{}^p$ calculated by the wind pressure theory), and hydrodynamic viscous force $F_{v1} = D_1\dot{X} + D_2\dot{X}\left|\dot{X}\right|$, where D_1 is the matrix of viscous damping coefficient,

$$F(t, X, \dot{X}) = F_{fk}(t) + F_d(t) + F_w(X, \dot{X}, t) + F_{v1}, \tag{26}$$

For the Spar type floating foundation, the nonlinear coupling between the heave and pitch motions is significant. Referring to the nonlinear heave–pitch coupling motion equations of the deepsea truss Spar platform [16,17], the restoring matrix can be written as follows,

$$K = \begin{bmatrix} f_1(\xi_1) & 0 & 0 \\ 0 & f_3(\xi_3) + \rho g A_w & -\frac{H_g}{2}\rho g A_w \xi_5 \\ 0 & \frac{1}{2}\rho g(\nabla + 2A_w\overline{GM})\xi_3 & \rho\nabla g\overline{GM} \end{bmatrix}, \quad (27)$$

where $f_i(\xi_i)$ $(i = 1,3)$ represents restoring stiffness of the mooring system in surge and heave of the floating VAWT. Aw is water plane area of the floating foundation, $\overline{GM}$ is the initial stability height of the floating VAWT, ∇ is the displacement, ρ is the density of seawater, and g is gravity acceleration. The hydrostatic restoring loads are calculated on the static equilibrium position of the floating VAWT.

3.1. Wind Pressure

The wind pressure and wind heeling moment acting on the tower and the blades (in the parked state) were calculated in accordance with the rules of the API [26], as follows,

$$F = 0.613 \times Cs \times Ch \times \overline{S} \times V_\infty{}^2, \quad (28)$$

$$M = F \times \overline{H}, \quad (29)$$

where Cs is the shape coefficient of the wind-bearing component; Ch is the height coefficient of the wind-bearing component; $\overline{S}$ is the frontal projected area of the wind-bearing component in the wind direction; $\overline{H}$ is the distance from the wind force acting points to the rotation center of the floating VAWT; and V_∞ is shown as Equation (1).

3.2. Hydrodynamic Loads

Hydrodynamic loads were calculated by using the potential theory and the Morison's equation [27]. The potential theory was used to calculate the wave loads, added mass and radiation damping of the large scale structure (including the upper buoyancy tank, upper mechanical tank, heaving plates, and bottom ballast tank). The first-order wave loads and second-order mean drift forces were considered in this paper.

The Morison equation was used to calculate the inertial load and viscous drag load of the slender structure (truss structures, $D/\lambda < 0.2$, where D is the diameter of the structure and λ is the wave length). The transverse hydrodynamic force per unit length can be written as follows,

$$dF = \rho\pi\frac{D^2}{4}\dot{u}_w + \rho\pi C_a\frac{D^2}{4}(\dot{u}_w - \dot{u}_b) + \frac{1}{2}\rho C_d D(u_w - u_b)|u_w - u_b|, \quad (30)$$

where C_a and C_d are the added mass and quadratic drag coefficients, respectively; u_w is the transverse wave particle velocity; and u_b is the local transverse body velocity.

The Wadam [28] was used to calculate the hydrodynamics in this paper. The wind turbine was in the parked state and the wind loads and mooring system were not considered in the hydrodynamic calculation, while the mass and shape of the blade were considered.

3.3. Damping and Mooring System

The heave plates were used to reduce the heave motions of the floating VAWT, and the viscous damping of floating foundation was important. In this paper, the viscous damping was obtained by the experiment of free decay model [29].

The mooring forces acting on the floating foundation in the directions of surge and heave were calculated by quasi-static catenary method, considering catenary mooring line and the fairlead set near the gravity center of the floating VAWT.

3.4. The Motions Calculation of the Floating VAWT

The motions calculation of the floating VAWT was carried out using the method of Runge–Kutta, and the computational procedures were as follows: give initial displacements and velocities of the floating VAWT to calculate the aerodynamic loads of the blades according to Equations (2)–(22); solve Equation (23) considering the wind and wave loads, and mooring loads; and calculate the new aerodynamic loads according to the new displacements and velocities of the floating VAWT. The coupling between aerodynamics and hydrodynamics was obtained by repeating the process above.

4. Results and Analysis

4.1. The Parameters of a 5 MW Floating VAWT

The H-type floating VAWT studied as an example is shown Figure 5. The upper wind turbine refers to the VAWT proposed by Collu et al. [15], and the floating foundation is Truss Spar type, which consists of buoyancy tank, the upper machinery tank, truss structure, heave plates and ballast tank. The specific parameters of the floating VAWT are shown in Table 1. The floating VAWT was located by four groups (one line for each group), the angle between each group of mooring lines was 90°. The lines were designed in the form of chain–cable–chain, and were connected with the floating foundation by the fairleads. The specific parameters of the mooring line are indicated in Table 2.

Table 1. Parameters of the floating VAWT. VAWT: Vertical Axis Wind Turbine.

Items	Values
Rated power	5.00 (MW)
Number of blade	2.00
Radius of rotor	37.00 (m)
Height of blade	78.75 (m)
Chord of blade	3.50 (m)
Hub height	72.00 (m)
Airfoil profile	NACA0018
Rated wind speed	15.00 (m/s)
Rated rotor speed	12.00 (rpm)
Mass of blades	76.75 (ton)
Draft of the foundation	65.00 (m)
Total water displacement	7475.10 (t)

Table 2. Parameters of the mooring line. EA: XXX.

Items	Chain	Cable
Length (m)	55	135
Diameter (mm)	147	84
Wet weight (Kg/m)	375.7	25.6
Dry weight (Kg/m)	432.2	29.5
Axial rigidity EA (N)	1.6×10^9	3.9×10^9

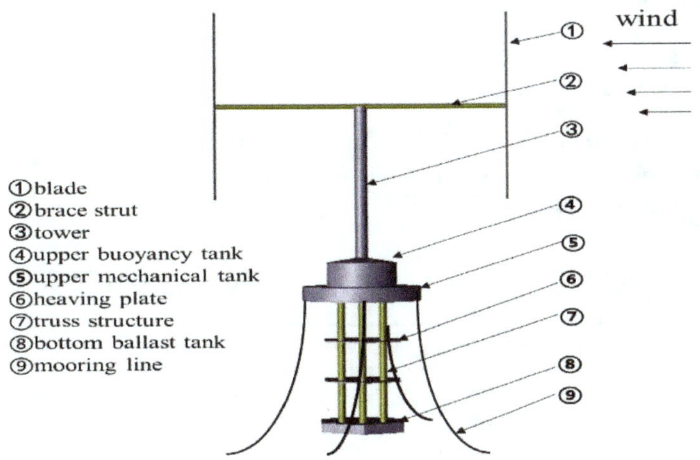

Figure 5. The H-Type floating VAWT system. VAWT: Vertical Axis Wind Turbine.

4.2. Control Strategy of Rotational Speed

The control strategy of rotor speed of the floating VAWT is as follows: for wind speeds less than 12 m/s, the rotational speed is designed to obtain the maximum power capture; the rotational speed is set to the rated rotational speed when wind speed is in the range of 12–15 m/s; and for wind speeds more than the rated wind speed, the rotational speed should be set to maintain the rated output power. Output power for different rotor speed and wind speed are shown in Figure 6, and rotor speed and output power for the different wind speed are shown in Figure 7.

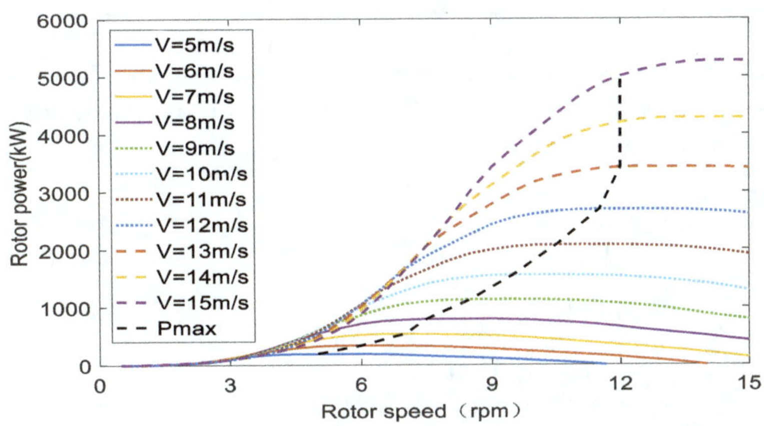

Figure 6. Power for different rotational speed and wind speed.

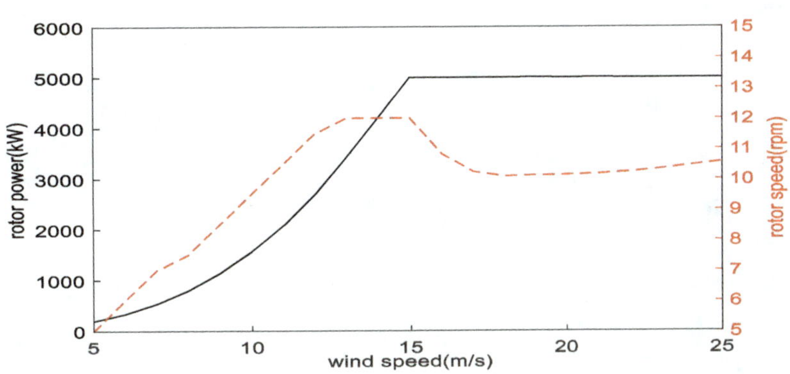

Figure 7. Power and rotational speed for different wind speed.

4.3. Results and Analysis

4.3.1. Free Decay Motions of the Floating VAWT

The free decay motions were carried out in this section. The wind turbine was in the parked state and wind and wave loads were not considered. The natural periods of surge, heave and pitch motions of the system are obtained as shown in Table 3.

Table 3. Natural periods of the floating VAWT.

DOFs	Natural Period (s)
surge	51.2
heave	20.5
pitch	29.3

4.3.2. Motions of the Floating VAWT Considering the Wind and Regular Wave Loads

Motions of the floating VAWT were calculated considering the wind loads and regular wave loads in this section. The parameters of wave and wind are shown in Table 4.

Table 4. Regular wave and wind cases (LC3).

Items	Wave Height (m)	Wave Period (s)	V_∞ (m/s)	Total Time (s)
LC3.1	2.10	9.74	5	3600
LC3.2	3.62	10.29	15	3600
LC3.3	6.02	11.38	25	3600

The time histories and the corresponding spectra (obtained by the Fast Fourier Transform) of the floating VAWT motions are shown in Figures 8–10, and the mean displacements and the corresponding standard deviations (STD) are shown in Table 5.

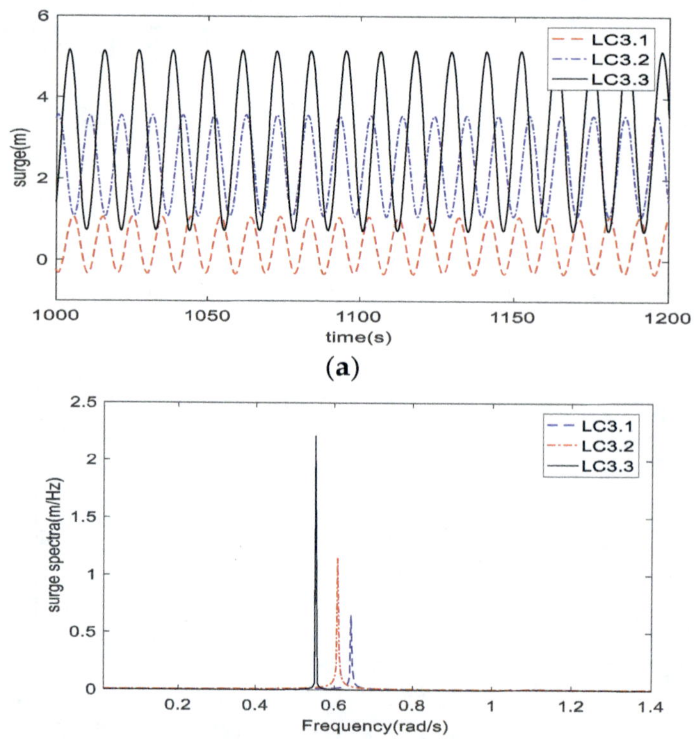

Figure 8. Surge responses of the floating VAWT (LC3): (**a**) time histories; and (**b**) response spectra.

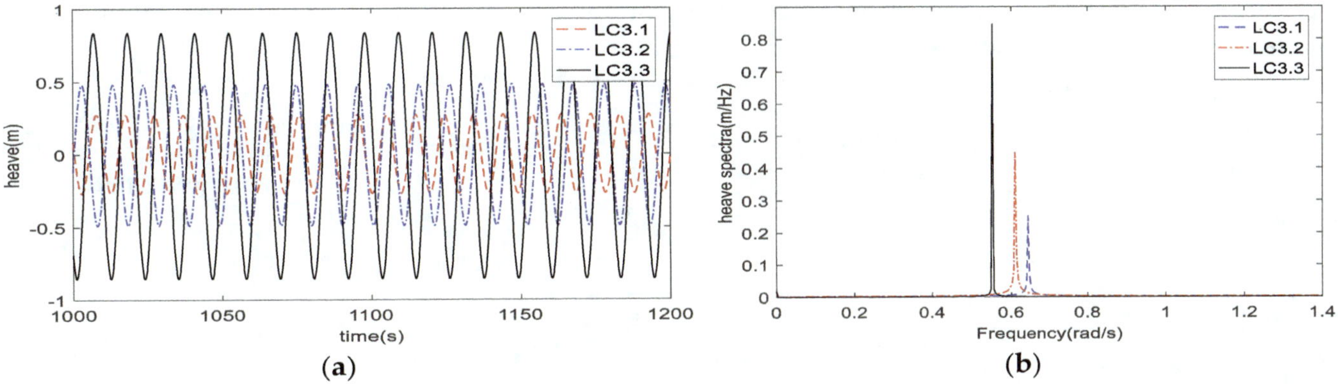

Figure 9. Heave responses of the floating VAWT (LC3): (**a**) time histories; and (**b**) response spectra.

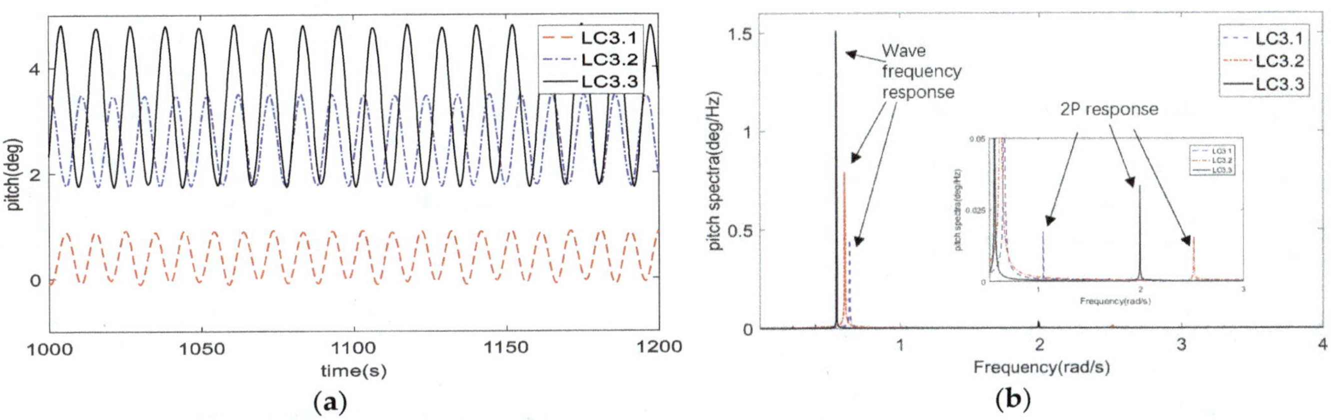

Figure 10. Pitch responses of the floating VAWT (LC3): (**a**) time histories; and (**b**) response spectra.

Table 5. The standard deviations and mean values (LC3). STD: standard deviations.

Case	STD			Mean Value		
	Surge (m)	Heave (m)	Pitch (deg)	Surge (m)	Heave (m)	Pitch (deg)
LC3.1	0.50	0.20	0.84	0.36	−0.00	0.40
LC3.2	0.89	0.35	1.49	2.32	0.006	2.62
LC3.3	1.55	0.60	2.64	2.95	0.011	3.27

Table 5 shows that, for the three cases, the max STDs of the surge, heave and pitch motions of the floating VAWT are 1.55 m and 0.60 m and 2.64 degrees, respectively; the mean values of heave are all small, and the max mean values of the surge and the pitch motions of the three cases are 2.95 m and 3.27 degrees, respectively. Therefore, the floating VAWT has good motion performance. Compared with LC3.2, the pitch STD of the floating VAWT in LC3.3 increases by approximately 77%, and the corresponding mean value increases by approximately 25%.

In Figures 8–10, we can find that the motion frequencies of the floating VAWT are dominated by the wave frequencies for the three cases. For the pitch frequencies of the floating VAWT, there are components of the rotor rotational frequency (2P response). However, they are not significant compared with the wave–frequency responses.

4.3.3. The Aerodynamics of the Floating VAWT

To analyze the effect of floating foundation's motion on the aerodynamics, two conditions are considered for LC3.2: (a) the foundation is floating, and the couple motions of floating foundation are considered in aerodynamics calculation; and (b) the wind turbine is considered to be set on a fixed foundation. The normal and tangential forces acting on the equator of one blade are calculated, and the results of time histories and spectra are shown in Figures 11 and 12.

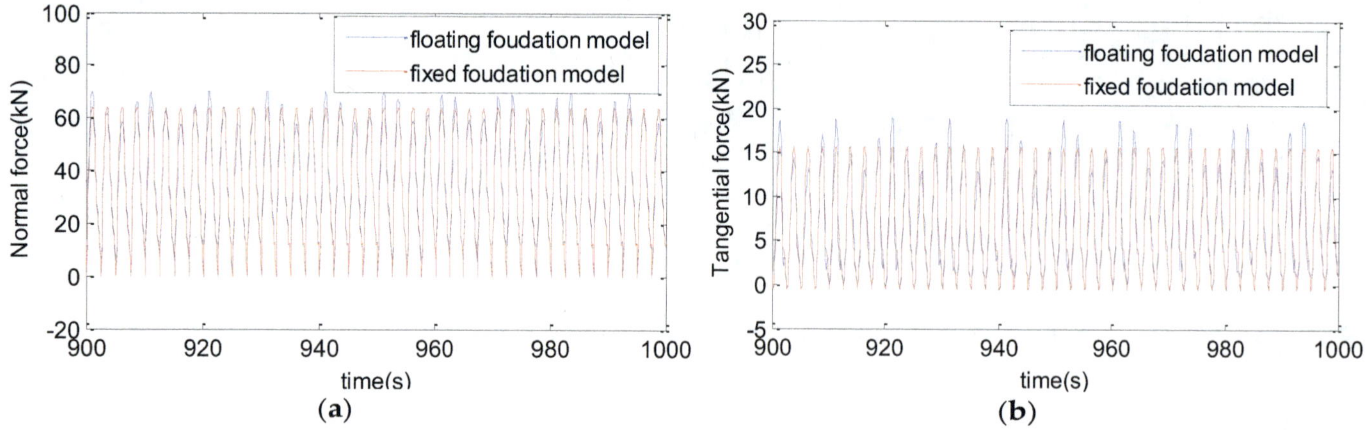

Figure 11. Time histories of aerodynamic force on the blade: (**a**) Normal force; and (**b**) Tangential force.

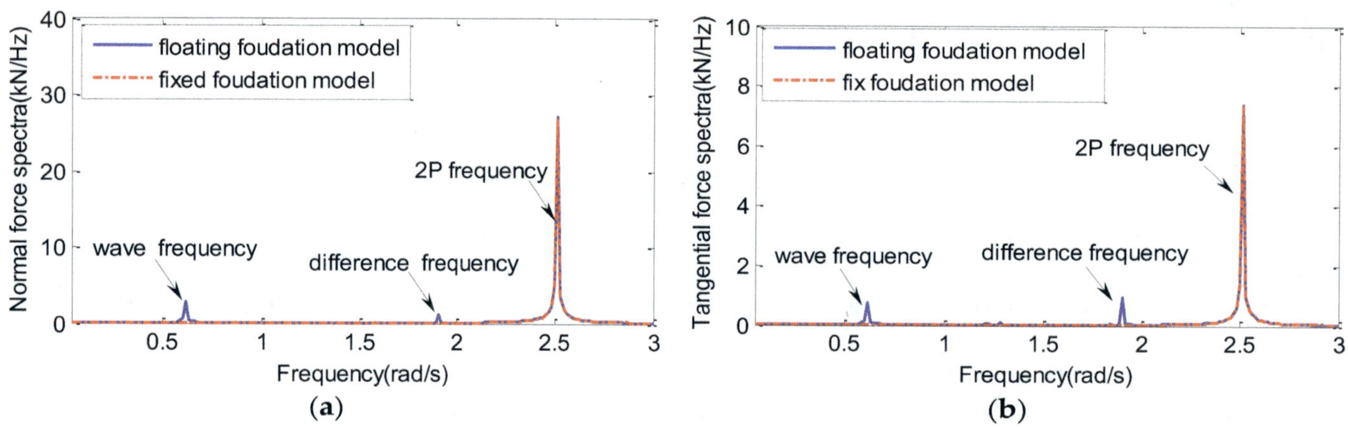

Figure 12. Spectra of blade aerodynamics: (**a**) Normal force spectra; and (**b**) Tangential force spectra.

Figures 11 and 12 show that the normal force is much bigger than tangential force. Compared with the fixed foundation case, the amplitudes of normal force and tangential force of the floating foundation cases increase up to 9.4% and 16.3%, respectively. As shown in Figure 12, the main frequency of aerodynamic force is 2P frequency. In addition, for the floating foundation model, there are frequency components of wave frequency and the difference frequency between 2P and wave frequency.

4.3.4. Motions of the Floating VAWT Considering the Wind and Irregular Wave Loads

Motions of the floating VAWT were calculated considering the wind loads and irregular wave (JONSWAP sea spectrum) loads in this section. The calculation parameters are shown in Table 6.

Table 6. Irregular wave and wind cases (LC4).

Items	Significant Wave Height (m)	Spectrum Peak Period (s)	V_∞(m/s)	Total Time (s)
LC4.1	2.1	9.74	5	3600
LC4.2	3.62	10.29	15	3600
LC4.3	6.02	11.38	25	3600

The time histories and spectra of pitch motions of LC4.2 and LC4.3 are presented in Figure 13. The max values, mean values and STDs of surge, heave and pitch responses of the floating VAWT are calculated, and the results are shown in Table 7.

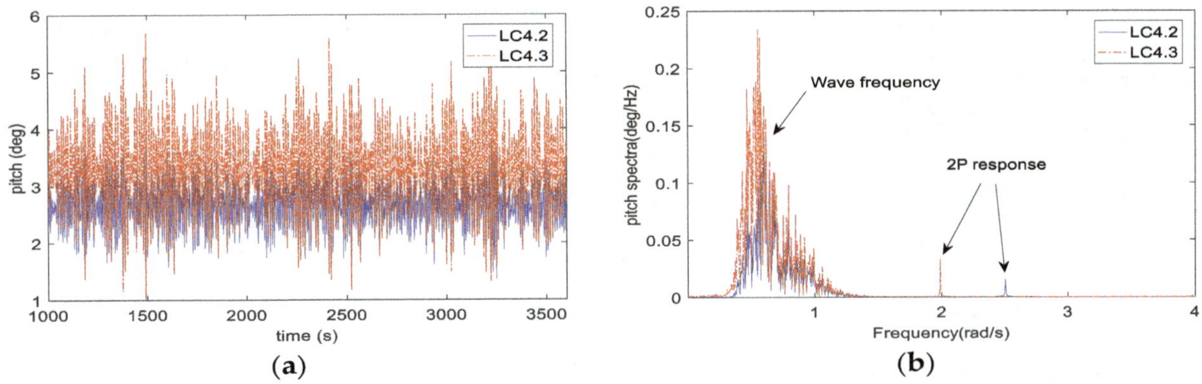

Figure 13. Pitch responses of the floating VAWT: (**a**) time histories; and (**b**) response spectra.

Table 7. Response statistics of the floating VAWT (cases LC4).

Items	Surge (m)			Heave(m)			Pitch(deg)		
	STD	Mean	Max	STD	Mean	Max	STD	Mean	Max
LC4.1	0.36	0.32	1.46	0.27	0.17	0.52	0.19	0.42	1.14
LC4.2	0.58	2.33	4.28	0.34	0.17	0.72	0.39	2.61	3.97
LC4.3	1.04	2.95	6.45	0.49	0.17	1.15	0.72	3.28	5.67

Figure 12 shows that the pitch motions of the floating VAWT increase with the increase of wave height and wind speed, and they are still dominated by the wave–frequency motions. Comparing Figure 12b with Figure 9b, the 2P frequency caused by the aerodynamic loads is more significant in the case LC4. Because of the effect of heave plates, the heave responses are small and the max values of heave is 1.15 m for LC4. In Table 7, we can find that the STDs of surge and pitch motions of the floating VAWT are small, which shows the floating VAWT has good motion performance in the wave and wind conditions.

4.3.5. Comparison with Φ-Type Floating VAWT

Liu et al. studied a 5 MW Φ-Darrieus type wind turbine supported by the same truss Spar type floating foundation and with the same mooring system, and its parameters are presented in Reference [11]. Compared with the Φ-type wind turbine (blade height is 129.56 cm, mass of two blades is 308 tons), the blade height and blades mass of this paper decreased by 39% and 75%, respectively. The two floating VAWTs have the same total mass.

Motions of the two floating VAWTs were compared considering the wind loads and regular wave loads, and the calculation parameters are shown in Table 8. The mean value and STDs of the surge, heave, and pitch motions of the two floating VAWTs are presented in Figures 14–16. The mean generator powers of the two floating VAWTs are shown in Figure 17.

Table 8. Comparison cases (LC5).

Case	Wind Turbine Type	Wave Height (m)	Wave Period (s)	V_∞ (m/s)
LC5.1	H-type Φ-type	2.10	9.74	5 5
LC5.2	H-type Φ-type	3.62	10.29	15 (rated wind speed) 14 (rated wind speed)
LC5.3	H-type Φ-type	6.02	11.38	25 25

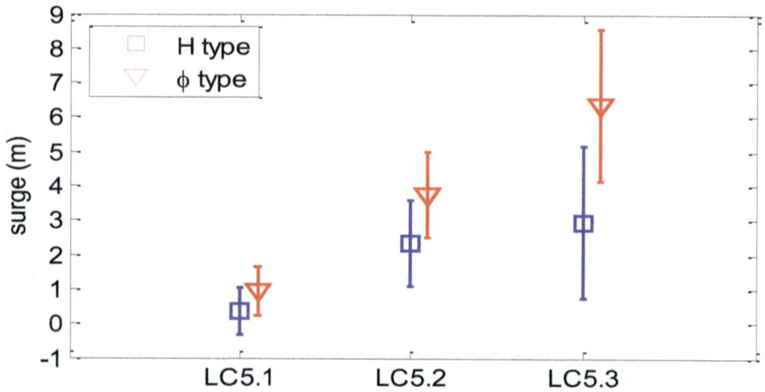

Figure 14. Surge comparison of two floating VAWTs (cases LC5).

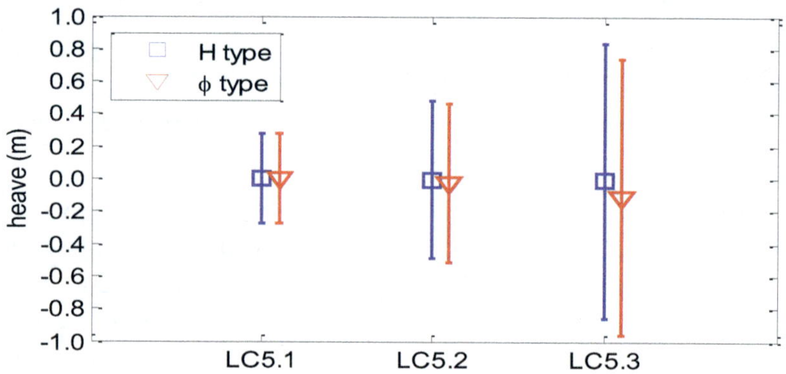

Figure 15. Heave comparison of two floating VAWTs (cases LC5).

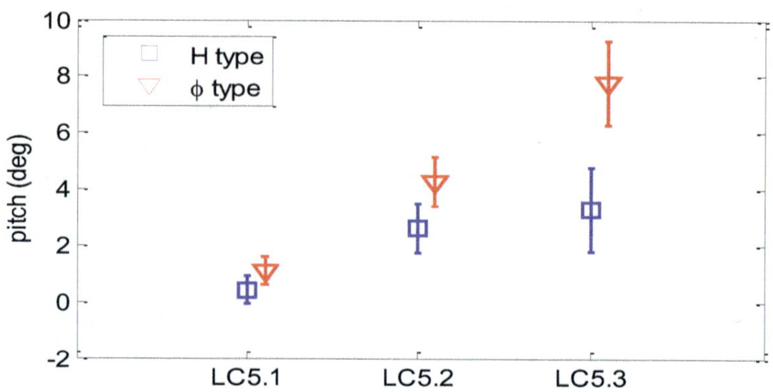

Figure 16. Pitch comparison of two floating VAWTs (cases LC5).

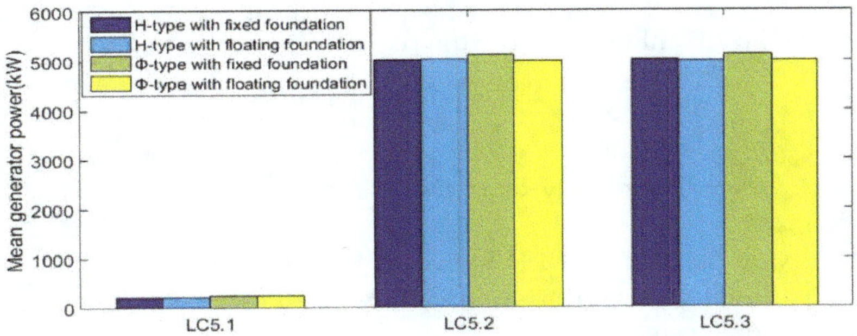

Figure 17. Mean generator power of two floating VAWTs (cases LC5).

Figures 14–16 show that compared with the Φ-type floating VAWT, the mean values of the H-type floating VAWT's motions are much smaller. Taking the rated wind speed case (LC5.2) as an example, the mean values of surge, heave and pitch of H-type floating VAWT decrease 38%, 74.2% and 38.6%, respectively. Therefore, H-type VAWT withstand smaller wind loads compared with the Φ-type VAWT. The STDs of motions of the two floating VAWTs are almost same, as they are supported by the same floating foundation, and the STD of the floating VAWT motions mainly depends on the wave loads. Figure 17 shows that the motion of floating VAWTs has insignificant effect on the mean generator power for both type wind turbines. It is a satisfactory result for the prospective development of a floating VAWT.

5. Conclusions

The surge–heave–pitch coupling nonlinear motion equations of the H-type floating VAWT were established considering the coupling between aerodynamic and hydrodynamic loads, and the motions of a 5 MW H-type floating VAWT were studied. The motion of the H-type floating VAWT was also compared with that of the Φ-type floating VAWT; they have the same floating foundation and total displacement. The main conclusions are as follows:

1. The natural periods of surge, heave, and pitch motions of the H-type floating VAWT are 51.2 s, 20.5 s, and 29.3 s, respectively. The small STDs of the LC3 and LC4 show the good motion performance of the H-type floating VAWT in the wave and wind conditions.
2. The effects of wind and wave loads on the motions of the H-type floating VAWT are assessed. For LC3, the motion frequencies of the floating VAWT are dominated by the wave frequencies, and there are small components of 2P frequency for the pitch motions. The 2P frequency is more significant in irregular wave conditions (LC4).
3. The effects of the floating foundation motions on the aerodynamics are also assessed. The results show that the normal force is much bigger than the tangential force. Compared with the fixed foundation case, the amplitudes of normal and tangential forces increase up to 9.4% and 16.3%, respectively, for the floating foundation case. Besides the 2P frequency, the frequencies of aerodynamics also include the wave frequency and the difference frequency between 2P and wave frequency for the floating foundation case. The effect of the floating foundation motion on the aerodynamics cannot be ignored.
4. The H-type VAWT withstand smaller wind loads compared with the Φ-type VAWT, and mean values of the H-type floating VAWT's motions are much smaller compared with the Φ-type floating VAWT. The STDs of motions of the two floating VAWTs are almost same, as they mainly depend on the wave loads.

Acknowledgments: The project was supported by the Natural Science Foundation of China (No. 51579176), the Natural Science Foundation of Tianjin (No. 16JCYBJC21200), and the Fund of the State Key Laboratory of Ocean Engineering, Shanghai Jiao Tong University (No. 1501).

Author Contributions: Ying Guo and Liqin Liu deduced the formulae and carried out numerical calculation; Xifeng Gao analyzed the data; and Wanhai Xu wrote the paper.

References

1. Paraschivoiu, I. *Wind Turbine Design with Emphasis on Darrieus Concept*; Polytechnic International Press: Montreal, QC, Canada, 2002.
2. Tjiu, W.; Marnoto, T.; Mat, S.; Ruslan, M.H.; Sopian, K. Darrieus vertical axis wind turbine for power generation II: Challenges in HAWT and the opportunity of multi-megawatt Darrieus VAWT development. *Renew. Energy* **2015**, *75*, 560–571. [CrossRef]
3. Collu, M.; Borg, M.; Shires, A.; Brennan, F.P. Progress on the development of a coupled model of dynamics for floating offshore vertical axis wind turbines. In Proceedings of the ASME 32nd International Conference on Ocean, Offshore and Arctic Engineering, Nantes, France, 9–14 June 2013. OMAE2013-54337.
4. Vita, L. Offshore Floating Vertical Axis Wind Turbines with Rotating Platform. Ph.D. Thesis, Danish Technical University, Copenhagen, Denmark, 2011.
5. Wang, K.; Hansen, M.O.L.; Moan, T. Dynamic analysis of a floating vertical axis wind turbine under emergency shutdown using hydrodynamic brake. *Energy Procedia* **2014**, *53*, 53–69. [CrossRef]
6. Cheng, Z.S.; Madsen, H.A.; Gao, Z.; Moan, T. A fully coupled method for numerical modeling and dynamic analysis of floating vertical axis wind turbines. *Renew. Energy* **2017**, *107*, 604–619. [CrossRef]
7. Owens, B.C.; Hurtado, J.E.; Paquette, J.A. Aero-elastic modelling of large offshore vertical-axis wind turbines: development of the offshore wind energy simulation toolkit. In Proceedings of the 54th AIAA/ASME/ASCE/A-HS/ASC Structures, Structural Dynamics, and Materials Conference, Boston, MA, USA, 8–11 April 2013.
8. Borg, M.; Collu, M. A comparison on the dynamics of a floating vertical axis wind turbine on three different floating support structures. *Energy Procedia* **2014**, *53*, 268–279. [CrossRef]
9. Cheng, Z.S.; Madesn, H.A.; Gao, Z.; Moan, T. Numerical study on aerodynamic damping of floating vertical axis wind turbines. *J. Phys. Conf. Ser.* **2016**, *753*, 102001. [CrossRef]
10. Cahay, M.; Luquiau, E.; Smadja, C.; Silvert, F. Use of a vertical wind turbine in an offshore floating wind farm. *J. Pet. Technol.* **2011**, *63*, 105–107.
11. Liu, L.Q.; Guo, Y.; Jin, W.C.; Yuan, R. Motion performances of a 5 MW VAWT supported by spar floating foundation with heave plates. In Proceedings of the ASME 36th International Conference on Ocean, Offshore and Arctic Engineering, Trondheim, Norway, 25–30 June 2017. OMAE2017-62625.
12. Paulsen, U.S.; Borg, M.; Madsen, H.A. Outcomes of the DeepWind conceptual design. *Energy Procedia* **2015**, *80*, 329–341. [CrossRef]
13. Bedon, G.; Paulsen, U.S.; Madsen, H.A.; Belloni, F.; Castelli, M.; Benini, E. Aerodynamic Benchmarking of the Deepwind Design. *Energy Procedia* **2015**, *75*, 677–682. [CrossRef]
14. Battisti, L.; Benini, E.; Brighenti, A.; Castelli, M.R.; Dell, A.S. Wind tunnel testing of the DeepWind demonstrator in design and tilted operating conditions. *Energy Procedia* **2016**, *111*, 484–497. [CrossRef]
15. Collu, M.; Borg, M.; Manuel, L. On the relative importance of loads acting on a floating vertical axis wind turbine system when evaluating the global system response. In Proceedings of the ASME 35th International Conference on Ocean, Offshore and Arctic Engineering, Busan, Korea, 19–24 June 2016. OMAE2016-54628.
16. Liu, L.Q.; Zhou, B.; Tang, Y.G. Study on the nonlinear dynamical behavior of deep sea spar platform by numerical simulation and model experiment. *J. Vib. Control* **2014**, *20*, 1528–1553. [CrossRef]
17. Keyvan, S.; Atilla, I.; Martin, J.D. Response analysis of a truss spar in the frequency domain. *J. Mar. Sci. Technol.* **2004**, *8*, 126–137.
18. Sandra, E. Evaluation of different turbine concepts for wind power. *Renew. Sustain. Energy Rev.* **2008**, *12*, 1419–1434.
19. International Electrochemical Commission. *Wind Turbines, Part 3: Design Requirements for Off-Shore Wind Turbines*; IEC61400-3; International Electrochemical Commission: Geneva, Switzerland, 2009.
20. Paraschivoiu, I. Aerodynamic loads and performance of the Darrieus rotor. *J. Energy* **1982**, *6*, 406–412. [CrossRef]
21. Sheldahl, R.E.; Klimas, P.C. *Aerodynamic Characteristics of Seven Symmetrical Airfoil Sections through 180-Degree Angle of Attack for Use in Aerodynamic Analysis of Vertical Axis Wind Turbines*; Sandia National Laboratories: Albuquerque, NM, USA, 1981.

22. Andrew, S. Development and Evaluation of an Aerodynamic Model for a Noval Vertical Axis Wind Turbine Concept. *Energies* **2013**, *6*, 2501–2520.
23. Berg, D.E. An improved double-multiple stream tube model for the Darrieus type vertical-axis wind turbine. In Proceedings of the Sixth Biennial Wind Energy Conference and Workshop, Minneapolis, NM, USA, 1–3 June 1983; pp. 231–233.
24. Morgan, C.A.; Gardner, P.; May, I.D.; Anderson, M.B. The Demonstration of a Stall Regulated 100 Kw Vertical Axis Wind Turbine. In Proceedings of the European Wind Energy Conference, Glasgow, UK, 10–13 July 1989.
25. Paraschivoiu, I.; Desy, P.; Massont, C. Blade Tip, Finite Aspect Ratio, and Dynamic Stall Effects on the Darrieus Rotor. *J. Propuls. Power* **1988**, *4*, 73–80. [CrossRef]
26. American Petroleum Institute. *Design and Analysis of Station keeping Systems for Floating Structures*; API RP-2SK; American Petroleum Institute: Washington, DC, USA, 2005.
27. Faltinsen, O.M. *Sea Loads on Ships and Offshore Structures*; Cambridge University Press: Cambridge, UK, 1990.
28. Pan, Z.Y.; Vada, T.; Finne, S. Benchmark study of numerical approaches for wave-current interaction problem of offshore floaters. In Proceedings of the ASME 35th International Conference on Ocean, Offshore and Arctic Engineering, Busan, Korea, 19–24 June 2016. OMAE2016-54411.
29. Liu, L.Q.; Guo, Y.; Zhao, H.X.; Tang, Y.G. Dynamic modeling, simulation and model tests research on the floating VAWT. *Chin. J. Theor. Appl. Mech.* **2017**, *49*, 299–307. (In Chinese)

12

Aerodynamic Performance of Wind Turbine Airfoil DU 91-W2-250 under Dynamic Stall

Shuang Li [1,2], Lei Zhang [1,3,4,*], Ke Yang [1,3,4], Jin Xu [1,2] and Xue Li [1,2]

1 Institute of Engineering Thermophysics, Chinese Academy of Sciences, Beijing 100190, China; lishuang@iet.cn (S.L.); yangke@iet.cn (K.Y.); xujin@iet.cn (J.X.); lixue@iet.cn (X.L.)
2 University of Chinese Academy of Sciences, Beijing 100049, China
3 Key Laboratory of Wind Energy Utilization, Chinese Academy of Sciences, Beijing 100190, China
4 National Research and Development Center of Wind Turbine Blade, Beijing 100190, China
* Correspondence: zhanglei@iet.cn

Abstract: Airfoils are subjected to the 'dynamic stall' phenomenon in significant pitch oscillations during the actual operation process of wind turbines. Dynamic stall will result in aerodynamic fatigue loads and further cause a discrepancy in the aerodynamic performance between design and operation. In this paper, a typical wind turbine airfoil, DU 91-W2-250, is examined numerically using the transition shear stress transport (SST) model under a Reynolds number of 3×10^5. The influence of a reduced frequency on the unsteady dynamic performance of the airfoil model is examined by analyzing aerodynamic coefficients, pressure contours and separation point positions. It is concluded that an increasingly-reduced frequency leads to lower aerodynamic efficiency during the upstroke process of pitching motions. The results show the movement of the separation point and the variation of flow structures in a hysteresis loop. Additionally, the spectrum of pressure signals on the suction surface is analyzed, exploring the level of dependence of pressure fluctuation on the shedding vortex and oscillation process. It provides a theoretical basis for the understanding of the dynamic stall of the wind turbine airfoil.

Keywords: wind turbine airfoil; dynamic stall; boundary layer separation; aerodynamic characteristics

---

1. Introduction

Wind turbines operate in diverse environments under complex control strategies, suffering unsteady loads including aerodynamic forces. An important source of these forces comes from the blades when the relative velocity with respect to blade sections varies in magnitude and direction. Generally, the airfoils on the blades are considered as operating in dynamic stall [1]. The unsteady factors that cause dynamic stall are, e.g., wind shear and yaw, and the angles of attack of airfoils periodically vary with the rotation of blades [2–4]. Then, the stall characteristics are dynamically dependent on the operating route of airfoils with dramatically different lift and drag forces from those in static operations. The main disadvantage of dynamic stall for the safety of blades is that it causes fatigue loads and a deviation between design and actual running conditions. With the trend of the large-scale development of wind turbines, dynamic stall will have increasingly greater influence on the unsteady aerodynamic characteristics of wind turbine blades.

Dynamic stall has been one of the major interesting research subjects and has been studied over the latest few decades. The dynamic stall phenomenon was first observed in the 1930s [5]. It was first systematically studied in the aerospace field and was studied in a wider field later, such as for maneuverable jets wings, gas turbines and wind turbines [6]. Previous studies primarily concentrating on dynamic stall consist of experimental and numerical simulation. Experiments are important because experimental data provide a verification basis for the numerical simulation. However, due to the limitation of experimental conditions and other reasons, it is necessary to employ numerical simulation to grasp a broader and more general conclusion. With the gradual maturity of the theory, there are many analytical models to predict the effects of dynamic stall and analyze the load of power machines nowadays; such as the Boeing–Vertol model [7], the ONERA model [8], the Beddoes–Leishman (B–L) semi-empirical model [9], etc. The B-L model has been adopted most widely to analyze dynamic stall. It has been modified and improved with the increasing attention on unsteady aerodynamics [10–13]. The modified model shows better agreement than the original one. Although the semi-empirical models have the advantage of using less computational resources, they depend on the accuracy of the empirical factors greatly, and it is not possible to obtain detailed and accurate flow field information. With the development of computer technology, the computational fluid dynamics (CFD) method is used to study dynamic stall for the purpose of obtaining detailed information of dynamic stall. Some researchers have been dedicated to finding more accurate turbulence models. For example, Martinat et al. [14] used three turbulence models to investigate the dynamic stall of a NACA 0012 airfoil, and the unsteady Reynolds-averaged Navier–Stokes (URANS) $k-\varepsilon$ Chien model provided the best results. Wang et al. [15] compared the renormalization-group (RNG) $k-\varepsilon$ model and transition shear stress transport (SST) $k-\omega$ model used to simulate dynamic stall of an oscillating NACA 0012 airfoil at a relatively low Reynolds number. It was found that the SST $k-\omega$-based detached eddy simulation (DES) approach was superior. Some other researchers have focused on how typical factors affecting dynamic stall. These factors include the Reynolds number, mean angle of pitch oscillation, amplitude of pitch oscillation, reduced frequency, the position of pitch axis, turbulence intensity, etc. For instance, Akbari et al. [16] studied the effects of reduced frequency, mean angle of attack, pitch axis location and Reynolds number on the dynamic stall of a NACA 0012. It was shown that the reduced frequency was the most influential parameter on dynamic stall. Gharali et al. [17] investigated the effects of the oscillations of the free-stream velocity on pitch oscillation by the SST $k-\omega$ model; Kim et al. [18] studied the effects of free-stream turbulence on the aerodynamic characteristics of pitching airfoils by large eddy simulations. Gandhi et al. [19] investigated the influence of reduced frequency on the dynamic stall characteristics of a NACA 0012 airfoil pitching in a turbulent wake with direct numerical simulations. In addition, McCroskey and his partners made two efforts for decades to investigate the details of the dynamic stall phenomena [20–22]. The above results are extensive and continuous. However, the problem is that most of the research objects are wings or symmetrical airfoils. The literature lacks observations of specific asymmetric and blunt tail edge wind turbine airfoils, such as Delft University of Technology (DU) series airfoils. On the other hand, the flow field contour is a general method to explain the phenomenon of dynamic stall used by many previous researchers. It is summarized that a typical dynamic stall process consists of four major stages: attached flow at low angles of attack, the development of the leading edge vortex (LEV), the shedding of the LEV from the suction surface and reattachment of the flow to the suction surface [23]. However, there is a little attention given to the relationship between the frequency of shedding vortices and the frequency of airfoil pitching motion. The interpretation of the "delay" is always analyzed by the lift and drag coefficients or the qualitative expression of the flow field contours. The quantitative expression of the separation of the airfoil boundary layer is not obviously given.

In order to avoid flow separation and reduce the dynamic stall effects, various devices have been proposed for flow control. Depending on the operating principle, they can be classified as active or passive [24]. Passive control includes vortex generators, microtabs, serrated trailing edges, and so on. Active control includes trailing-edge flaps, synthetic jets, and so on. Johnson et al. [25] overviewed 15 different devices for wind turbine control. Lei Zhang et al. [26,27] made long-term efforts to investigate the effects of vortex generators. Some other researchers focused on microtabs [28,29] and obtained some good results. T.K. Barlas et al. [30] and Unai Fernandez-Gamiz et al. [31] reviewed some devices and discussed their effects. The flow control devices have good effects on avoiding flow separation, but the blades have a higher cost for their production. Based on the above, cases with different reduced frequencies (the most important influential factor) are compared regarding the aerodynamic forces and lift-to-drag ratio. In addition, the delay effect of dynamic stall is deeply analyzed from the perspective of two aspects: the pressure field and spectrum analysis.

2. Method

2.1. Numerical Method

The airfoil DU 91-W2-250 was developed by TU Delft, which has a relative thickness of 25% with a trailing edge gap of 0.65% chord [32]. The present study is based on a 2D simulation with ceiling and bottom boundary similarity in a wind tunnel. With this kind of set, the numerical method is easily validated by the authors' previous measurements [33]. In order to simulate the dynamic stall phenomenon, a rotational zone whose radius is three-times that of the model chord is placed at the pitching center, as shown in Figure 1. The rotational zone and airfoil model oscillate like a rigid body, and the movement is defined with a user-defined function (UDF) in ANSYS Fluent [22]. The rest of grid is connected to the rotational grid by interfaces. The block interfaces between the dynamic and static zones are non-conformal because, as the airfoil moves, the O-grid attached to the airfoil rotates rigidly (as a solid body) together with the airfoil.

The Reynolds number is 3×10^5, which confirmed the turbulence that appeared in the previous study. Flows under such a relatively low Reynolds number are highly non-linear [34]. Therefore, turbulence should be considered in the simulation. There are three main turbulence simulation methods, i.e., direct numerical simulation (DNS), Reynolds-averaged Navier–Stokes (RANS) and large eddy simulation (LES). A turbulence model should present a satisfactory representation of an important phenomenon while introducing the least amount of complexity [35]. In the present study, considering numerical simulation cost and acceptable accuracy, the unsteady Reynolds-averaged Navier–Stokes (URANS) method is seem to be the most suitable one to simulate the dynamic stall phenomenon [15]. Consequently, in the CFD software ANSYS Fluent, the transition SST $k - \omega$ model, which is capable of capturing the transition process, was chosen. Some researchers have also discussed that the accuracy of this model is satisfactory [15,17,36–40].

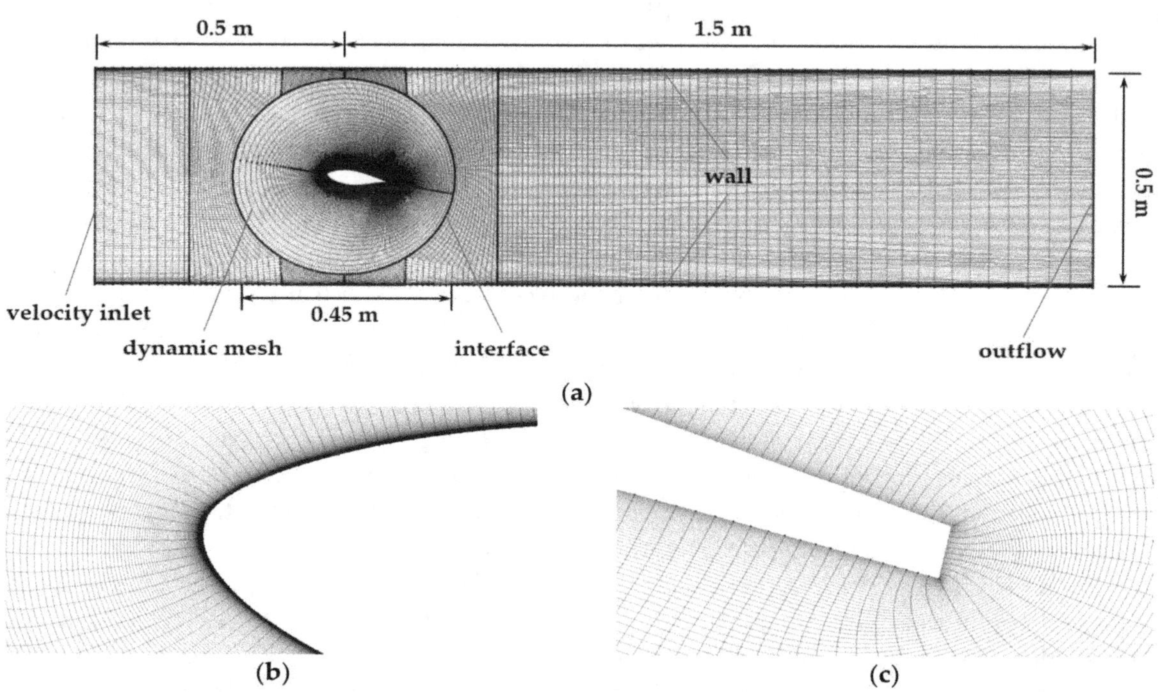

Figure 1. (**a**) Mesh topology of the whole domain; (**b**) mesh details near the leading edge; (**c**) mesh details near the trailing edge.

2.2. Oscillation of the Airfoil

The movement of the airfoil model is described as Equation (1) in a sinusoidal mode, as shown in Figure 2.

$$\alpha_t = \alpha_0 + \alpha_{amp} \sin(\omega t), \tag{1}$$

where,

α_t is the angle of attack at flow time t,
α_0 is the mean angle of attack,
α_{amp} is the pitch oscillation amplitude,
ω is the oscillation angular frequency.

The angular frequency ω is related to the reduced frequency k, which is an important parameter used to describe the 'degree of unsteadiness', as defined by Equation (2).

$$k = \frac{\omega c}{2U_\infty} \tag{2}$$

where U_∞ is the free-stream velocity and c is the chord length of the airfoil. It can be obtained from dimensionless Navier-Stokes (N-S) equations. J. G. Leishman has the empirical view that when $0 \leq k \leq 0.5$, the flow can be considered as quasi-steady flow, which means for some cases that the unsteady effects may be neglected completely [1]. Therefore, k = 0.026, 0.502 and 0.106 are set to be computed to find out whether the previous empirical view applies to the pitching motion of the DU 91-W2-250 airfoil. They correspond to the oscillating period T = 0.6 s, 0.3 s and 0.15 s, respectively.

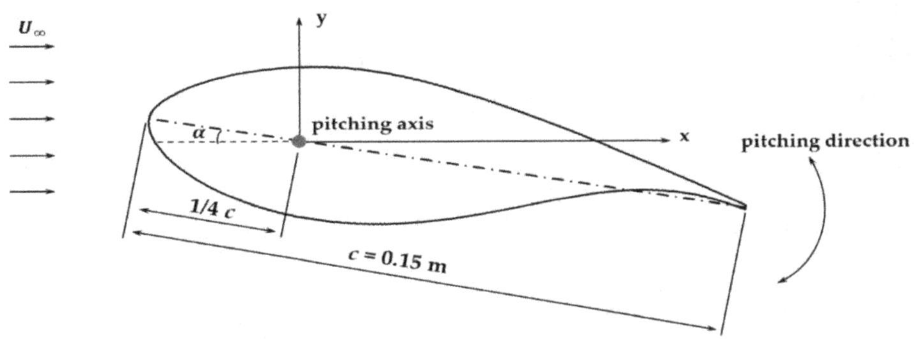

Figure 2. Schematic of the pitching movement of the airfoil model.

2.3. Grid Sensitivity

The independence of the solution from the grid is examined by different grid densities in Table 1. Only the circumferential grid densities are changed, but the radial grid densities are the same. The growth rate of grid is set as 1.05, and the maximum grid is 0.3c. It is important that the mesh near the wall be properly sized to ensure accurate simulation and no wall functions are used [41]. The height of the first mesh cell off the wall is 10^{-5} m, so that the non-dimensional wall-distance y^+ [42] is less than 1.0 at a Reynolds number of 3×10^5. It is computed by using flat-plate boundary layer theory [43]. The lift and drag coefficients during the 4th cycle of pitching motion are shown in Figure 3. The comparison shows that the numerical results for grid cases G3 and G4 do not differ considerably from other grid densities. In order to observe more detailed characteristics on the suction surface of the airfoil, the grid nodes on the suction side are increased to 200 and keep the same set of grids as G3 on the pressure side of the airfoil.

Table 1. Cases for grid sensitivity examination.

Grid	Suction Surface	Pressure Surface	Trailing Edge	Total Grids
G1	50 [1]	50	11	25,753
G2	100	100	11	45,387
G3	150	150	11	67,238
G4	200	200	11	89,070

[1] The number of cells on the indicated surfaces.

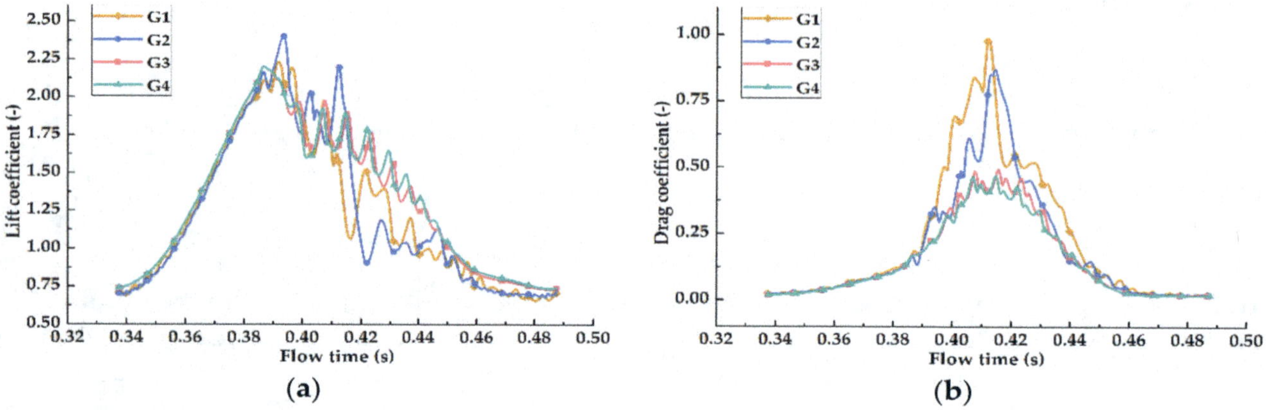

Figure 3. Comparison of computed (**a**) C_l and (**b**) C_d with different grid densities during the 4th cycle of pitching motion.

2.4. Time Step and Periodic Repeatability

To specify the optimum time step in the unsteady simulations, a series case with different time steps was simulated. A characteristic time T_c defined by Equation (3) is employed as base of the time step:

$$T_c = \frac{c}{U_\infty} \tag{3}$$

It has the same significance as the Courant, Friedrichs and Levy criterion (CFL) number in numerical simulations. The time steps of 0.1T$_c$, 0.05T$_c$, 0.025T$_c$, 0.0125T$_c$ and 0.00625T$_c$ (denoted as dt1 to dt5, respectively) are used to examine the time step sensitivity.

As shown in Figure 4, when the time step dt4 = 6.25×10^{-5} s, the time cost and accuracy are acceptable for the present study. This is consistent with dynamic stall cases applied by other researchers [17,38].

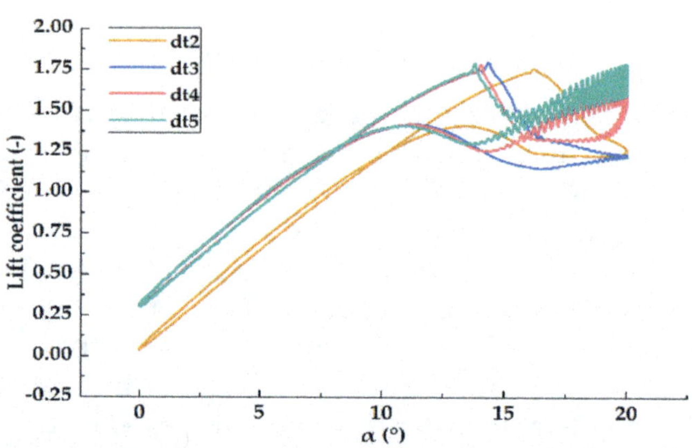

Figure 4. Comparison of computed C_l with different time steps.

Periodic repeatability is also examined, as shown in Figure 5. It shows that values tend to be well reproducible from the third cycle. That is why the other cases are computed only for 4 periods.

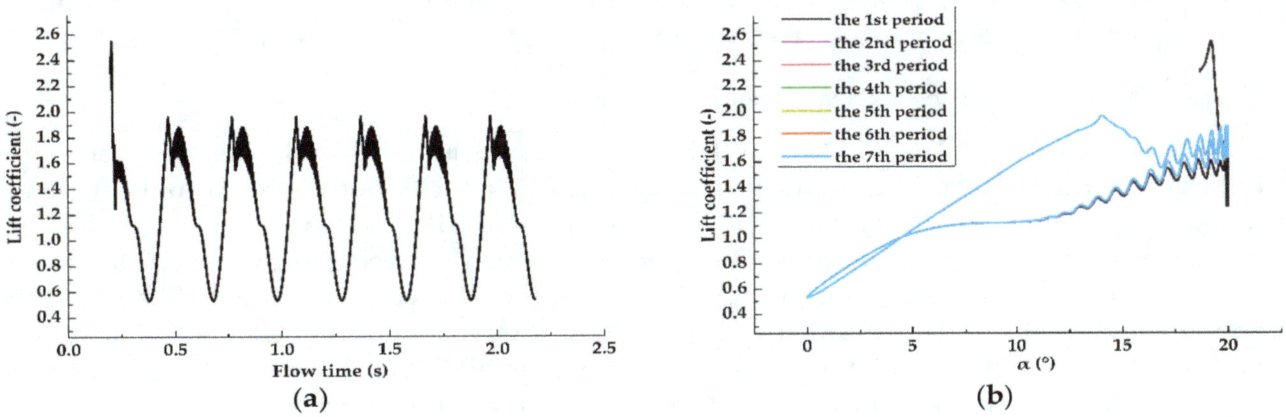

Figure 5. (**a**) history of the lift coefficient C_l; (**b**) the repeatability of the lift coefficient C_l with $k = 0.052$, $\alpha_0 = 10°$, $\alpha_{amp} = 10°$.

2.5. Comparison of the Static Result with Experiment and Vortex Panel Method

Due to the lack of experimental data for the dynamic stall of the DU 91-W2-250 airfoil at present, the static experimental data, which have been obtained from Jingyan Bai et al. [33], are used to examine the accuracy of the numerical method. The data computed by the software Xfoil and Rfoil, which are currently widely accepted, are also compared to each other, as shown in Figure 6.

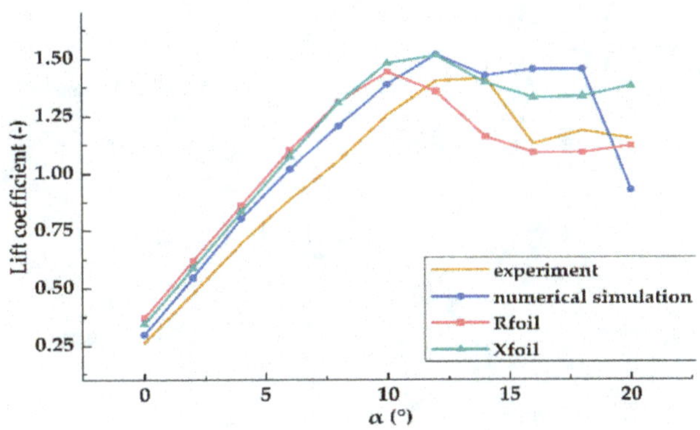

Figure 6. Comparison between the numerical results, Rfoil, Xfoil and experimental data [25].

Generally, Xfoil and Rfoil predict acceptable lift and drag coefficients for thin airfoils in attached flow. It is shown that the numerical simulation is more consistent with the data obtained by Xfoil. Because of the constraints of the experimental equipment and conditions in 2014, a large discrepancy appears between the experimental and CFD results. The accuracy of the zero angle of attack and the stiffness of the support structure may be the main factors, which usually cause the shift of the lift curve along the angle-axis and the variation of the slope of the linear section on the lift curve, respectively. Moreover, CFD greatly overestimates the lift coefficient at 18 and 20 with respect to that of Xfoil, Rfoil and the experiment. This is because the airfoil goes into a deep stall and the flow becomes fully separated at high angles of attack. Raffel et al. [44] showed that the 3D effects of separation flows should be more significant than those without separations at smaller angles of attack. Therefore, in the following analysis, static data calculated by Xfoil are considered as baseline data in the static condition.

3. Results and Discussions

In this part, the results of the cases defined in Section 2.2 are analyzed. The performance of airfoil DU 91-W2-250 under dynamic stall is illustrated with respect to three aspects, i.e.: (a) the influence of reduced frequency; (b) flow structures and separation points; and (c) spectrum characteristics.

3.1. Effects of Reduced Frequency on Aerodynamic Coefficients

Three cases when k = 0.026, 0.502 and 0.106 were simulated, and the results are shown in Figure 7. The dynamic characteristics of the airfoil undergoing pitch oscillation were strongly dependent on the reduced frequency. The hysteresis loop increased with the increasingly-reduced frequency. In a loop, as can be seen, the maximum lift coefficient in upstroke and the corresponding angle of attack were much higher than those in the static condition. Previous researchers have observed this phenomenon already as mentioned in Section 1. Contrarily, the lift coefficient in downstroke was smaller than that in upstroke and sometimes even smaller than that in the static condition. On the lift curves, the intersection point in a loop was named the "reunion point" by the authors. An increase in the reduced frequency made the angle of reunion point smaller. This also means that the effect of delay was more significant when the frequency reduction was greater. On the other hand, the increasingly-reduced frequency would cause the increase of the drag coefficient. This effect is clearly revealed in the lift-drag-ratio curve in Figure 7c.

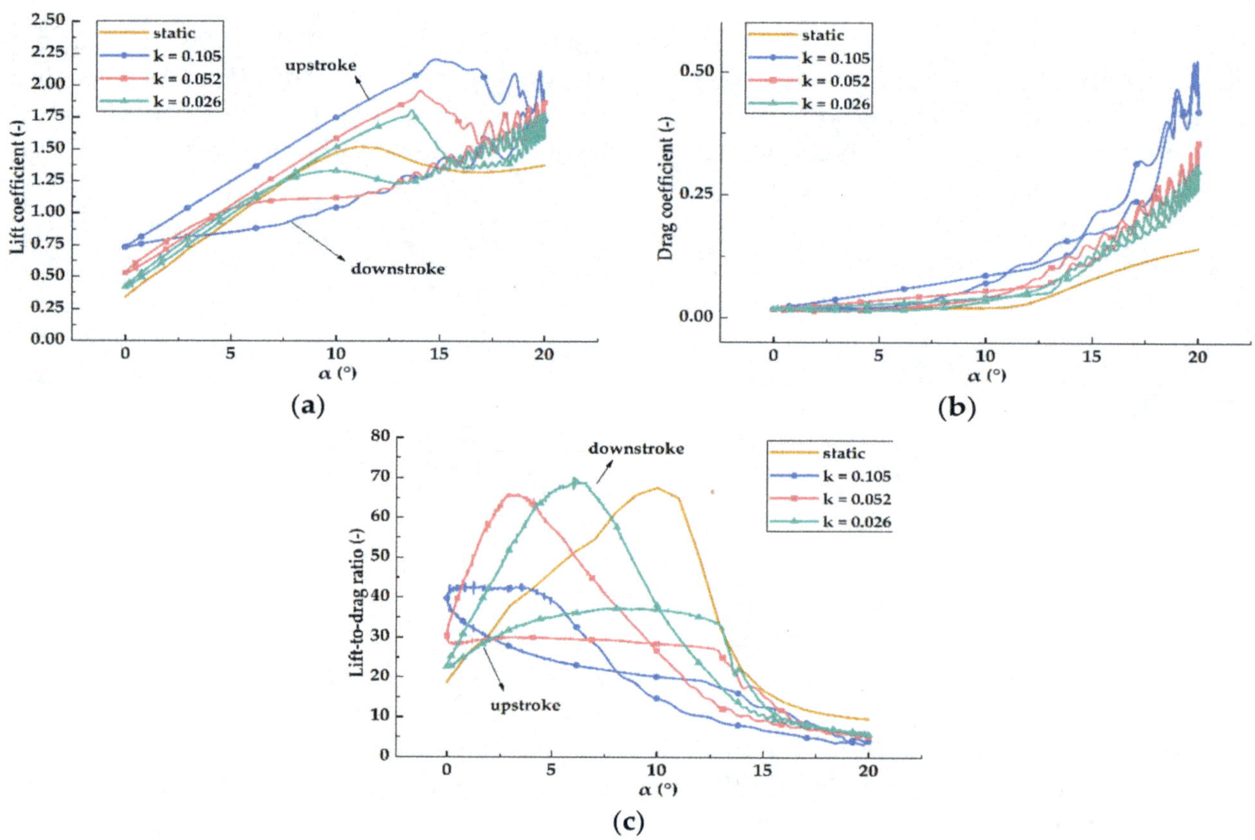

Figure 7. Comparison of the different reduced frequencies $k = 0.105, 0.052, 0.026$ of the computed (**a**) lift coefficient C_l, (**b**) drag coefficient C_d and (**c**) lift-to-drag ratio C_l/C_d with $\alpha_0 = 10°$, $\alpha_{amp} = 10°$.

The lift-to-drag ratio is an important parameter that indicates the aerodynamic efficiency of the airfoils. The aerodynamic performance of the airfoils is better with a larger lift-to-drag ratio. The angle of attack corresponding to the maximum lift-to-drag ratio is called a favorable angle of attack. In general, from the minimum angle to the favorable angle of attack, lift increases fast, but drag increases slowly. Therefore, the lift-to-drag ratio increases. Then, from the favorable angle to a critical stall angle of attack, the increase of lift is slow, and the drag increases rapidly. This makes the lift-to-drag ratio relatively small. After the critical stall angle of attack, the lift-to-drag ratio sharply decreases with the sharply increasing pressure drag. The lift-to-drag ratio of the airfoil during a complete hysteresis loop was quite different with a regular pattern. The peak of the lift-to-drag ratio curve in upstroke decreased and even disappeared with the increasingly-reduced frequency because of the faster increase of the drag coefficient. This effect was amplified in the lift-to-drag ratio and showed the tremendous difference between static and dynamic stall. Instead, the lift-to-drag ratio in the downstroke process showed a different form. Its shape was closer to the static form, which also had a peak. The angle of attack corresponding to the peak was smaller with increasingly-reduced frequency. This also shows the degree of the delay effect. Therefore, the aerodynamic efficiency was lower in dynamic stall process though the lift in upstroke was higher than that in the static condition.

3.2. *Flow Development and Separation Points on the Suction Surface of the Airfoil*

The reason for the variation of forces and lift-to-drag ratio can be found in flow field contours. One case was selected to investigate the characteristics as an example. The pressure fields and corresponding streamlines at different angles of attack through a full pitching cycle in the case with $k = 0.052$, $\alpha_0 = 10°$, $\alpha_{amp} = 10°$ are demonstrated in Figure 8. It was observed that the development trend of flow field structures of the DU 91-W2-250 was qualitatively in good agreement with the conclusions of other papers with different shapes of airfoils [15,45,46]. The flow was fully attached to the airfoil from the angles of attack $\alpha_t = 0°$~$10.31°$. The hysteresis of stall angles of attack in lift curves, which is the most obvious feature of dynamic stall, is observed clearly here. The lift dropped suddenly at about 14° in the hysteresis loop, while the static stall angle was 12° in the experiment. It was manifested as reversed flow generated near the trailing edge, as shown in Figure 8c. The reverse flow was due to the adverse pressure gradient caused by the rapid movement of the airfoil surface. As the angle of attack increased, the clockwise-rotating vortex on the suction surface moved forward to the leading edge, and the strength of the recirculation increased. There was a small counterclockwise-rotating vortex behind the trailing edge vortex in Figure 8c, implying that the vortex near the trailing edge started to move downstream slowly. This would lead to a decrease in the lift coefficient compared with the lift coefficient curves in Figure 7a. The effects of vortices rotating in two opposite directions on the lift coefficient are shown in Figure 9 in more detail. At Point A in Figure 9a, the clockwise-rotating vortex appeared. With the growth of the vortex, the lift coefficient was still increasing. The growing vortex increased the slope of the lift curves until the maximum value of C_l when the angle of attack was 14.07° at Point B; see Figure 9b. The reason that C_l decreased was the generation of the counterclockwise-rotating vortex, as shown in Figure 9c. C_l decreased with the growth of the counterclockwise-rotating vortex until the counterclockwise-rotating vortex traveled into the wake and the clockwise-rotating vortex appeared again; see Figure 9e. This process would repeat until the trend of the lift coefficient curve became flat when $\alpha_t = 10.31°$ in the downstroke process. The counterclockwise-rotating vortex did not appear from this moment. In other words, the growth and shedding of vortices in opposite directions led to the fluctuations of the lift coefficient curve. The lift coefficient C_l increased with the growth of the vortex, and it decreased with the shedding of the vortex. What needs to be noted here is that the range of the angles of attack in this case was from 0°–20°, so the airfoil did not experience a deep stall. Vortices on the suction surface of airfoil did not develop to reach the leading edge, but only reached 30% of the chord from the leading edge. Therefore, in this case with $k = 0.052$, $\alpha_0 = 10°$, $\alpha_{amp} = 10°$, the LEV observed in other papers [15] did not appear, but it appeared when $k = 0.052$, $\alpha_0 = 15°$, $\alpha_{amp} = 10°$. After the maximum angle of 20°, the airfoil was performing the downstroke motion. It is known that the lift coefficient of the downstroke process was much smaller than the upstroke process from Section 3.1. This is because the vortex still existed near the trailing edge and the pressure of pressure surface was lower at the same angle of attack by comparing Figure 8b,h and Figure 8c,g. As the airfoil attack angle became smaller, vortices eventually moved into the wake. The last vortex left the trailing edge at $\alpha_t = 6.61°$, and the flow completely reattached to the airfoil at $\alpha_t = 4.73°$. They were smaller than the angles respectively in upstroke with a similar structure. Therefore, the hysteresis of flow attachment during the downstroke process was considerable.

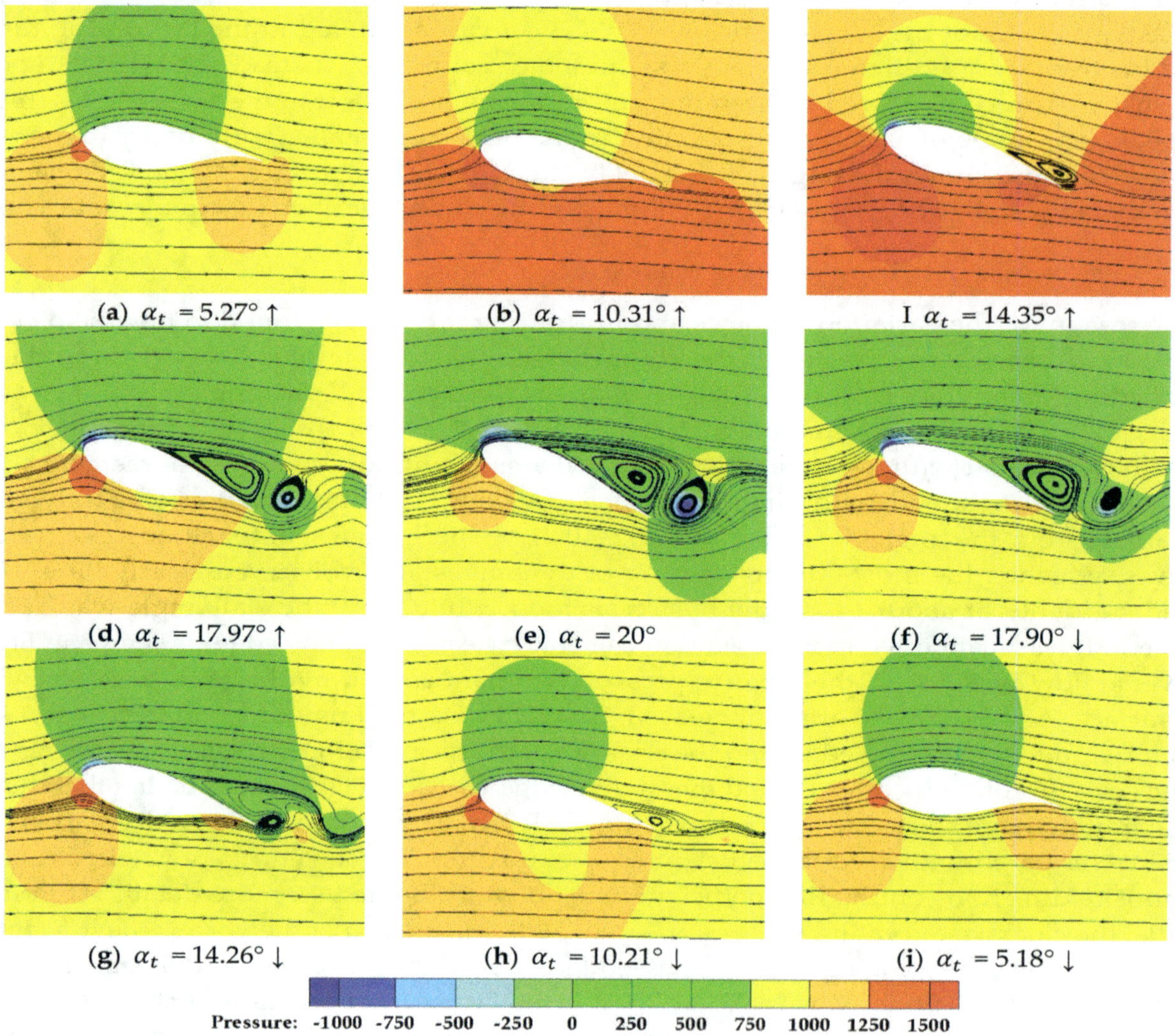

Figure 8. Pressure fields and streamlines for different angles of attack in the pitching cycle with $k = 0.052$, $\alpha_0 = 10°$, $\alpha_{amp} = 10°$.

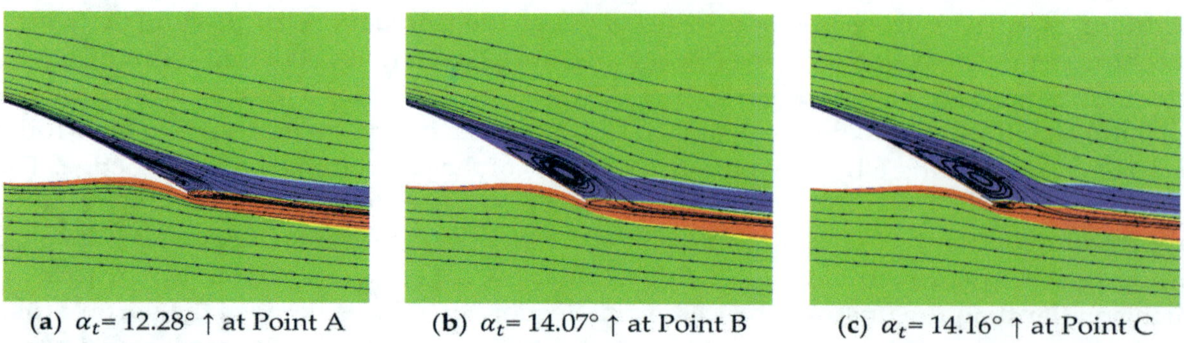

Figure 9. *Cont.*

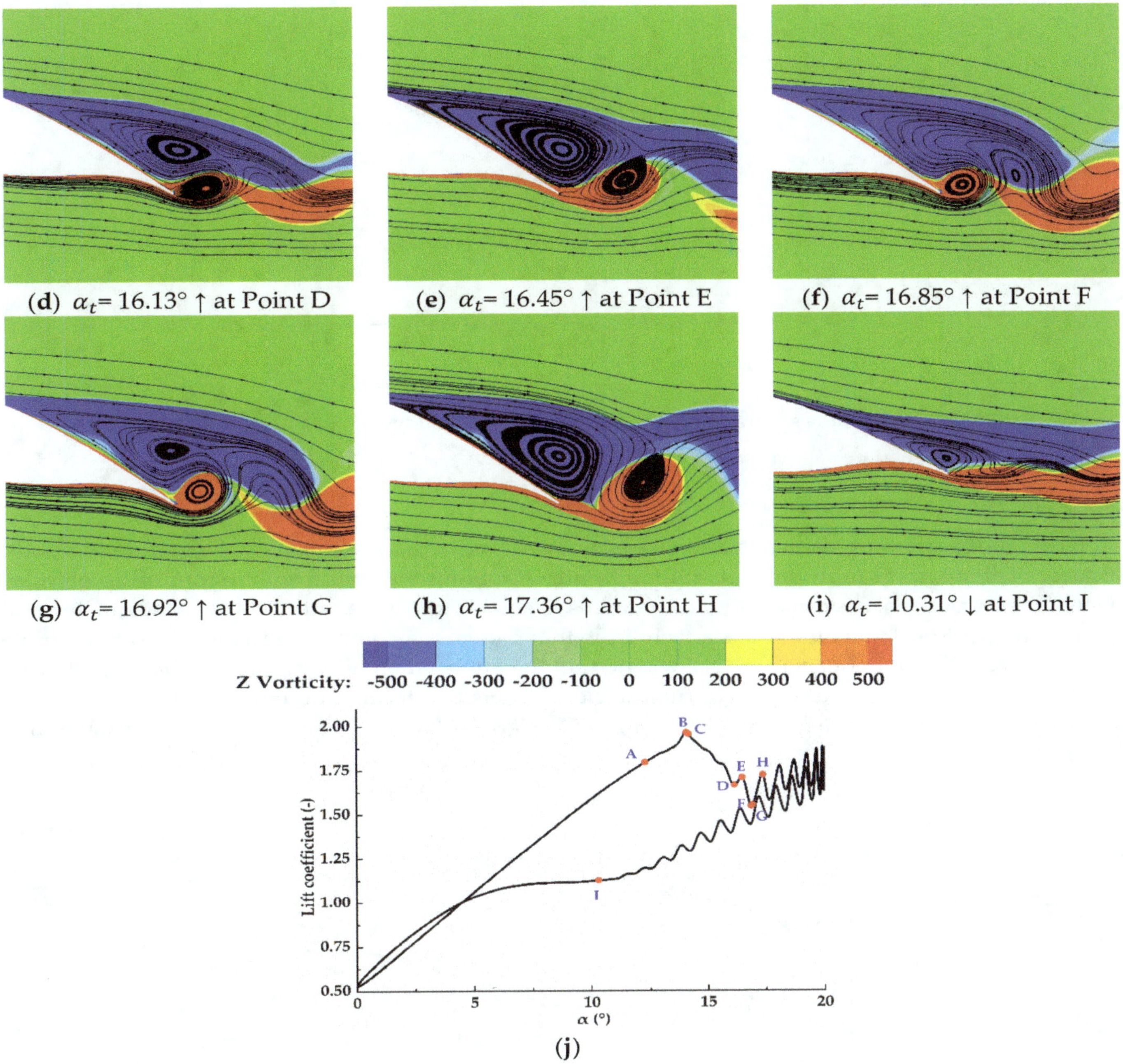

Figure 9. (**a–i**) Vorticity fields and streamlines for different points of angles of attack corresponding to (**j**) the lift coefficient curve.

It is not convenient to show separation situations at every angle of attack with contours. In order to examine dynamic stall delay quantitatively, the pressure coefficient and the wall shear stress around the airfoil were recorded at each time step. The separation points appeared when the wall shear stress abruptly became zero or the pressure signal showed a plateau. Figure 10 shows the position, i.e., x/c, from the leading edge of the separation point at each angle of attack. The delay to stall angle is observed clearly here, as well.

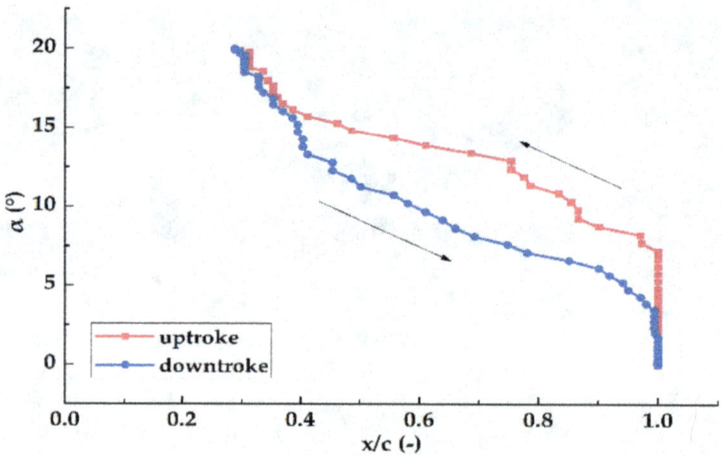

Figure 10. Positions of the separation point during dynamic stall with $k = 0.052$, $\alpha_0 = 10°$, $\alpha_{amp} = 10°$.

3.3. *Frequency Spectrum Analysis of the Pressure Coefficient*

The above pressure contours and streamlines can only be observed from a qualitative point of view. This requires multiple consecutive figures to obtain the characteristics of multiple cycles. Instead, spectrum analysis can capture the features of shedding vortices around the airfoil quickly and quantitatively. Spectrum analysis is a technique that converts a complex signal into a set of simple signals with different frequencies. It is a way to find the information (e.g., amplitude, power, intensity or phase, etc.) of a signal at different frequencies. In this paper, nine points are set on the suction surface of the airfoil to collect time series of pressure signals. The positions of the points are shown in Figure 11.

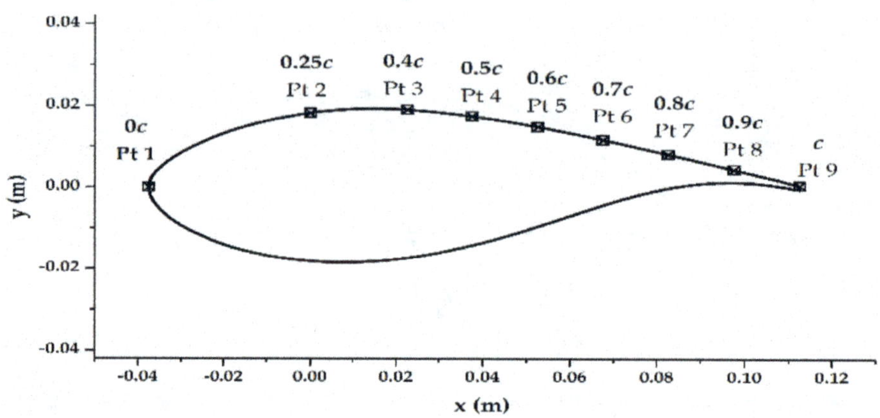

Figure 11. Positions of points to gather the history of pressure signals.

The fast fourier transform (FFT) of the pressure signals was performed with the software MATLAB under different calculations. The spectra of pressure signals were generated by two continuous period signals. Figure 12 shows an example of the time series signals and energy spectrum at Point 6. Figure 12b is the result of Figure 12a after FFT. It shows that the grid in the boundary layer was able to capture small-scale vortices properly, with the turbulence model in the present study.

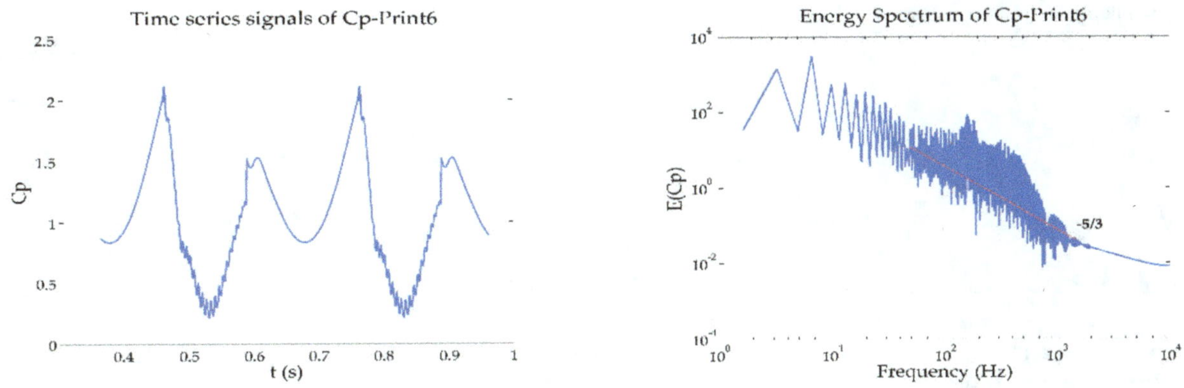

(a) (b)

Figure 12. (**a**) time series signals and (**b**) energy spectrum of the pressure coefficient of Point 9 for two periods with $k = 0.052$, $= 10°$, $\alpha_{amp} = 10°$.

In order to show the differences between each frequency more clearly, the spectra of surface pressure at all nine points on one airfoil are shown in Figure 13 with the semi-log-x function. The abscissa value at the first peak corresponds to the pitch oscillation frequency $f = 3.3333$, which can be computed by Equation (4):

$$f = \frac{1}{T} = \frac{1}{2\pi/\omega} = \frac{\omega}{2\pi} = \frac{2U_\infty k}{2\pi c} = \frac{U_\infty k}{\pi c} \tag{4}$$

The abscissa value at the second peak corresponds to the vortices' shedding mode. The flow field around the pitching airfoil consists of large-scale motions and small-scale turbulence motions. The large-scale one has the same time scale as the unsteady pitch oscillations.

By observing the spectra of pressure at the different points on the airfoil in the same case, i.e., $k = 0.052$, $\alpha_0 = 10°$, $\alpha_{amp} = 10°$, it can be seen that most of the energy of the leading edge point, Pt, is concentrated in the first order frequency pitch oscillation. This is because that the pressure signals of the leading edge point basically were affected by the airfoil pitch oscillation and unaffected by shedding vortices. With the position of the points moving backward, the energy of pressure at the second order frequency became higher. This means that the pressure signals begin to be affected not only by the airfoil pitch oscillation, but also by the shedding of vortices. Due to the proximity of the location, the features of Pt 4 and Pt 5 were similar. This was the same for Pt 6 and Pt 7. It is worth noting that the energy of pressure at high frequency $f_v = 163.333$ at the trailing edge point Pt 9 was much higher than that of other points. This means that there were severe high-frequency fluctuations in the pressure signal at the trailing edge during pitching motion cycles.

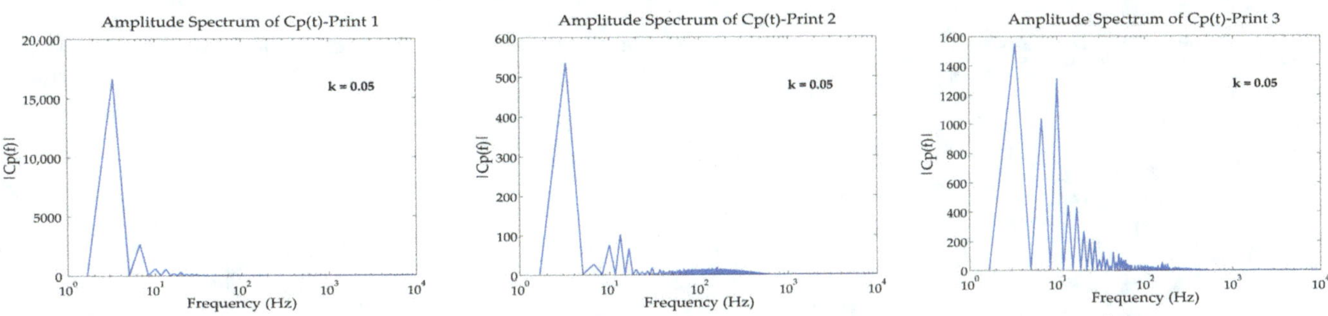

(a) (b) (c)

Figure 13. *Cont.*

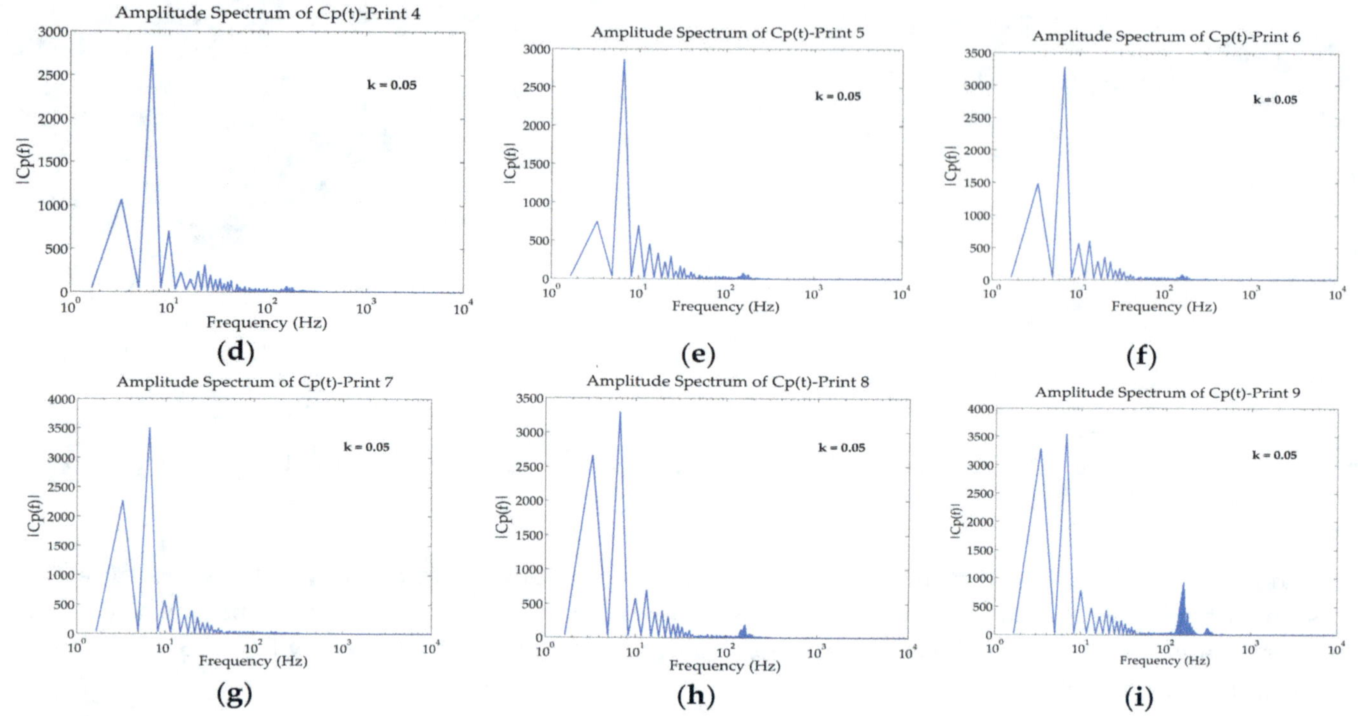

Figure 13. (**a–i**) Energy spectra of the pressure coefficient at different points on the suction surface of the airfoil in the semi-log-x coordinate system with $k = 0.052$, $\alpha_0 = 10°$, $\alpha_{amp} = 10°$.

By comparing the pressure spectra of the cases with different reduced frequencies, as shown in Figure 14, it can clearly be seen that the vortex shedding frequency increased with the increase of the reduced frequency. At the same time, pressure fluctuations at each point on the airfoil showed different characteristics dependent on reduced frequencies. At the leading edge of the airfoil, the proportion of energy for pressure at the first order of frequency became lower with the increasingly-reduced frequency. This shows that the proportion of pressure variation affected by pitch oscillation and shedding vortex changed. This implies that the increasingly-reduced frequency intensified the unsteady effects. Instead, the change of this proportion was reversed at the trailing edge point. Pitch oscillation played a dominant role in pressure fluctuations with increasingly-reduced frequency at the trailing edge point. The pitch oscillation dominated the fluctuations of pressure signals with $k = 0.1$, but did not with $k = 0.05$. This phenomenon provides new evidence of J.G. Leishman's conclusion that when k is smaller than 0.5, the flow can be considered as quasi-steady [1]. Besides the high-frequency fluctuations in the pressure signal, at the trailing edge during pitching motion, cycles had higher energy, but lower frequency with increasingly-reduced frequency; see Figure 14i.

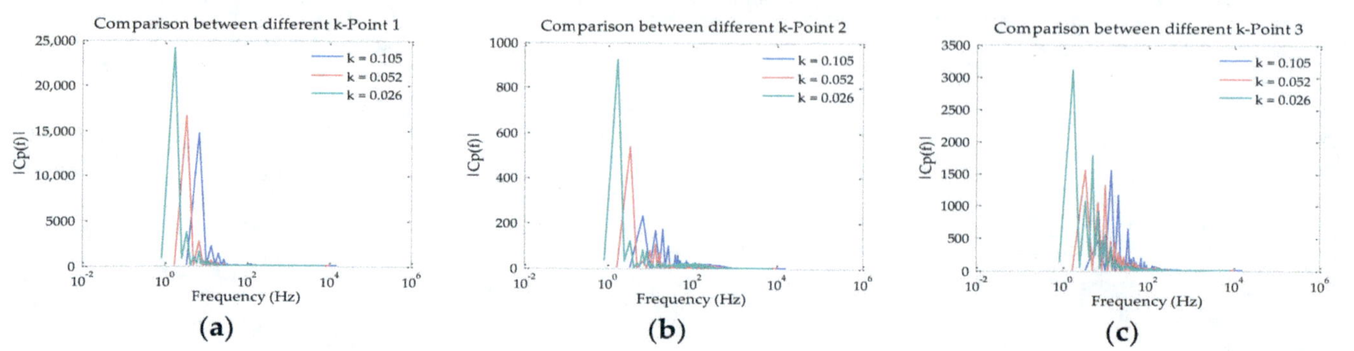

Figure 14. *Cont.*

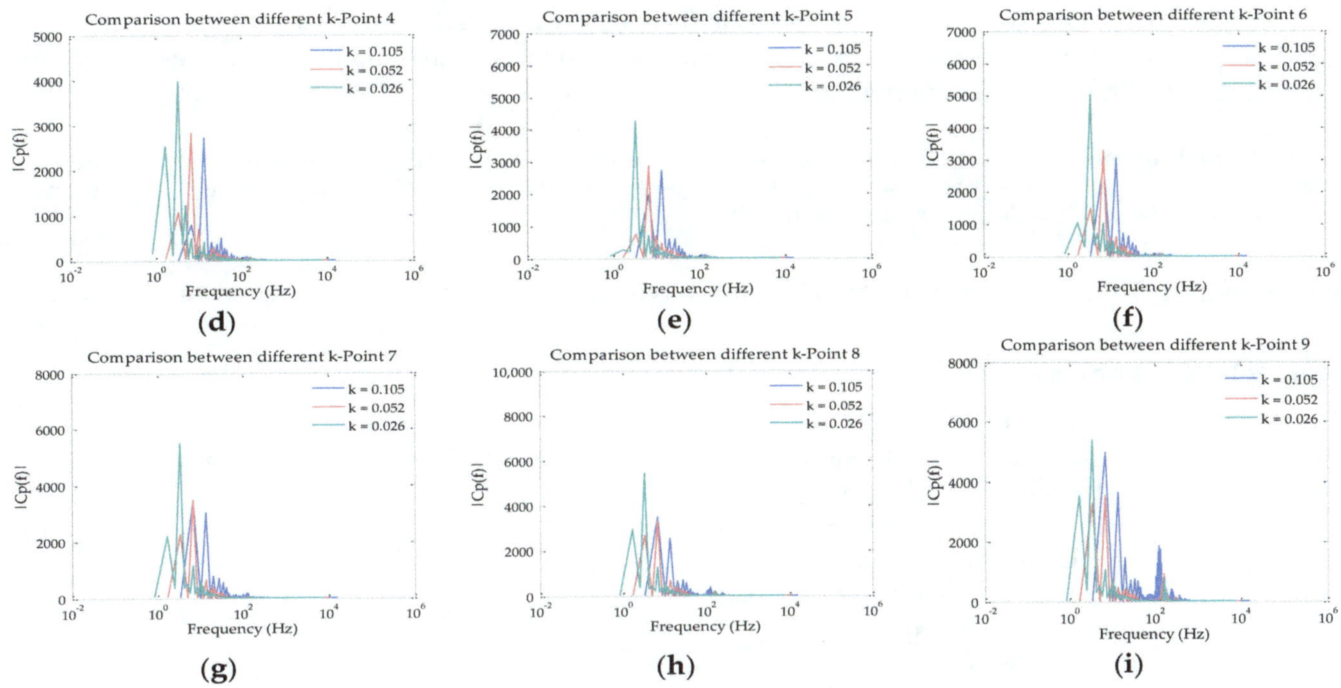

Figure 14. (**a–i**) Comparison of the amplitude spectrum of the pressure coefficient at typical points with different reduced frequencies k.

4. Conclusions

In this paper, in order to better understand the effects of dynamic stall on the aerodynamic characteristics, a CFD method was employed to investigate the influence of the pitch oscillations of the DU 91-W2-250 airfoil. Cases with different reduced frequency were studied by comparing aerodynamic forces and the lift-to-drag ratio. In addition, the hysteresis effect of dynamic stall was analyzed from the perspective of two aspects: the pressure field and spectrum analysis.

The effects of reduced frequency on dynamic stall were analyzed qualitatively and quantitatively. Aerodynamic forces and the lift-to-drag ratio during pitching motion were compared. During the hysteresis loop, the aerodynamic efficiency of the upstroke was lower than that of the downstroke. The effect of delay was more significant with the increased reduction of the frequency. Increasingly-reduced frequency led to lower aerodynamic efficiency during the upstroke process of pitching motions. The result that increasingly-reduced frequency aggravates dynamic stall effects was the same as the conclusions of other observations. Flow field structures and positions of separation points were used to explain the delay of lift stall from other aspects. There were a clockwise-rotating vortex and a counterclockwise-rotating vortex near the trailing edge of airfoil during pitching motions. The growth of the clockwise-rotating vortex increased the value of the lift coefficient, while the growth of the counterclockwise-rotating vortex decreased the value of the lift coefficient.

Frequency spectrum analysis of the pressure coefficient was used as a new method to find the relationship of shedding vortices and pitching motion. The vortex shedding frequency increased with the increase of the reduced frequency. The fluctuations of the pressure signals were the results of pitching motion and vortex shedding. The dominant factor of the pressure signal fluctuations was different at the leading edge and trailing edge. It was also different with different reduced frequencies. Pitch oscillation played a dominant role in pressure fluctuations with increasingly-reduced frequency at the trailing edge point.

In this paper, the dynamic stall characteristics of the DU 91-W2-250 airfoil at $Re = 3 \times 10^5$ were analyzed systematically. However, due to the large amount of computing resources and time cost, many problems need to be further studied in the project. For example, a conclusion of this paper was that increasingly-reduced frequency caused angles of attack of stall to increase. However, it is not known yet when will this trend would stop, i.e., what is the biggest reduced frequency to maintain this trend. In addition, dynamic experiments of the DU series airfoil should be carried out as soon as possible to obtain dynamic data for the purpose of verification.

Author Contributions: S.L. carried out numerical calculation, analyzed the data, and wrote the paper. L.Z. supervised the work, provided theoretical guidance and experimental data. K.Y. provided subject inspiration and writing comments. J.X. and X.L. provided writing help.

Acknowledgments: The authors are grateful to Xingxing Li for several helpful discussions and suggestions.

Nomenclature/Abbreviations

c	chord of airfoil (m)
C_f	skin friction coefficient, $\tau_w/0.5\rho U_\infty{}^2$
C_l	lift coefficient, $L/0.5\rho U_\infty{}^2 c$
C_p	pressure coefficient, $(p - p_\infty)/0.5\rho U_\infty{}^2$
f	pitch oscillation frequency (Hz)
f_v	frequency of shedding vortices
k	reduced frequency, $\omega c/2U_\infty$
L	lift force (N)
p	pressure at one point (Pa)
t	time (s)
T	time period (s)
T_c	characteristic time (s)
U	velocity (m/s)
U_∞	free-stream velocity (m/s)
y	height of the first grid in the boundary layer of the airfoil (m)
y^+	dimensionless wall distance
α	angle of attack (deg)
α_t	angle of attack at flow time t (deg)
α_0	mean angle of attack (deg)
α_{amp}	amplitude of angle of attack (deg)
Δt	time step (s)
ρ	density (kg/m^3)
τ_{wx}	wall shear stress in the x direction (Pa)
ω	angular frequency (rad/s)
2D	two-dimensional
CFD	computational fluid dynamics
CFL	Courant, Friedrichs and Levy criterion
DES	detached eddy simulation
DNS	direct numerical simulation
dt1, dt2 ...	time step size 1, time step size 2 ...
DU	Delft University of Technology
FFT	fast fourier transform
G1, G2 ...	gird case 1, grid case 2 ...
LEV	leading edge vortex
N-S	Navier-Stokes
Pt1, Pt2 ...	point1, point2 ...
RANS	Reynolds-averaged Navier–Stokes
RNG	renormalization-group
SST	shear stress transport
UDF	user-defined function
URANS	unsteady Reynolds-averaged Navier–Stokes

References

1. Leishman, J.G. *Principles of Helicopter Aerodynamics*; Cambridge University Press: Cambridge, UK, 2006; p. 78, ISBN 9781107013353.
2. Holierhoek, J.; De Vaal, J.; Van Zuijlen, A.; Bijl, H. Comparing different dynamic stall models. *Wind Energy* **2013**, *16*, 139–158. [CrossRef]
3. Choudhry, A.; Leknys, R.; Arjomandi, M.; Kelso, R. An insight into the dynamic stall lift characteristics. *Exp. Therm. Fluid Sci.* **2014**, *58*, 188–208. [CrossRef]
4. Xiong, L.; Xianmin, Z.; Gangqiang, L.; Yan, C.; Zhiquan, Y. Dynamic response analysis of the rotating blade of horizontal axis wind turbine. *Wind Eng.* **2010**, *34*, 543–559. [CrossRef]
5. Ham, N.D. Aerodynamic loading on a two-dimensional airfoil during dynamic stall. *AIAA J.* **1968**, *6*, 1927–1934. [CrossRef]
6. Galvanetto, U.; Peiró, J.; Chantharasenawong, C. An assessment of some effects of the nonsmoothness of the leishman–beddoes dynamic stall model on the nonlinear dynamics of a typical aerofoil section. *J. Fluids Struct.* **2008**, *24*, 151–163. [CrossRef]
7. Tarzanin, F. Prediction of control loads due to blade stall. *J. Am. Helicopter Soc.* **1972**, *17*, 33–46. [CrossRef]
8. Tran, C.; Petot, D. Semi-Empirical Model for the Dynamic Stall of Airfoils in View of the Application to the Calculation of Responses of a Helicopter Blade in Forward Flight. *Vertica* **1981**, *5*, 35–53.
9. Leishman, J.G.; Beddoes, T. A semi-empirical model for dynamic stall. *J. Am. Helicopter Soc.* **1989**, *34*, 3–17. [CrossRef]
10. Larsen, J.W.; Nielsen, S.R.; Krenk, S. Dynamic stall model for wind turbine airfoils. *J. Fluids Struct.* **2007**, *23*, 959–982. [CrossRef]
11. Hansen, M.H.; Gaunaa, M.; Madsen, H.A. *A Beddoes–Leishman Type Dynamic Stall Model in State-Space and Indicial Formulation*; Technical University of Denmark: Lyngby, Denmark, 2004; ISBN 8755030890.
12. Gupta, S.; Leishman, J.G. Dynamic stall modelling of the s809 aerofoil and comparison with experiments. *Wind Energy* **2010**, *9*, 521–547. [CrossRef]
13. Sheng, W.; Galbraith, R.A.M.; Coton, F.N. A modified dynamic stall model for low mach numbers. *J. Sol. Energy Eng.* **2008**, *130*, 653. [CrossRef]
14. Martinat, G.; Braza, M.; Harran, G.; Sevrain, A.; Tzabiras, G.; Hoarau, Y.; Favier, D. *Dynamic Stall of a Pitching and Horizontally Oscillating Airfoil*; Springer: Berlin, Germany, 2009; pp. 395–403.
15. Wang, S.; Ingham, D.B.; Ma, L.; Pourkashanian, M.; Tao, Z. Numerical investigations on dynamic stall of low reynolds number flow around oscillating airfoils. *Comput. Fluids* **2010**, *39*, 1529–1541. [CrossRef]
16. Akbari, M.H.; Price, S.J. Simulation of dynamic stall for a naca 0012 airfoil using a vortex method. *J. Fluids Struct.* **2003**, *17*, 855–874. [CrossRef]
17. Gharali, K.; Johnson, D.A. Dynamic stall simulation of a pitching airfoil under unsteady freestream velocity. *J. Fluids Struct.* **2013**, *42*, 228–244. [CrossRef]
18. Kim, Y.; Xie, Z.T. Modelling the effect of freestream turbulence on dynamic stall of wind turbine blades. *Comput. Fluids* **2016**, *129*, 53–66. [CrossRef]
19. Gandhi, A.; Merrill, B.; Peet, Y.T. Effect of Reduced Frequency on Dynamic Stall of a Pitching Airfoil in a Turbulent Wake. In Proceedings of the 55th AIAA Aerospace Sciences Meeting, Grapevine, TX, USA, 9–13 January 2017; p. 0720. [CrossRef]
20. Martin, J.; Empey, R.; McCroskey, W.; Caradonna, F. An experimental analysis of dynamic stall on an oscillating airfoil. *J. Am. Helicopter Soc.* **1974**, *19*, 26–32. [CrossRef]
21. Mccroskey, W.J. Unsteady airfoils. *Annu. Rev. Fluid Mech.* **2003**, *14*, 285–311. [CrossRef]
22. McCroskey, W.J. *The Phenomenon of Dynamic Stall*; NASA Ames Research Center: Mountain View, CA, USA, 1981. [CrossRef]
23. Wernert, P.; Geissler, W.; Raffel, M.; Kompenhans, J. Experimental and numerical investigations of dynamic stall on a pitching airfoil. *AIAA J.* **1996**, *34*, 982–989. [CrossRef]
24. Aramendia, I.; Fernandez-Gamiz, U.; Ramos-Hernanz, J.A.; Sancho, J.; Lopez-Guede, J.M.; Zulueta, E. Flow control devices for wind turbines. In *Energy Harvesting and Energy Efficiency*; Springer: Berlin, Germany, 2017; pp. 629–655.
25. Johnson, S.J.; Baker, J.P.; Van Dam, C.P.; Berg, D. An overview of active load control techniques for wind turbines with an emphasis on microtabs. *Wind Energy* **2010**, *13*, 239–253. [CrossRef]

26. Zhang, L.; Li, X.; Yang, K.; Xue, D. Effects of vortex generators on aerodynamic performance of thick wind turbine airfoils. *J. Wind Eng. Ind. Aerodyn.* **2016**, *156*, 84–92. [CrossRef]
27. Yang, K.; Zhang, L.; Jianzhong, X.U. Simulation of aerodynamic performance affected by vortex generators on blunt trailing-edge airfoils. *Sci. China Technol. Sci.* **2010**, *53*, 1–7. [CrossRef]
28. Tsai, K.-C.; Pan, C.-T.; Cooperman, A.M.; Johnson, S.J.; Van Dam, C. An innovative design of a microtab deployment mechanism for active aerodynamic load control. *Energies* **2015**, *8*, 5885–5897. [CrossRef]
29. Fernandez-Gamiz, U.; Zulueta, E.; Boyano, A.; Ramos-Hernanz, J.A.; Lopez-Guede, J.M. Microtab design and implementation on a 5 mw wind turbine. *Appl. Sci.* **2017**, *7*, 536. [CrossRef]
30. Barlas, T.K.; Van Kuik, G. Review of state of the art in smart rotor control research for wind turbines. *Prog. Aerosp. Sci.* **2010**, *46*, 1–27. [CrossRef]
31. Fernandez-Gamiz, U.; Zulueta, E.; Boyano, A.; Ansoategui, I.; Uriarte, I. Five megawatt wind turbine power output improvements by passive flow control devices. *Energies* **2017**, *10*, 742. [CrossRef]
32. Timmer, W.; Van Rooij, R. Summary of the delft university wind turbine dedicated airfoils. *J. Sol. Energy Eng.* **2003**, *125*, 488–496. [CrossRef]
33. Bai, J.Y.; Zhang, L.; Xingxing, L.I.; Yang, K. Analyzing the effect of wind tunnel wall on the aerodynamic performance of airfoils. *Sci. Sin.* **2016**, *46*, 124707. (In Chinese)
34. Poirel, D.; Métivier, V.; Dumas, G. Computational aeroelastic simulations of self-sustained pitch oscillations of a naca0012 at transitional reynolds numbers. *J. Fluids Struct.* **2011**, *27*, 1262–1277. [CrossRef]
35. Blazek, J. *Computational Fluid Dynamics: Principles and Applications*, 2nd ed.; Elsevier: New York, NY, USA, 2005; pp. 1–4.
36. Hand, B.; Kelly, G.; Cashman, A. Numerical simulation of a vertical axis wind turbine airfoil experiencing dynamic stall at high reynolds numbers. *Comput. Fluids* **2017**, *149*, 12–30. [CrossRef]
37. Liu, X.; Lu, C.; Liang, S.; Godbole, A.; Chen, Y. Vibration-induced aerodynamic loads on large horizontal axis wind turbine blades. *Appl. Energy* **2017**, *185*, 1109–1119. [CrossRef]
38. Wang, S.; Ingham, D.B.; Ma, L.; Pourkashanian, M.; Tao, Z. Turbulence modeling of deep dynamic stall at relatively low reynolds number. *J. Fluids Struct.* **2012**, *33*, 191–209. [CrossRef]
39. Gharali, K.; Johnson, D.A. Numerical modeling of an s809 airfoil under dynamic stall, erosion and high reduced frequencies. *Appl. Energy* **2012**, *93*, 45–52. [CrossRef]
40. Karbasian, H.R.; Esfahani, J.A.; Barati, E. Effect of acceleration on dynamic stall of airfoil in unsteady operating conditions. *Wind Energy* **2016**, *19*, 17–33. [CrossRef]
41. *Ansys Fluent 14 User's Guide*; Ansys Inc.: Canonsburg, PA, USA, 2011.
42. Durbin, P.A.; Reif, B.P. *Statistical Theory and Modeling for Turbulent Flows*; John Wiley & Sons: Hoboken, NJ, USA, 2011; ISBN 1119957524.
43. White, F.M. *Fluid mechanics*, 7th ed.; McGraw-Hill: New York, NY, USA, 2011; ISBN 978-0-07-352934-9.
44. Raffel, M.; Favier, D.; Berton, E.; Rondot, C.; Nsimba, M.; Geissler, W. Micro-piv and eldv wind tunnel investigations of the laminar separation bubble above a helicopter blade tip. *Meas. Sci. Technol.* **2005**, *17*, 1–13. [CrossRef]
45. Barakos, G.N.; Drikakis, D. Computational study of unsteady turbulent flows around oscillating and ramping aerofoils. *Int. J. Numer. Methods Fluids* **2010**, *42*, 163–186. [CrossRef]
46. Martinat, G.; Braza, M.; Hoarau, Y.; Harran, G. Turbulence modelling of the flow past a pitching naca0012 airfoil at 10^5 and 10^6 reynolds numbers. *J. Fluids Struct.* **2008**, *24*, 1294–1303. [CrossRef]

PERMISSIONS

All chapters in this book were first published in MDPI; hereby published with permission under the Creative Commons Attribution License or equivalent. Every chapter published in this book has been scrutinized by our experts. Their significance has been extensively debated. The topics covered herein carry significant findings which will fuel the growth of the discipline. They may even be implemented as practical applications or may be referred to as a beginning point for another development.

The contributors of this book come from diverse backgrounds, making this book a truly international effort. This book will bring forth new frontiers with its revolutionizing research information and detailed analysis of the nascent developments around the world.

We would like to thank all the contributing authors for lending their expertise to make the book truly unique. They have played a crucial role in the development of this book. Without their invaluable contributions this book wouldn't have been possible. They have made vital efforts to compile up to date information on the varied aspects of this subject to make this book a valuable addition to the collection of many professionals and students.

This book was conceptualized with the vision of imparting up-to-date information and advanced data in this field. To ensure the same, a matchless editorial board was set up. Every individual on the board went through rigorous rounds of assessment to prove their worth. After which they invested a large part of their time researching and compiling the most relevant data for our readers.

The editorial board has been involved in producing this book since its inception. They have spent rigorous hours researching and exploring the diverse topics which have resulted in the successful publishing of this book. They have passed on their knowledge of decades through this book. To expedite this challenging task, the publisher supported the team at every step. A small team of assistant editors was also appointed to further simplify the editing procedure and attain best results for the readers.

Apart from the editorial board, the designing team has also invested a significant amount of their time in understanding the subject and creating the most relevant covers. They scrutinized every image to scout for the most suitable representation of the subject and create an appropriate cover for the book.

The publishing team has been an ardent support to the editorial, designing and production team. Their endless efforts to recruit the best for this project, has resulted in the accomplishment of this book. They are a veteran in the field of academics and their pool of knowledge is as vast as their experience in printing. Their expertise and guidance has proved useful at every step. Their uncompromising quality standards have made this book an exceptional effort. Their encouragement from time to time has been an inspiration for everyone.

The publisher and the editorial board hope that this book will prove to be a valuable piece of knowledge for researchers, students, practitioners and scholars across the globe.

LIST OF CONTRIBUTORS

Jiufa Cao
School of Hydraulic Energy and Power Engineering, Yangzhou University, Yangzhou 225009, China

Weijun Zhu
School of Hydraulic Energy and Power Engineering, Yangzhou University, Yangzhou 225009, China
Department of Wind Energy, Technical University of Denmark, 2800 Kgs. Lyngby, Denmark

Wenzhong Shen and Jens Nørkær Sørensen
Department of Wind Energy, Technical University of Denmark, 2800 Kgs. Lyngby, Denmark

Chengyong Zhu and Tongguang Wang
Jiangsu Key Laboratory of Hi-Tech Research for Wind Turbine Design, Nanjing University of Aeronautics and Astronautics, Nanjing 210016, China

Zhenye Sun and Jiufa Cao
School of Hydraulic Energy and Power Engineering, Yangzhou University, Yangzhou 225127, China

Wei Jun Zhu
School of Hydraulic Energy and Power Engineering, Yangzhou University, Yangzhou 225127, China
Technical University of Denmark, 2800 Lyngby, Denmark

Emre Barlas
Goldwind Energy ApS, 8382 Hinnerup, Denmark

Wenlin Yu
Department of Civil Engineering, Nanjing University of Aeronautics and Astronautics, 29 Yudao Road, Nanjing 210016, China

Jiufa Cao
Jiangsu Key Laboratory of Hi-Tech Research for Wind Turbine Design, Nanjing University of Aeronautics and Astronautics, Nanjing 210016, China

Shitang Ke
Department of Civil Engineering, Nanjing University of Aeronautics and Astronautics, 29 Yudao Road, Nanjing 210016, China
Jiangsu Key Laboratory of Hi-Tech Research for Wind Turbine Design, Nanjing University of Aeronautics and Astronautics, Nanjing 210016, China

Junwei Yang
College of Electrical, Energy and Power Engineering, Yangzhou University, Yangzhou 225127, China
New Energy Research Center, Yangzhou University, Yangzhou 225009, China
College of Hydraulic Science and Engineering, Yangzhou University, Yangzhou 225009, China

Hua Yang, Weijun Zhu and Nailu Li
College of Electrical, Energy and Power Engineering, Yangzhou University, Yangzhou 225127, China
New Energy Research Center, Yangzhou University, Yangzhou 225009, China

Yiping Yuan
Jiangsu Key Laboratory of Hi-Tech Research for Wind Turbine Design, Nanjing University of Aeronautics and Astronautics, Nanjing 210016, China

Kurt Schaldemose Hansen
Department of Wind Energy, Technical University of Denmark, Lyngby Campus, 2800 Kgs. Lyngby, Denmark

Maarten Paul van der Laan
Department of Wind Energy, Technical University of Denmark, Risø Campus, 4000 Roskilde, Denmark

Zhenzhou Zhao, Ruixin Wang, Bofeng Xu, Yuan Zheng and Siyue Qian
College of Energy and Electrical Engineering, Hohai University, Nanjing 210098, Jiangsu, China

Wenzhong Shen
Department of Wind Energy, Technical University of Denmark, 2800 Lyngby, Denmark

Han Yang, Jin Chen and Xiaoping Pang
State Key Laboratory of Mechanical Transmissions, Chongqing University, Chongqing 400044, China

Mingyue Zhou, Matias Sessarego and Wen Zhong Shen
Fluid Mechanics Section, Department of Wind Energy, Technical University of Denmark, Nils Koppels Allé, Building 403, 2800 Lyngby, Denmark

Sebastian Ehrich and Carl Michael Schwarz
ForWind, University of Oldenburg, Institute of Physics, Küpkersweg 70, 26129 Oldenburg, Germany

Bernhard Stoevesandt
Fraunhofer IWES, Küpkersweg 70, 26129 Oldenburg, Germany

Hamid Rahimi and Joachim Peinke
ForWind, University of Oldenburg, Institute of Physics, Küpkersweg 70, 26129 Oldenburg, Germany
Fraunhofer IWES, Küpkersweg 70, 26129 Oldenburg, Germany

Ying Guo, Liqin Liu, Xifeng Gao and Wanhai Xu
State Key Laboratory of Hydraulic Engineering Simulation and Safety, Tianjin University, Tianjin 300072, China

Shuang Li, Jin Xu and Xue Li
Institute of Engineering Thermophysics, Chinese Academy of Sciences, Beijing 100190, China
University of Chinese Academy of Sciences, Beijing 100049, China

Lei Zhang and Ke Yang
Institute of Engineering Thermophysics, Chinese Academy of Sciences, Beijing 100190, China
Key Laboratory of Wind Energy Utilization, Chinese Academy of Sciences, Beijing 100190, China
National Research and Development Center of Wind Turbine Blade, Beijing 100190, China

Index

A

Actuator Disc, 1-2, 6, 18-19, 42-43, 61-62, 111-113, 115, 117, 129-131

Aerodynamic Drag, 93

Aerodynamics, 1-2, 19-20, 24, 31, 40-43, 87, 89, 109, 133-135, 140, 145-146, 172, 191-192, 195, 205, 207, 214, 217-218, 221, 225, 239

Amplitude Modulation, 173-174, 184, 189-190

Atmospheric Boundary Layer, 3, 20, 63, 108, 112

Atmospheric Stability, 21, 43, 131

B

Blade Length, 72-73, 155-156, 160, 163-164, 166, 171, 193-194

Blade Model, 73

C

Capital Cost, 156, 159, 171

Cell Size, 195

Circumferential Angle, 79

Climate Change, 155

Complex Terrain, 111-112, 114, 117, 129-131

Composite Materials, 157-158

Computational Fluid Dynamics, 20, 23, 42, 62, 64-65, 77, 111, 129-130, 152, 174, 192-193, 205, 225, 238, 240

Computer-aided Design, 113

Conceptualization, 61, 87, 109, 189, 204

Conjunction, 25

Constant Value, 139

Convergence, 8, 26, 69, 142, 183

D

Data Collection, 40, 109

Discretization, 7, 26, 192

Dissipation, 77, 96, 118

E

Eigenvalues, 85-86

F

Fatigue, 90, 137, 192-193, 196, 200, 202, 204-206, 224

Feedback, 9

Finite Element Method, 66, 86

Flow Characteristics, 111

Flow Control, 109, 225, 239-240

Flow Field, 6, 8, 22-24, 47, 64-66, 87, 91, 96-97, 110-112, 114-115, 122-123, 125-126, 129, 175, 189, 192, 194-195, 225, 231, 235, 237

Force Coefficients, 36, 135, 142, 210

Fourier Transform, 216, 234, 238

Friction Coefficient, 159, 238

G

Genetic Algorithm, 1, 3, 21, 88, 172

I

Immersed Boundary Method, 3, 20, 91, 110

Induction Factor, 19, 134, 142, 146, 151

Integer, 89

K

Kinetic Energy, 6, 96

L

Laminar-turbulent Transition, 195

Layout Optimization, 2, 19, 89

Linear Equations, 23

Local Variables, 41

Lookup Tables, 195

Loops, 37-39

N

Negative Pressure, 79-80, 87

O

Operating Expenses, 159

Oscillations, 7, 107, 204, 224-225, 235, 237, 240

P

Parameterization, 68

Perturbation, 23-25

Plane Of Rotation, 73

Power Coefficient, 73, 132-134, 137, 139-140, 149-152, 157, 160, 183, 207

Precipitation, 69, 89

R

Relative Thickness, 1, 3, 13, 17-18, 226

Resolution, 26, 32, 64-68, 86, 88, 178, 195, 203-204

Rotor Blades, 1, 3, 6, 41, 108, 192-193, 195, 200

S

Sampling Frequency, 94

Sea Level, 68-69

Shear Stress, 224-225, 233, 238

Speed Ratio, 8, 132, 152, 157, 160-161, 170

Speed Sites, 155-156

Standard Deviation, 96, 122-124, 126-130, 192, 196-198, 200-202, 204

T

Tangential Force, 135, 142-143, 152, 180, 210, 218, 221

Thermodynamics, 65

Tilt Angle, 42, 160, 177, 194

Time Scale, 66, 235

Tip Loss Correction, 21

Topography, 131

Topology, 3, 12, 51, 226

Turbine Optimization, 2, 15, 18, 155-156, 158, 171-172

Turbine Performance, 2, 20, 40, 65, 89, 131, 180, 205

Turbulence, 2-3, 6, 8-9, 11, 18, 20-24, 26, 40-42, 45, 51, 64, 77-79, 87-91, 94-97, 100-102, 105-108, 110, 112-113, 115-116, 118, 122-131, 174, 178, 185, 190, 193-195, 204-205, 225-226, 234-235, 239-240

U

Unsteady Aerodynamics, 31, 40-41, 195, 225

V

Variables, 1, 3, 13, 15, 41, 133, 156, 158, 161-165, 175

Velocity, 2, 6-7, 9-10, 12-13, 22-27, 30-32, 36-37, 40, 42-43, 45, 50-52, 59, 73, 90, 93-94, 96-97, 104, 106-107, 122-129, 132, 134-135, 142, 159-160, 175-176, 178, 180-181, 185, 194-197, 199, 209-210, 213, 224-225, 227, 238-239

Velocity Gradient, 23

Viscosity, 3, 20, 23-24, 26, 40, 42, 113, 140

Vorticity, 33-34, 122-125, 233

W

Wake Field, 52

Wind Farm, 1-2, 11-12, 15, 19, 21-22, 42-43, 61-62, 89, 91, 111-118, 120-121, 129-130, 174, 207, 222

Wind Shear Exponent, 185

Wind Tunnel Experiment, 90, 97

Wind Turbine Blades, 6, 19-20, 40-41, 72, 89-90, 156, 161, 171-172, 175, 224, 239-240

Wind Turbines, 1-3, 7-13, 15-23, 40-43, 47, 50-53, 56, 58-66, 88, 90-91, 96, 108, 110-114, 116-117, 119, 121, 129-133, 152-156, 159, 163, 169-174, 189-190, 193, 205-207, 221-222, 224, 239-240

www.ingramcontent.com/pod-product-compliance
Lightning Source LLC
LaVergne TN
LVHW081013210926
848485LV00004B/855